W9-AUD-567

ALGEBRA

Revised Edition

THE HISTORY OF MATHEMATICS

ALGEBRA

SETS, SYMBOLS, AND THE LANGUAGE OF THOUGHT

Revised Edition

John Tabak, Ph.D.

ALGEBRA: Sets, Symbols, and the Language of Thought, Revised Edition

Facts On File, Inc.
An imprint of Infobase Learning
132 West 31st Street
New York NY 10001

Library of Congress Cataloging-in-Publication Data

Tabak, John.
Algebra : sets, symbols, and the language of thought / John Tabak—Rev. ed.
p. cm.—(The history of mathematics)
Includes bibliographical references and index.
ISBN 978-0-8160-7944-5 (alk. paper)
1. Algebra—History. I. Title.
QA151.T33 2011
512.009—dc22 2010021597

Facts On File books are available at special discounts when purchased in bulk quantities for businesses, associations, institutions, or sales promotions. Please call our Special Sales Department in New York at (212) 967-8800 or (800) 322-8755.

You can find Facts On File on the World Wide Web at http://www.infobaselearning.com

Text design by David Strelecky
Composition by Hermitage Publishing Services
Illustrations by Dale Williams
Photo research by Elizabeth H. Oakes
Cover printed by Yurchak Printing, Inc., Landisville, Pa.
Book printed and bound by Yurchak Printing, Inc., Landisville, Pa.
Date printed: May 2011
Printed in the United States of America

10 9 8 7 6 5 4 3 2 1

This book is printed on acid-free paper.

T 17568

To Diane Haber; teacher; mathematician, and inspirator

CONTENTS

PREFACE

Of all human activities, mathematics is one of the oldest. Mathematics can be found on the cuneiform tablets of the Mesopotamians, on the papyri of the Egyptians, and in texts from ancient China, the Indian subcontinent, and the indigenous cultures of Central America. Sophisticated mathematical research was carried out in the Middle East for several centuries after the birth of Muhammad, and advanced mathematics has been a hallmark of European culture since the Renaissance. Today, mathematical research is carried out across the world, and it is a remarkable fact that there is no end in sight. The more we learn of mathematics, the faster the pace of discovery.

Contemporary mathematics is often extremely abstract, and the important questions with which mathematicians concern themselves can sometimes be difficult to describe to the interested nonspecialist. Perhaps this is one reason that so many histories of mathematics give so little attention to the last 100 years of discovery—this, despite the fact that the last 100 years have probably been the most productive period in the history of mathematics. One unique feature of this six-volume History of Mathematics is that it covers a significant portion of recent mathematical history as well as the origins. And with the help of in-depth interviews with prominent mathematicians—one for each volume—it is hoped that the reader will develop an appreciation for current work in mathematics as well as an interest in the future of this remarkable subject.

Numbers details the evolution of the concept of number from the simplest counting schemes to the discovery of uncomputable numbers in the latter half of the 20th century. Divided into three parts, this volume first treats numbers from the point of view of computation. The second part details the evolution of the concept of number, a process that took thousands of years and culminated in what every student recognizes as "the real number line," an extremely important and subtle mathematical idea. The third part of this volume concerns the evolution of the concept of the infinite. In particular, it covers Georg Cantor's discovery (or creation, depending on one's point of view) of transfinite numbers and his efforts to place set theory at the heart of modern mathematics. The most important ramifications of Cantor's work, the attempt to axiomatize mathematics carried out by David Hilbert and Bertrand Russell, and the discovery by Kurt Gödel and Alan Turing that there are limitations on what can be learned from the axiomatic method, are also described. The last chapter ends with the discovery of uncomputable numbers, a remarkable consequence of the work of Kurt Gödel and Alan Turing. The book concludes with an interview with Professor Karlis Podnieks, a mathematician of remarkable insights and a broad array of interests.

Probability and Statistics describes subjects that have become central to modern thought. Statistics now lies at the heart of the way that most information is communicated and interpreted. Much of our understanding of economics, science, marketing, and a host of other subjects is expressed in the language of statistics. And for many of us statistical language has become part of everyday discourse. Similarly, probability theory is used to predict everything from the weather to the success of space missions to the value of mortgage-backed securities.

The first half of the volume treats probability beginning with the earliest ideas about chance and the foundational work of Blaise Pascal and Pierre Fermat. In addition to the development of the mathematics of probability, considerable attention is given to the application of probability theory to the study of smallpox and the misapplication of probability to modern finance. More than most branches of mathematics, probability is an applied discipline, and its uses and misuses are important to us all. Statistics is the subject of the second half of the book. Beginning with the earliest examples of statistical thought, which are found in the writings of John Graunt and Edmund Halley, the volume gives special attention to two pioneers of statistical thinking, Karl Pearson and R. A. Fisher, and it describes some especially important uses and misuses of statistics, including the use of statistics in the field of public health, an application of vital interest. The book concludes with an interview with Dr. Michael Stamatelatos, director of the Safety and Assurance Requirements Division in the Office of Safety and Mission Assurance at NASA, on the ways that probability theory, specifically the methodology of probabilistic risk assessment, is used to assess risk and improve reliability.

Geometry discusses one of the oldest of all branches of mathematics. Special attention is given to Greek geometry, which set the standard both for mathematical creativity and rigor for many centuries. So important was Euclidean geometry that it was not until the 19th century that mathematicians became willing to consider the existence of alternative and equally valid geometrical systems. This 19th-century revolution in mathematical, philosophical, and scientific thought is described in some detail, as are some alternatives to Euclidean geometry, including projective geometry, the non-Euclidean geometry of Nikolay Ivanovich Lobachevsky and János Bolyai, the higher (but finite) dimensional geometry of Riemann, infinite-dimensional geometric ideas, and some of the geometrical implications of the theory of relativity. The volume concludes with an interview with Professor Krystyna Kuperberg of Auburn University about her work in geometry and dynamical systems, a branch of mathematics heavily dependent on ideas from geometry. A successful and highly insightful mathematician, she also discusses the role of intuition in her research.

Mathematics is also the language of science, and mathematical methods are an important tool of discovery for scientists in many disciplines. *Mathematics and the Laws of Nature* provides an overview of the ways that mathematical thinking has influenced the evolution of science—especially the use of deductive reasoning in the development of physics, chemistry, and population genetics. It also discusses the limits of deductive reasoning in the development of science.

In antiquity, the study of geometry was often perceived as identical to the study of nature, but the axioms of Euclidean geometry were gradually supplemented by the axioms of classical physics: conservation of mass, conservation of momentum, and conservation of energy. The significance of geometry as an organizing principle in nature was briefly subordinated by the discovery of relativity theory but restored in the 20th century by Emmy Noether's work on the relationships between conservation laws and symmetries. The book emphasizes the evolution of classical physics because classical insights remain the most important insights in many branches of science and engineering. The text also includes information on the relationship between the laws of classical physics and more recent discoveries that conflict with the classical model of nature. The main body of the text concludes with a section on the ways that probabilistic thought has sometimes supplanted older ideas about determinism. An interview with Dr. Renate Hagedorn about her work at the European Centre for Medium-Range Weather Forecasts (ECMWF), a leading center for research into meteorology and a place where many of the concepts described in this book are regularly put to the test, follows.

Of all mathematical disciplines, algebra has changed the most. While earlier generations of geometers would recognize—if not immediately understand—much of modern geometry as an extension of the subject that they had studied, it is doubtful that earlier generations of algebraists would recognize most of modern algebra as in any way related to the subject to which they devoted their time. *Algebra* details the regular revolutions in thought that have occurred in one of the most useful and vital areas of contemporary mathematics: Ancient proto-algebras, the concepts of algebra that originated in the Indian subcontinent and in the Middle East, the "reduction" of geometry to algebra begun by René Descartes, the abstract algebras that grew out of the work of Évariste Galois, the work of George Boole and some of the applications of his algebra, the theory of matrices, and the work of Emmy Noether are all described. Illustrative examples are also included. The book concludes with an interview with Dr. Bonita Saunders of the National Institute of Standards and Technology about her work on the Digital Library of Mathematical Functions, a project that mixes mathematics and science, computers and aesthetics.

New to the History of Mathematics set is *Beyond Geometry*, a volume that is devoted to set-theoretic topology. Modern mathematics is often divided into three broad disciplines: analysis, algebra, and topology. Of these three, topology is the least known to the general public. So removed from daily experience is topology that even its subject matter is difficult to describe in a few sentences, but over the course of its roughly 100-year history, topology has become central to much of analysis as well as an important area of inquiry in its own right.

The term *topology* is applied to two very different disciplines: set-theoretic topology (also known as general topology and point-set topology), and the very different discipline of algebraic topology. For two reasons, this volume deals almost exclusively with the former. First, set-theoretic topology evolved along lines that were, in a sense, classical, and so its goals and techniques, when viewed from a certain perspective, more closely resemble those of subjects that most readers have already studied or will soon encounter. Second, some of the results of set-theoretic topology are incorporated into elementary calculus courses. Neither of these statements is true for algebraic topology, which, while a very important branch of mathematics, is based on ideas and techniques that few will encounter until the senior year of an undergraduate education in mathematics.

The first few chapters of *Beyond Geometry* provide background information needed to put the basic ideas and goals of set-theoretic topology into context. They enable the reader to better appreciate the work of the pioneers in this field. The discoveries of Bolzano, Cantor, Dedekind, and Peano are described in some detail because they provided both the motivation and foundation for much early topological research. Special attention is also given to the foundational work of Felix Hausdorff.

Set-theoretic topology has also been associated with nationalism and unusual educational philosophies. The emergence of Warsaw, Poland, as a center for topological research prior to World War II was motivated, in part, by feelings of nationalism among Polish mathematicians, and the topologist R. L. Moore at the University of Texas produced many important topologists while employing a radical approach to education that remains controversial to this day. Japan was also a prominent center of topological research, and so it remains. The main body of the text concludes with some applications of topology, especially dimension theory, and topology as the foundation for the field of analysis. This volume contains an interview with Professor Scott Williams, an insightful thinker and pioneering topologist, on the nature of topological research and topology's place within mathematics.

The five revised editions contain a more comprehensive chronology, valid for all six volumes, an updated section of further resources, and many new color photos and line drawings. The visuals are an important part of each volume, as they enhance the narrative and illustrate a number of important (and very visual) ideas. The History of Mathematics should prove useful as a resource. It is also my hope that it will prove to be an enjoyable story to read—a tale of the evolution of some of humanity's most profound and most useful ideas.

ACKNOWLEDGMENTS

The author is deeply appreciative of Frank Darmstadt, executive editor, for his many helpful suggestions, and to Elizabeth Oakes, for her fine photo research.

Thanks, also, to Penelope Pillsbury and the staff of the Brownell Library, Essex Junction, Vermont, for their extraordinary help with the many questions that arose during the preparation of this book. Finally, thanks to Dr. Bonita Saunders of the National Institute of Standards and Technology (NIST) for the generous way that she shared her time and expertise.

INTRODUCTION

algebra n.

1. a generalization of arithmetic in which letters representing numbers are combined according to the rules of arithmetic
2. any of various systems or branches of mathematics or logic concerned with the properties and relationships of abstract entities (as complex numbers, matrices, sets, vectors, groups, rings, or fields) manipulated in symbolic form under operations often analogous to those of arithmetic

—*from* Merriam-Webster's Collegiate Dictionary, 11th Edition
(© 2003 by Merriam-Webster, Incorporated [www.Merriam-Webster.com])

Algebra is one of the oldest of all branches of mathematics. Today it is also one of the most abstract. Consider the following important algebraic theorem first proved by the French mathematician Augustin-Louis Cauchy (1789–1857):

> Let the letter G denote a finite group. Let N represent the number of elements in G. Let p represent a prime number. If p divides N, then G has an element of order p.

To many people without a background in mathematics, it is not clear what this statement means or even if it means anything at all. But Cauchy's theorem reveals an important property shared by every *group* that is comprised of a finite collection of objects. The theorem is important because groups are central to modern algebra.

When contemporary mathematicians speak of algebra, they usually use the term in a way that is very different from what most of us understand the word *algebra* to mean. To be sure, all branches of mathematics have become increasingly abstract, but those mathematicians who study geometry, for example, continue to study properties that most people would—with some prompting—still recognize as geometric, and probability, no matter how advanced the treatment, still concerns itself with the odds that some event will occur. But modern algebra is completely divorced from the subject most learn about in school.

For thousands of years, algebra consisted solely of expanding the list of problem-solving algorithms—a list of procedures similar in concept to the quadratic formula. But much of contemporary algebra focuses on identifying and describing the logical structures upon which mathematics is built. It is now clear that identifying and exploiting these structures is just as important for mathematical and scientific progress as the development of new algorithms. *Algebra, Revised Edition* describes the history of both strands of algebraic thought.

Chapters 1 and 2 describe some of the earliest progress in algebra. Mathematicians in Mesopotamia, Egypt, China, and Greece all contributed to this early period—although there was more progress in Mesopotamia, China, and Greece than in Egypt.

Chapter 3 describes the research that was conducted in present-day India, the Mideast, and North Africa. From the Indian subcontinent came important breakthroughs in algebraic techniques and a more inclusive concept of what constitutes a solution to an algebraic equation. Mathematicians in the Mideast and North Africa were the first to adopt a logically rigorous approach to algebra.

Chapter 4 describes how European mathematicians, after absorbing the work of their predecessors, began to use letters to represent numbers. They sought an abstract visual language that would generalize arithmetic and enable them to develop a theory of equations.

Chapter 5 describes how mathematicians of the European Enlightenment further developed the language of algebra. Coordinate systems were first widely adopted at this time and served as a bridge between the formerly separate branches of geometry and algebra. The result was a new field of mathematics, analytic geometry, and progress in both geometry and algebra was accelerated.

Chapters 6 and 8 describe the beginnings of what is now often called "modern" algebra. The origins of modern algebra date to the 19th century. It was during this time that mathematicians began to notice the same logical structures embedded in many different mathematical systems. Mathematical structures called groups were the first to be described and exploited, but the identification and study of other logical structures soon followed. This type of mathematical research continues to occupy the attention of many mathematicians today. Much of the material in chapter 8 is new to this edition.

Finally, chapter 7 describes Boolean algebra, which was developed in the first half of the 19th century by George Boole. Boolean algebra is important, in part, because its results are used in the design of computer chips. Perhaps more important from the point of view of a history of algebra is Boole's insistence that the proper subject matter of mathematics is the relationships that exist among abstract symbols. His heightened sense of rigor was a revelation to his contemporaries, and many, including the mathematician and philosopher Bertrand Russell, now assert that modern mathematics began with the work of George Boole.

In addition to an expanded chronology, an updated glossary, and new additions to the section on further reading, the second edition contains an interview with Dr. Bonita Saunders, a research mathematician at the National Institute of Standards and Technology (NIST), about her work in creating the Digital Library of Mathematical Functions. Her work is an elegant blend of algebra, geometry, and computer graphics with applications to the physical sciences.

Algebra has become the language of engineering, science, and mathematics. It is doubtful that many of the basic concepts upon which these subjects are founded could now be expressed without algebraic notation. Modern life depends on algebra, and developing an appreciation for the history of algebra is an important part of understanding how that dependence has evolved and why algebra is so important today.

1

THE FIRST ALGEBRAS

How far back in time does the history of algebra begin? Some scholars begin the history of algebra with the work of the Greek mathematician Diophantus of Alexandria (ca. third century C.E.). It is easy to see why Diophantus is always included. His works contain problems that most modern readers have no difficulty recognizing as algebraic.

Other scholars begin much earlier than the time of Diophantus. They believe that the history of algebra begins with the mathematical texts of the Mesopotamians. The Mesopotamians were a people who inhabited an area that is now inside the country of Iraq. Their written records begin about 5,000 years ago in the city-state of Sumer. The Mesopotamians were one of the first, perhaps *the* first, of all literate civilizations, and they remained at the forefront of the world's mathematical cultures for well over 2,000 years. Since the 19th century, when archaeologists began to unearth the remains of Mesopotamian cities in search of clues to this long-forgotten culture, hundreds of thousands of their clay tablets have been recovered. These include a number of mathematics tablets. Some tablets use mathematics to solve scientific and legal problems—for example, the timing of an eclipse or the division of an estate. Other tablets, called problem texts, are clearly designed to serve as "textbooks."

Mesopotamia: The Beginnings of Algebra

We begin our history of algebra with the Mesopotamians. Not everyone believes that the Mesopotamians knew algebra. That they were a mathematically sophisticated people is beyond doubt.

Ruins of a massive mud-brick Mesopotamian temple called a ziggurat. For 3,000 years, Mesopotamia was one of the most mathematically advanced civilizations on the planet. Remnants of Mesopotamian mathematics remain in the way we measure angles (in degrees) and time (in minutes and seconds).

They solved a wide variety of mathematical problems. The difficulty in determining whether the Mesopotamians knew any algebra arises not in what the Mesopotamians did—because their mathematics is well documented—but in how they did it. Mesopotamian mathematicians solved many important problems in ways that were quite different from the way we would solve those same problems. Many of the problems that were of interest to the Mesopotamians *we would solve with algebra.*

Although they spent thousands of years solving equations, the Mesopotamians had little interest in a general theory of equations. Moreover, there is little algebraic language in their methods of solution. Mesopotamian mathematicians seem to have learned mathematics simply by studying individual problems. They moved from one problem to the next and thereby advanced from the simple to the complex in much the same way that students today might learn to play the piano. An aspiring piano student might begin with "Old McDonald" and after much practice master the works of Frédéric Chopin. Ambitious piano students can learn the theory of music as they progress in their musical studies, but there is no necessity to do so—not if their primary interest is in the area of performance. In a similar way, Mesopotamian students began with simple arithmetic and advanced to problems that we would solve with, for example, the quadratic formula. For this reason Mesopotamian mathematics is sometimes called protoalgebra or arithmetic algebra or numerical algebra. Their work is an important first step in the development of algebra.

It is not always easy to appreciate the accomplishments of the Mesopotamians and other ancient cultures. One barrier to our appreciation emerges when we express their ideas in our notation. When we do so it can be difficult for us to see why they had to work so hard to obtain a solution. The reason for their difficulties, however, is not hard to identify. Our algebraic notation is so powerful that it makes problems that were challenging to them appear almost trivial to us. Mesopotamian problem texts, the equivalent of our school textbooks, generally consist of one or more problems that are communicated in the following way: First, the problem is stated; next, a step-by-step algorithm or method of solution is described; and, finally, the presentation concludes with the answer to the problem. The algorithm does not contain "equals signs" or other notational conveniences. Instead it consists of one terse phrase or sentence after another. The lack of symbolic notation is one important reason the problems were so difficult for them to solve.

The Mesopotamians did use a few terms in a way that would roughly correspond to our use of an abstract notation. In particular they used the words *length* and *width* as we would use the variables x and y to represent unknowns. The product of the length and width they called *area.* We would write the product of x and y as xy. Their use of the geometric words *length*, *width*, and *area*, however, does not indicate that they were interpreting their work geometrically. We can be sure of this because in some problem texts the reader is advised to perform operations that involve multiplying *length* and *width* to obtain *area* and then adding (or subtracting) a *length* or a *width* from an *area*. Geometrically, of course, this makes no sense. To see the difference between the brief, to-the-point algebraic symbolism that we use and the very wordy descriptions of algebra used by all early mathematical cultures, and the Mesopotamians in particular, consider a simple example. Suppose we wanted to add the difference $x - y$ to the product xy. We would write the simple phrase

$$xy + x - y$$

In this excerpt from an actual Mesopotamian problem text, the short phrase $xy + x - y$ is expressed this way:

> Length, width. I have multiplied length and width, thus obtaining the area. Next I added to the area the excess of the length over the width.
>
> *(Van der Waerden, B. L.* Geometry and Algebra in Ancient Civilizations. *New York: Springer-Verlag, 1983. Page 72. Used with permission)*

Despite the lack of an easy-to-use symbolism, Mesopotamian methods for solving algebraic equations were extremely advanced for their time. They set a sort of world standard for at least 2,000 years.

Translations of the Mesopotamian algorithms, or methods of solution, can be difficult for the modern reader to appreciate. Part of the difficulty is associated with their complexity. From our point of view, Mesopotamian algorithms sometimes appear unnecessarily complex given the relative simplicity of the problems that they were solving. The reason is that the algorithms contain numerous separate procedures for what the Mesopotamians perceived to be different types of problems; each type required a different method. Our understanding is different from that of the Mesopotamians: We recognize that many of the different "types" of problems perceived by the Mesopotamians can be solved with just a few different algorithms. An excellent example of this phenomenon is the problem of solving second-degree equations.

Mesopotamians and Second-Degree Equations

There is no better example of the difference between modern methods and ancient ones than the difference between our approach and their approach to solving *second-degree equations.* (These are equations involving a polynomial in which the highest exponent appearing in the equation is 2.) Nowadays we understand that all second-degree equations are of a single form:

$$ax^2 + bx + c = 0$$

where a, b, and c represent numbers and x is the unknown whose value we wish to compute. We solve all such equations with a single very powerful algorithm—a method of solution that most students learn in high school—called the quadratic formula. The quadratic formula allows us to solve these problems without giving much thought to either the size or the sign of the numbers represented by the letters a, b, and c. For a modern reader it hardly matters. The Mesopotamians, however, devoted a lot of energy to solving equations of this sort, because for them there was not one form of a second-degree equation but several. Consequently, there could not be one method of solution. Instead the Mesopotamians required several algorithms for the several different types of second-degree equations that they perceived.

The reason they had a more complicated view of these problems is that they had a narrower concept of number than we do. They did not accept negative numbers as "real," although they must have run into them at least occasionally in their computations. The price they paid for avoiding negative numbers was a more complicated approach to what we perceive as essentially a single problem. The approach they took depended on the values of a, b, and c.

Today, we use negative numbers, irrational numbers, and even imaginary numbers. We accept all such numbers as solutions to second-degree equations, but all of this is a relatively recent historical phenomenon. Because we have such a broad idea of number we are able to solve all second-degree algebraic equations with the quadratic formula, a one-size-fits-all method of solution. By contrast the Mesopotamians *perceived* that there were three basic types of second-degree equations. In our notation we would write these equations like this:

$$x^2 + bx = c$$
$$x^2 + c = bx$$
$$x^2 = bx + c$$

where, in each equation, b and c represent positive numbers. This approach avoids the "problem" of the appearance of negative numbers in the equation. The first job of any scribe or mathematician was to reduce or "simplify" the given second-degree equation to one of the three types listed. Once this was done, the appropriate algorithm could be employed for that type of equation and the solution could be found.

In addition to second-degree equations the Mesopotamians knew how to solve the much easier first-degree equations. We call these linear equations. In fact, the Mesopotamians were advanced enough that they apparently considered these equations too simple to warrant much study. We would write a first-degree equation in the form

$$ax + b = 0$$

where a and b are numbers and x is the unknown.

They also had methods for finding accurate approximations for solutions to certain third-degree and even some fourth-degree equations. (Third- and fourth-degree equations are polynomial equations in which the highest exponents that appear are 3 and 4, respectively.) They did not, however, have a general method for finding the precise solutions to third- and fourth-degree equations. Algorithms that enable one to find the exact solutions to equations of the third and fourth degrees were not developed until about 450 years ago. What the Mesopotamians discovered instead were methods for developing *approximations* to the solutions. From a practical point of view an accurate approximation is usually as good as an exact solution, but from a mathematical point of view the two are quite different. The distinctions that we make between exact and approximate solutions were not important to the Mesopotamians. They seemed satisfied as long as their approximations were accurate enough for the applications that they had in mind.

The Mesopotamians and Indeterminate Equations

In modern notation an indeterminate equation—that is, an equation with many different solutions—is usually easy to recognize. If we have one equation and more than one unknown then the equation is generally indeterminate. One of the most famous examples of an indeterminate equation from Mesopotamia can be expressed in our notation as

$$x^2 + y^2 = z^2$$

The fact that that we have three variables but only one equation is a good indicator that this equation is probably indeterminate.

Cuneiform tablet, Plimpton 322—this tablet is the best known of all Mesopotamian mathematical tablets; its meaning is still a subject of scholarly debate.

And so it is. Geometrically we can interpret this equation as the Pythagorean theorem, which states that for a right triangle the square of the length of the hypotenuse (here represented by z^2) equals the sum of the squares of the lengths of the two remaining sides. The Mesopotamians knew this theorem long before the birth of Pythagoras, however, and their problem texts are replete with exercises involving what we call the Pythagorean theorem.

The Pythagorean theorem is usually encountered in high school or junior high in a problem in which the length of two sides of a

CLAY TABLETS AND ELECTRONIC CALCULATORS

The positive square root of the positive number *a*–usually written as $\sqrt{a}$–is the positive number with the property that if we multiply it by itself we obtain *a*. Unfortunately, writing the square root of *a* as $\sqrt{a}$ does not tell us what the number is. Instead, it tells us what $\sqrt{a}$ does: If we square $\sqrt{a}$ we get *a*.

Some square roots are easy to write. In these cases the square root sign, $\sqrt{}$, is not really necessary. For example, 2 is the square root of 4, and 3 is the square root of 9. In symbols we could write $2 = \sqrt{4}$ and $3 = \sqrt{9}$ but few of us bother.

The situation is a little more complicated, however, when we want to know the square root of 2, for example. How do we find the square root of 2? It is not an especially easy problem to solve. It is, however, equivalent to finding the solution of the second-degree equation

$$x^2 - 2 = 0$$

Notice that when the number $\sqrt{2}$ is substituted for *x* in the equation we obtain a true statement. Unfortunately, this fact does not convey much information about the size of the number we write as $\sqrt{2}$.

The Mesopotamians developed an algorithm for computing square roots that yields an accurate approximation for any positive square root. (As the Mesopotamians did, we will consider only positive square roots.) For definiteness, we will apply the method to the problem of calculating $\sqrt{2}$.

The Mesopotamians used what we now call a recursion algorithm to compute square roots. A recursion algorithm consists of several steps. The output of one step becomes the input for the next step. The more often one repeats the process–that is, the more steps one takes–the right triangle are given and the student has to find the length of the third side. The Mesopotamians solved problems like this as well, but the indeterminate form of the problem—with its three unknowns rather than one—can be a little more challenging. The indeterminate version of the problem consists of identifying what we now call Pythagorean triples. These are solutions to the equation given here that involve only whole numbers.

There are infinitely many Pythagorean triples, and Mesopotamian mathematicians exercised considerable ingenuity and

closer one gets to the exact answer. To get started, we need an "input" for the first step in our algorithm. We can begin with a guess; they did. Almost any guess will do. After we input our initial guess we just repeat the process over and over again until we are as close as we want to be. In a more or less modern notation we can represent the Mesopotamian algorithm like this:

$$OUTPUT = 1/2(INPUT + 2/INPUT)$$

(If we wanted to compute $\sqrt{5}$, for example, we would only have to change 2/INPUT into 5/INPUT. Everything else stays the same.)

If, at the first step, we use 1.5 as our input, then our output is $1.41\bar{6}$ because

$$1.41\bar{6} = 1/2(1.5 + 2/1.5)$$

At the end of the second step we would have

$$1.414215\ldots = 1/2(1.41\bar{6} + 2/1.41\bar{6})$$

as our estimate for $\sqrt{2}$. We could continue to compute more steps in the algorithm, but after two steps (and with the aid of a good initial guess) our approximation agrees with the actual value of $\sqrt{2}$ up to the millionth place–an estimate that is close enough for many practical purposes.

What is especially interesting about this algorithm from a modern point of view is that it may be the one that your calculator uses to compute square roots. The difference is that instead of representing the algorithm on a clay tablet, the calculator represents the algorithm on an electronic circuit! This algorithm is as old as civilization.

mathematical sophistication in finding solutions. They then compiled these whole number solutions in tables. Some simple examples of Pythagorean triples include (3, 4, 5) and (5, 12, 13), where in our notation, taken from a preceding paragraph, $z = 5$ in the first triple and $z = 13$ in the next triple. (The numbers 3 and 4 in the first triple, for example, can be placed in either of the remaining positions in the equation and the statement remains true.)

The Mesopotamians did not indicate the method that they used to find these Pythagorean triples, so we cannot say for certain how they found them. Of course a few correct triples could be attributed to lucky guesses. We can be sure, however, that the Mesopotamians had a method worked out because their other solutions to the problem of finding Pythagorean triples include (2,700, 1,771, 3,229), (4,800, 4,601, 6,649), and (13,500, 12,709, 18,541).

The search for Pythagorean triples occupied mathematicians in different parts of the globe for millennia. A very famous generalization of the equation we use to describe Pythagorean triples was proposed by the 17th-century French mathematician Pierre de Fermat. His conjecture about the nature of these equations, called Fermat's last theorem, occupied the attention of mathematicians right up to the present time and was finally solved only recently; we will describe this generalization later in this volume. Today the mathematics for generating all Pythagorean triples is well known but not especially easy to describe. That the mathematicians in the first literate culture in world history should have solved the problem is truly remarkable.

Egyptian Algebra

Little is left of Egyptian mathematics. The primary sources are a few papyri, the most famous of which is called the Ahmes papyrus, and the first thing one notices about these texts is that the Egyptians were not as mathematically adept as their neighbors and contemporaries the Mesopotamians—at least there is no indication of a higher level of attainment in the surviving records. It would be tempting to concentrate exclusively on the Mesopotamians, the Chinese, and the Greeks as sources of early algebraic thought. We include the Egyptians because Pythagoras, who is an important figure in our story, apparently received at least some of his mathematical education in Egypt. So did Thales, another very early and very important figure in Greek mathematics. In addition, certain other peculiar characteristics of Egyptian mathematics, especially their penchant for writing all fractions as sums of what are called unit fractions, can be found in several cultures throughout the region and even as far away as China. (A *unit fraction* is a fraction with a 1 in the numerator.) None of these commonalities proves that Egypt was the original source of a lot of commonly held mathematical ideas and practices, but there are indications that this could be true. The Greeks, for example, claimed that their mathematics originated in Egypt.

Egyptian arithmetic was considerably more primitive than that of their neighbors the Mesopotamians. Even multiplication was not treated in a general way. To multiply two numbers together they used a method that consisted of repeatedly doubling one of the numbers and then adding together some of the intermediate steps. For example, to compute 5×80, first find 2×80 and then double the result to get 4×80. Finally, 1×80 would be added to 4×80 to get the answer, 5×80. This method, though it works, is awkward.

Egyptian algebra employed the symbol *heap* for the unknown. Problems were phrased in terms of "heaps" and then solved. To paraphrase a problem taken from the most famous of Egyptian mathematical texts, the Ahmes papyrus: If 1 heap and 1/7 of a heap together equal 19, what is the value of the heap? (In our notation we would write the corresponding equation as $x + x/7 = 19$.) This type of problem yields what we would call a linear equation. It is not the kind of exercise that attracted much attention from Mesopotamian mathematicians, who were concerned with more difficult problems, but the Egyptians apparently found them challenging enough to be worth studying.

What is most remarkable about Egyptian mathematics is that it seemed to be adequate for the needs of the Egyptians for thousands of years. Egyptian culture is famous for its stunning architecture and its high degree of social organization and stability. These were tremendous accomplishments, and yet the Egyptians seem to have accomplished all of this with a very simple mathematical system, a system with which they were apparently quite satisfied.

Chinese Algebra

The recorded history of Chinese mathematics begins in the Han dynasty, a period that lasted from 206 B.C.E. until 220 C.E. Records from this time are about 2,000 years younger than many Mesopotamian mathematics texts. What we find in these earliest of records of Chinese mathematics is that Chinese mathematicians had already developed an advanced mathematical culture. It would be interesting to know when the Chinese began to develop their mathematics and how their ideas changed over time, but little is known about mathematics in China before the founding of the Han dynasty. This lack of knowledge is the result of a deliberate act. The first emperor of China, Qin Shi Huang, who died in the year 210 B.C.E., ordered that all books be burned. This was done. The book burners were diligent. As a consequence, little information is available about Chinese mathematical thought before 206 B.C.E.

One of the first and certainly the most important of all early surviving Chinese mathematical texts is *Nine Chapters on the Mathematical Art*, or the *Nine Chapters* for short. (It is also known as *Arithmetic in Nine Sections.*) The mathematics in the *Nine Chapters* is already fairly sophisticated, comparable with the mathematics of Mesopotamia. The *Nine Chapters* has more than one author and is based on a work that survived, at least in part, the book burning campaign of the emperor Qin Shi Huang. Because it was extensively rewritten and enlarged, knowing what the original text was like is difficult. In any case, it is one of the earliest extant Chinese mathematical texts. It is also one of the best known. It was used as a math text for generations, and it served as an important source of inspiration for Chinese mathematicians.

In its final form the *Nine Chapters* consists of 246 problems on a wide variety of topics. There are problems in taxation,

Temple built during the Han dynasty—Chinese mathematics flourished during the Han dynasty.

surveying, engineering, and geometry and methods of solution for determinate and indeterminate equations alike. The tone of the text is much more conversational than that adopted by the Mesopotamian scribes. It is a nice example of what is now known as rhetorical algebra. (*Rhetorical algebra* is algebra that is expressed with little or no specialized algebraic notation.) Everything—the problem, the solution, and the algorithm that is used to obtain the solution—is expressed in words and numbers, not in mathematical symbols. There are no "equals" signs, no *x*'s to represent unknowns, and none of the other notational tools that we use when we study algebra. Most of us do not recognize what a great advantage algebraic notation is until after we read problems like those in the *Nine Chapters.* These problems make for fairly difficult reading for the modern reader precisely because they are expressed without the algebraic symbolism to which we have become accustomed. Even simple problems require a lot of explanatory prose when they are written without algebraic notation. The authors of the *Nine Chapters* did not shy away from using as much prose as was required.

Aside from matters of style, Mesopotamian problem texts and the *Nine Chapters* have a lot in common. There is little in the way of a general theory of mathematics in either one. Chinese and Mesopotamian authors are familiar with many algorithms that work, but they express little interest in *proving* that the algorithms work as advertised. It is not clear why this is so. Later Mesopotamian mathematicians, at least, had every opportunity to become familiar with Greek mathematics, in which the idea of proof was central. The work of their Greek contemporaries had little apparent influence on the Mesopotamians. Some historians believe that there was also some interaction between the Chinese and Greek cultures, if not directly, then at least by way of India. If this was the case, then Chinese mathematics was not overly influenced by contact with the Greeks, either. Perhaps the Chinese approach to mathematics was simply a matter of taste. Perhaps Chinese mathematicians (and their Mesopotamian counterparts) had little interest in exploring the mathematical landscape in the way that the Greeks did. Or perhaps the Greek approach was unknown to the authors of the *Nine Chapters*.

Another similarity between Mesopotamian and Chinese mathematicians lay in their use of approximations. As with the Mesopotamians, Chinese mathematicians made little distinction between exact results and good approximations. And as with their Mesopotamian counterparts, Chinese mathematicians developed a good deal of skill in obtaining accurate approximations for square roots. Even the method of conveying mathematical knowledge used by the authors of the *Nine Chapters* is similar to that of the Mesopotamian scribes in their problem texts. Like the Mesopotamian texts, the *Nine Chapters* is written as a straightforward set of problems. The problems are stated, as are the solutions, and an algorithm or "rule" is given so the reader can solve the given problem for himself or herself. The mathematics in the *Nine Chapters* is not higher mathematics in a modern sense; it is, instead, a highly developed example of "practical" mathematics.

The authors of the *Nine Chapters* solved many determinate equations (see the sidebar Rhetorical Algebra for an example). They were at home manipulating positive whole numbers, fractions, and

RHETORICAL ALGEBRA

The following problem is an example of Chinese rhetorical algebra taken from the *Nine Chapters.* This particular problem is representative of the types of problems that one finds in the *Nine Chapters;* it is also a good example of rhetorical algebra, which is algebra that is expressed without specialized algebraic notation.

In this problem the authors of the *Nine Chapters* consider three types or "classes" of corn measured out in standard units called measures. The corn in this problem, however, is not divided into measures; it is divided into "bundles." The number of measures of corn in one bundle depends on the class of corn considered. The goal of the problem is to discover how many measures of corn constitute one bundle for each class of corn. The method of solution is called the Rule. Here are the problem and its solution:

> There are three classes of corn, of which three bundles of the first class, two of the second and one of the third make 39 measures. Two of the first, three of the second and one of the third make 34 measures. And one of the first, two of the second and three of the third make 26 measures. How many measures of grain are contained in one bundle of each class?
>
> Rule. Arrange the 3, 2, and 1 bundles of the three classes and the 39 measures of their grains at the right.
>
> Arrange other conditions at the middle and at the left. With the first class in the right column multiply currently the middle column, and directly leave out.
>
> Again multiply the next, and directly leave out.
>
> Then with what remains of the second class in the middle column, directly leave out.
>
> Of the quantities that do not vanish, make the upper the *fa,* the divisor, and the lower the *shih,* the dividend, i.e., the dividend for the third class.
>
> To find the second class, with the divisor multiply the measure in the middle column and leave out of it the dividend for the third class. The remainder, being divided by the number of bundles of the second class, gives the dividend for the third class. To find the second class, with the divisor multiply the measure in the middle column and leave out of it the dividend for the third class. The remainder, being divided by the number

(continues)

RHETORICAL ALGEBRA
(continued)

of bundles of the second class, gives the dividend for the second class.

To find the first class, also with the divisor multiply the measures in the right column and leave out from it the dividends for the third and second classes. The remainder, being divided by the number of bundles of the first class, gives the dividend for the first class.

Divide the dividends of the three classes by the divisor, and we get their respective measures.

(*Mikami, Yoshio. The Development of Mathematics in China and Japan. New York: Chelsea Publishing, 1913*)

The problem, which is the type of problem often encountered in junior high or high school algebra classes, is fairly difficult to read, but only because the problem–and especially the solution–are expressed rhetorically. In modern algebraic notation we would express the problem with three variables. Let x represent a bundle for the first class of corn, y represent a bundle for the second class of corn, and z represent a bundle for the third class of corn. In our notation the problem would be expressed like this:

$$3x + 2y + z = 39$$
$$2x + 3y + z = 34$$
$$x + 2y + 3z = 26$$

The answer is correctly given as 9 1/4 measures of corn in the first bundle, 4 1/4 measures of corn in the second bundle, and 2 3/4 measures of corn in the third bundle.

Today this is not a particularly difficult problem to solve, but at the time that the *Nine Chapters* was written this problem was for experts only. The absence of adequate symbolism was a substantial barrier to mathematical progress.

even negative numbers. Unlike the Mesopotamians, the Chinese accepted the existence of negative numbers and were willing to work with negative numbers to obtain solutions to the problems that interested them. In fact, the *Nine Chapters* even gives rules for dealing with negative numbers. This is important because negative numbers can arise during the process of solving many different algebraic problems even when the final answers are positive. When one refuses to deal with negative numbers, one's work becomes much harder. In this sense the Chinese methods for solving algebraic equations were more adaptable and "modern" than were the methods used by the Mesopotamians, who strove to avoid negative numbers.

In addition to their work on determinate equations, Chinese mathematicians had a deep and abiding interest in indeterminate equations, equations for which there are more unknowns than there are equations. As were the Mesopotamians, Chinese mathematicians were also familiar with the theorem of Pythagoras and used the equation (which we might write as $x^2 + y^2 = z^2$) to pose indeterminate as well as determinate problems. They enjoyed finding Pythagorean triples just as the Mesopotamians did, and they compiled their results just as the Mesopotamians did.

The algebras that developed in the widely separated societies described in this chapter are remarkably similar. Many of the problems that were studied are similar. The approach to problem solving—the emphasis on algorithms rather than a theory of equations—was a characteristic that all of these cultures shared. Finally, not one of the cultures developed a specialized set of algebraic symbols to express their ideas. All these algebras were rhetorical. There was one exception, however. That was the algebra that was developed in ancient Greece.

2

GREEK ALGEBRA

Greek mathematics is fundamentally different from the mathematics of Mesopotamia and China. The unique nature of Greek mathematics seems to have been present right from the outset in the work of Thales of Miletus (ca. 625 B.C.E.–ca. 546 B.C.E.) and Pythagoras of Samos (ca. 582 B.C.E.–ca. 500 B.C.E.). In the beginning, however, the Greeks were not solving problems that were any harder than those of the Mesopotamians or the Chinese. In fact, the Greeks were not interested in problem solving at all—at least not in the sense that the Mesopotamian and Chinese mathematicians were. Greek mathematicians for the most part did not solve problems in taxation, surveying, or the division of food. They were interested, instead, in questions about the nature of number and form.

It could be argued that Chinese and Mesopotamian mathematicians were not really interested in these applications, either—that they simply used practical problems to express their mathematical insights. Perhaps they simply preferred to express their mathematical ideas in practical terms. Perhaps, as it was for their Greek counterparts, it was the mathematics and not the applications that provided them with their motivation.

There is, however, no doubt about how the Greeks felt about utilitarian mathematics. The Greeks did not—would not—express their mathematical ideas through problems involving measures of corn or the division of estates or any other practical language. They must have known, just as the Mesopotamian and Chinese mathematicians knew, that all of these fields are rich sources of mathematical problems. To the Greeks this did not matter. The

The ruins of the Lyceum, the school where Aristotle taught. Mathematics was an integral part of Greek philosophical thought. (History of Macedonia Blog)

Greeks were interested in mathematics for the sake of mathematics. They expressed their ideas in terms of the properties of numbers, points, curves, planes, and geometric solids. Most of them had no interest in applications of their subject, and in case anyone missed the point they were fond of reciting the story about the mathematician Euclid of Alexandria, who, when a student inquired about the utility of mathematics, instructed his servant to give the student a few coins so that he could profit from his studies.

Another important difference between Greek mathematicians and the mathematicians of other ancient cultures was the distinction that the Greeks made between exact and approximate results. This distinction is largely absent from other mathematical cultures of the time. In a practical sense, exact results are generally no more useful than good approximations. Practical problems involve measurements, and measurements generally involve some uncertainty. For example, when we measure the length of a line segment our measurement removes some of our uncertainty about the "true" length of the segment, but some uncertainty remains. This uncertainty is our margin of error. Although we can further reduce our uncertainty with better measurements, we cannot eliminate all uncertainty. As a consequence, any computations that depend on this measurement must also reflect our initial imprecision about the length of the segment. Our methods may be exact in the sense that if we had exact data then our solution would be exact as well. Unfortunately, exact measurements are generally not available.

The Greek interest in precision influenced not only the way they investigated mathematics; it also influenced what they investigated. It was their interest in exact solutions that led to one of the most profound discoveries in ancient mathematics.

The Discovery of the Pythagoreans

Pythagoras of Samos was one of the first Greek mathematicians. He was extremely influential, although, as we will soon see, we cannot attribute any particular discoveries to him. As a young man Pythagoras is said to have traveled widely. He apparently received his mathematics education in Egypt and Mesopotamia. He may have traveled as far east as India. Eventually he settled on the southeastern coast of what is now Italy in the Greek city of Cortona. (Although we tend to think of Greek civilization as situated within the boundaries of present-day Greece, there was a time when Greek cities were scattered throughout a much larger area along the Mediterranean Sea.)

Pythagoras was a mystic as well as a philosopher and mathematician. Many people were attracted to him personally as well as to his ideas. He founded a community in Cortona where he and his many disciples lived communally. They shared property, ideas, and credit for those ideas. No Pythagorean took individual credit for a discovery, and as a consequence we cannot be sure which of the discoveries attributed to Pythagoras were his and which were his disciples'. For that reason we discuss the contributions of the Pythagoreans rather than the contributions of Pythagoras himself. There is, however, one point about Pythagoras about which we can be sure: Pythagoras did not discover the Pythagorean theorem. The theorem was known to Mesopotamian mathematicians more than 30 generations before Pythagoras's birth.

At the heart of Pythagorean philosophy was the maxim "All is number." There is no better example of this than their ideas about music. They noticed that the musical tones produced by a string could be described by whole number ratios. They investigated music with an instrument called a monochord, a device consisting of one string stretched between two supports. (The supports may have been attached to a hollow box to produce a richer, more harmonious sound.) The Pythagorean monochord had a third support that was slid back and forth under the string. It could be placed anywhere between the two end supports.

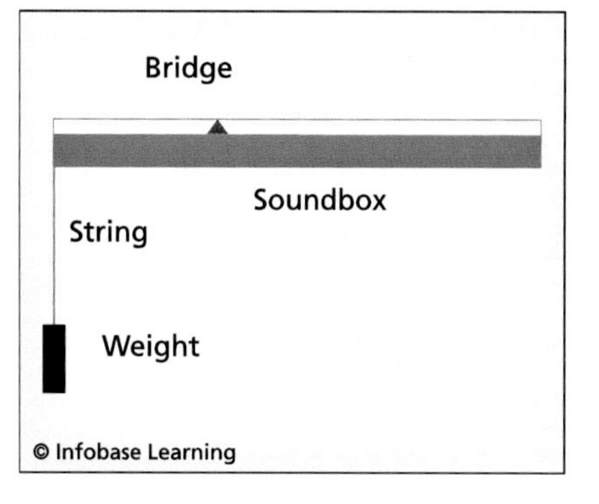

A monochord, a device used by the Pythagoreans to investigate the relationships that exist between musical pitches and mathematical ratios

The Pythagoreans discovered that when the third support divided the length of the string into certain whole number ratios, the sounds produced by the two string segments were harmonious or consonant. This observation indicated to them that music could be described in terms of certain numerical ratios. They identified these ratios and listed them. The ratios of the lengths of the two string segments that they identified as consonant were 1:1, 1:2, 2:3, and 3:4. The ratio 1:1, of course, is the unison: Both string segments are vibrating at the same pitch. The ratio 1:2 is what musicians now call an octave. The ratio 2:3 is the perfect fifth, and the ratio 3:4 is the perfect fourth. The identification of these whole number ratios was profoundly important to the Pythagoreans. They believed that everything—all human creations, the natural world, and mathematics—could be expressed via whole number ratios.

The Pythagoreans worshipped numbers. It was part of their beliefs that certain numbers were invested with special properties. The number 4, for example, was the number of justice and retribution. The number 1 was the number of reason. When they referred to "numbers," however, they meant *only* what we would call positive, whole numbers, that is, the numbers belonging to the sequence 1, 2, 3, . . . (Notice that the consonant tones of the monochord were produced by dividing the string into simple *whole number ratios.*) They did not recognize negative numbers, the number 0, or any type of fraction as a number. Quantities that we might describe with a fraction they would describe as a ratio between two whole numbers, and although we might not make a distinction between a ratio and a fraction, we need to recognize that they did. They only recognized ratios.

To the Pythagoreans the number 1 was the generator of all numbers—by adding 1 to itself often enough they could obtain every number (or at least every number as they understood the concept). What we would use fractions to represent, they described as ratios of sums of the number 1. A consequence of this concept of number—coupled with their mystical belief that "all is number"—is that everything in the universe can be generated from the number 1. *Everything*, in the Pythagorean view, was in the end a matter of whole number arithmetic. This idea, however, was incorrect, and their discovery that their idea of number was seriously flawed is one of the most important and far-reaching discoveries in the history of mathematics.

To understand the flaw in the Pythagorean idea of number we turn to the idea of commensurability. We say that two line segments—we call them L_1 and L_2—are commensurable, if there is a third line segment—we call it L_3—with the property that the lengths L_1 and L_2 are whole number multiples of length L_3. In this sense L_3 is a "common measure" of L_1 and L_2. For example, if segment L_1 is 2 units long and L_2 is 3 units long then we can take L_3 to be 1 unit long, and we can use L_3 to measure (evenly) the lengths of both L_1 and L_2. The idea of commensurability agrees with our intuition. It agrees with our experience. Given two line segments we can always measure them and then find a line segment whose length evenly divides both. This idea is at the heart of the Pythagorean concept of number, and that is why it came as such a shock to discover that there existed pairs of line segments that were incommensurable, that is, that there exist pairs of segments that *share no common measure!*

The discovery of incommensurability was a fatal blow to the Pythagorean idea of number; that is why they are said to have tried to hide the discovery. Happily, knowledge of this remarkable fact spread rapidly. Aristotle (384–332 B.C.E.) wrote about the concept and described what is now a standard proof. Aristotle's teacher, Plato (ca. 428–347 B.C.E.), described himself as having lived as an animal lives—that is, he lived without reasoning—until he learned of the concept.

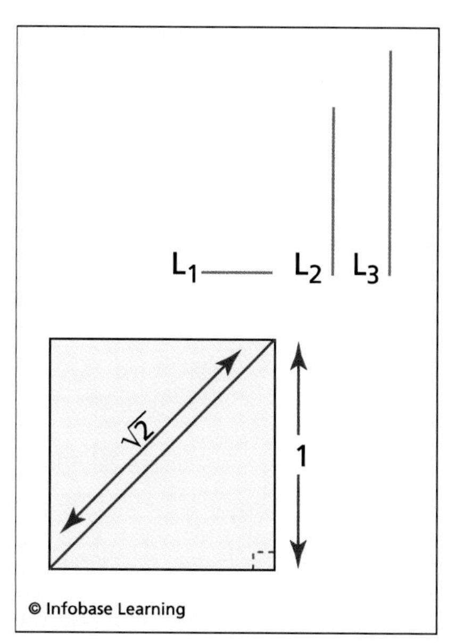

Two lengths are commensurable if they are whole number multiples of a third length. For example, segments L_2 *and* L_3 *are commensurable because* $L_2 = 2L_1$ *and* $L_3 = 3L_1$*. Segment* L_1 *is called a common measure of* L_2 *and* L_3*. Not every pair of lengths is commensurable. The side of a square and its diagonal, for example, share no common measure; these segments are called incommensurable.*

It is significant that the Greeks so readily accepted the proof of the concept of incommensurability because that acceptance shows just how early truly abstract reasoning began to dominate Greek mathematical thinking. They were willing to accept a mathematical result that violated their worldview, their everyday experience, and their sense of aesthetics. They were willing to accept the idea of incommensurability because it was a logical consequence of other, previously established, mathematical results. The Greeks often expressed their understanding of the concept by saying that the length of a diagonal of a square is incommensurable with the length of one of its sides.

Incommensurability is a perfect example of the kind of result that distinguished Greek mathematical thought from the mathematical thought of all other ancient cultures. In a practical sense incommensurability is a "useless" concept. We can always find a line segment whose length is so close to the length of the diagonal of the square as to be indistinguishable from the diagonal, and we can always

THE INCOMMENSURABILITY OF √2

The proof that the length of a diagonal of a square whose sides are 1 unit long is incommensurable with the length of a side of the square is one of the most famous proofs in the history of mathematics. The proof itself is only a few lines long. (Note that a square whose side is 1 unit long has a diagonal that is $\sqrt{2}$ units long. This is just a consequence of the Pythagorean theorem.) In modern notation the proof consists of demonstrating that there do not exist natural numbers a and b such that $\sqrt{2}$ equals a/b. The following *nonexistence proof* requires the reader to know the following two facts:

1. If a^2 is divisible by 2 then $a^2/2$ is even.
2. If b^2 (or a^2) is divisible by 2 then b (or a) is even.

We begin by assuming the opposite of what we intend to prove: We suppose that $\sqrt{2}$ *is* commensurable with 1–that is, we suppose that $\sqrt{2}$ can be written as a fraction a/b where a and b are positive whole numbers. We also assume–and this is critical–that the fraction a/b is expressed in *lowest terms.* In particular, this means that a and b cannot both be even numbers. It is okay if one is even. It is okay if neither is even, but both cannot be even or our fraction would not be in lowest terms. (Notice that if we could find integers such that $\sqrt{2} = a/b$, and if the fraction were not in lowest terms we could certainly reduce it to lowest terms. There is, therefore, no harm in assuming that it is in lowest terms from the outset.) Here is the proof:

choose this segment with the additional property that its length and the length of a side of the square share a common measure. In a practical sense, commensurable lengths are always sufficient.

In a theoretical sense, however, the discovery of incommensurability was an important insight into mathematics. It showed that the Pythagorean idea that everything could be expressed in terms of whole number ratios was flawed. It showed that the mathematical landscape is more complex than they originally perceived it to be. It demonstrated the importance of rigor (as opposed to intuition) in the search for mathematical truths. Greek mathematicians soon moved away from Pythagorean concepts and toward a geometric view of mathematics and the world around them. How much of this was due to the discoveries of the Pythagoreans and how much was due to the success of later generations of geometers is not clear. In

Suppose $a/b = \sqrt{2}$.

Now solve for b to get
$a/\sqrt{2} = b$

Finally, square both sides.
$a^2/2 = b^2$

This completes the proof. Now we have to read off what the last equation tells us. First, a^2 is evenly divisible by 2. (The quotient is b^2.) Therefore, by fact 2, a is even. Second, since $a^2/2$ is even (this follows by fact 1) b^2–which *is* $a^2/2$–is also even. Fact 2 enables us to conclude that b is even as well. Since both a and b are even our assumption that a/b is in lowest terms cannot be true. This is the contradiction that we wanted. We have proved that a and b do not exist.

This proof resonated through mathematics for more than 2,000 years. It showed that intuition is not always a good guide to truth in mathematics. It showed that the number system is considerably more complicated than it first appeared. Finally, and perhaps unfortunately, mathematicians learned from this proof to describe $\sqrt{2}$ and other similar numbers in terms of what they are *not:* $\sqrt{2}$ is not expressible as a fraction with whole numbers in the numerator and denominator. Numbers like $\sqrt{2}$ came to be called irrational numbers. A definition of irrational numbers in terms of what they *are* would have to wait until the late 19th century and the work of the German mathematician Richard Dedekind.

any case Greek mathematics does not turn back toward the study of algebra as a separate field of study for about 700 years.

Geometric Algebra

The attempt by the Pythagoreans to reduce mathematics to the study of whole number ratios was not successful, and Greek mathematics soon shifted away from the study of number and ratio and toward the study of geometry. The Greeks did not study geometry only as a branch of knowledge; they used it as a tool to study everything from astronomy to the law of the lever. Geometry became the language that the Greeks used to describe and understand the world about them. It should come as no surprise, then, that the Greeks also learned to use the language of geometry to express

Greek ruins in Alexandria, Egypt. Most of Greek mathematics was created outside of present-day Greece. Alexandria occupies an especially important place in the history of Greek mathematics.

ideas that we learn to express algebraically. We call this geometric algebra, and it is an important part of the mathematical legacy of the ancient Greeks. Today the principal source of Greek ideas about geometric algebra is the set of books entitled *Elements* by Euclid of Alexandria, who lived in Alexandria, Egypt, in the third century B.C.E.

Little is known about Euclid. We do not know when he was born or when he died. We know that the institution where Euclid worked—it was called the Museum—was home to many of the most successful Greek mathematicians of the time. We know that many of the mathematicians who lived and worked at the school were born elsewhere. Perhaps the same can be said of Euclid.

Euclid is best remembered for having written one of the most popular textbooks of all time. Called *Elements*, it has been translated into most of the world's major languages over the last 2,000 years. In recent years it has fallen out of favor as a textbook, but

ALGEBRA MADE VISIBLE

Today one of the first ideas that students learn as they begin to study algebra is that "multiplication distributes over addition." This is called the distributive law and in symbols it looks like this:

$$x(y + z) = xy + xz$$

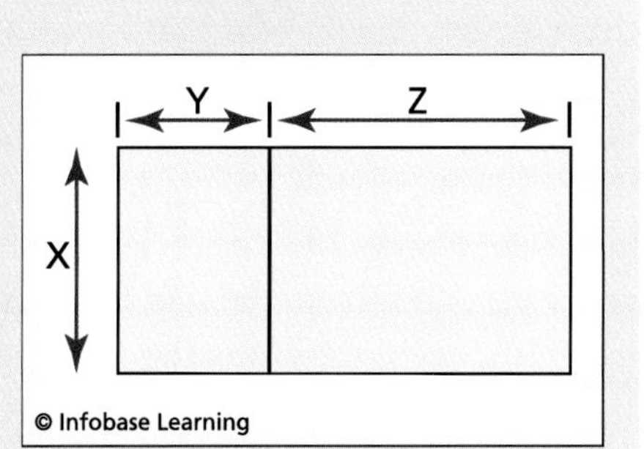

Diagram of Euclid's proof that xy + xz = x(y + z)

Though most of us eventually succeed in learning this rule, few of us could give a reason why it might be true. The very first proposition that Euclid proves in book II of the *Elements* is exactly this statement, but it is expressed in the language of geometrical algebra. More than 2,000 years ago Euclid expresses the distributive law in the following words:

> If there be two straight lines, and one of them be cut into any number of segments whatever, the rectangle contained by the two straight lines is equal to the rectangles contained by the uncut straight line and each of the segments.

(Euclid. Elements. Translated by Sir Thomas L. Heath. Great Books of the Western World, vol. 11. Chicago: Encyclopaedia Britannica, 1952.)

See the pictorial version of Euclid's statement. Notice that the illustration shows three rectangles, two smaller ones and a large one. (The large rectangle is made of the four outside line segments. The smaller rectangles lie inside the large one.) All three rectangles have the same height. We use x to represent the height of each of the rectangles. The rectangle on the left has length y and the rectangle on the right has length z. The length of the largest rectangle is $y + z$. Now we recall the formula for the area of a rectangle: Area = length × width. Finally, we can express the idea that the area of the largest rectangle equals the area of the two smaller rectangles by using the algebraic equation given. When the distributive law is expressed geometrically the reason that it is true is obvious.

many high school treatments of plane geometry are still only simplified versions of parts of Euclid's famous work. To describe the *Elements* solely as a textbook, however, is to misrepresent its impact. The type of geometry described in Euclid's textbook—now called Euclidean geometry, though it was not Euclid's invention—dominated mathematical thought for 2,000 years. We now know that there are other kinds of geometry, but as late as 200 years ago many mathematicians and philosophers insisted that Euclidean geometry was the single true geometry of the universe. It was not until the 19th century that mathematicians began to realize that Euclidean geometry was simply one kind of geometry and that other, equally valid geometries exist.

The *Elements* was written in 13 brief books. Of special interest to us is the very brief book II, which lays out the foundations of geometric algebra. In book II we see how thoroughly geometric thinking pervaded all of Greek mathematics including algebra. For example, when we speak of unknowns, x, y, and z, we generally assume that these variables represent numbers. Part of learning elementary algebra involves learning the rules that enable us to manipulate these symbols as if they were numbers. Euclid's approach is quite different. In Euclid's time "variables" were not numbers. Euclid represented unknowns by line segments, and in his second book he establishes the rules that allow one to manipulate segments in the way that we would manipulate numbers. What we represent with equations, Euclid represented with pictures of triangles, rectangles, and other forms. Geometric algebra is algebra made visible.

Much of the geometry that one finds in the *Elements* is performed with a straightedge and compass. This is constructible mathematics in the sense that the truth of various mathematical statements can be demonstrated through the use of these implements. Though it would be hard to imagine simpler implements, the Greeks used these devices successfully to investigate many important mathematical ideas. But as with any set of techniques, the use of the straightedge and compass has its limitations. Although it is not immediately apparent, certain classes of problems cannot be solved by using straightedge and compass techniques. In fact, some of the most famous mathematical problems from antiquity are famous precisely because they *cannot* be solved with a straightedge and compass.

There are three classical geometry problems that are very important in the history of algebra. Their importance in geometry is that they remained unsolved for more than 2,000 years. They were not unsolved because they were neglected. These problems attracted some of the best mathematical minds for generation after generation. Interesting mathematical ideas and techniques were discovered as individuals grappled with these problems and searched for solutions, but in the end none of these mathematicians could solve any of the three problems as originally stated, nor could they show that solutions did not exist. The problems are as follows:

Problem 1: Given an arbitrary angle, divide the angle into three equal parts, *using only a straightedge and compass.*

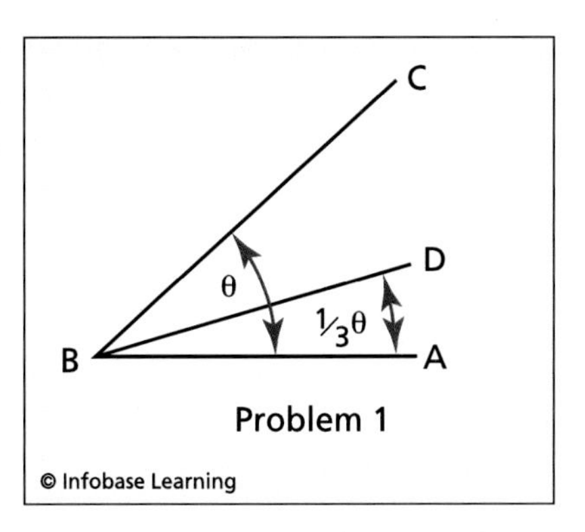

Trisecting the angle: Given angle ABC, *use a straightedge and compass to construct angle* ABD *so that the measure of angle* ABD *is one-third that of the measure of angle* ABC.

Problem 2: Given a circle, construct a square having the same area as the circle, *using only a straightedge and compass.*

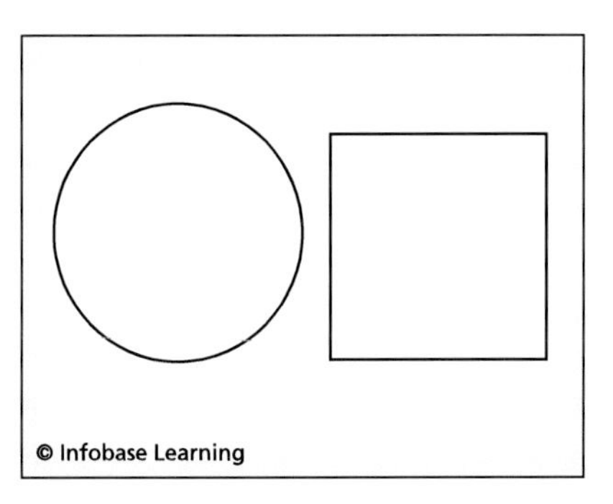

Squaring the circle: Given a circle and using only a straightedge and compass, construct a square of equal area.

Problem 3: Given a cube, find the length of the side of a new cube whose volume is twice that of the original cube. Do this *using only a straightedge and compass.*

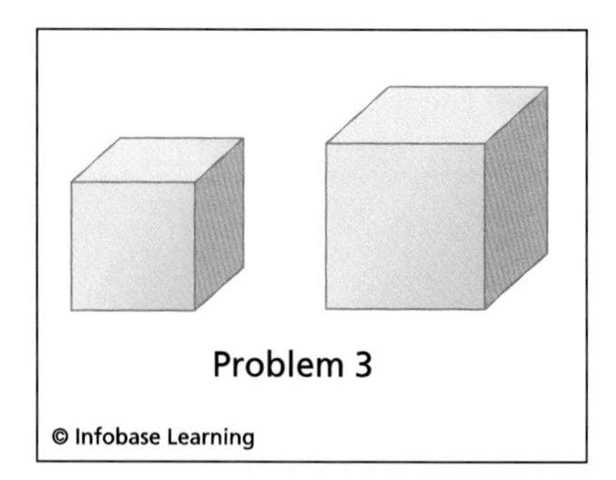

Doubling the cube: Given a cube and using only a straightedge and compass, construct a second cube that has precisely twice the volume of the original cube.

Notice that each problem has the same restriction: using only a straightedge and compass. This is critical. It is also critical to remember that the Greeks were not interested in approximate solutions to these problems. The Greeks could have easily constructed highly accurate *approximations* to a third of an angle, a squared circle, and a cube with a volume approximately twice as large as the given cube—and all with only a straightedge and compass. But approximations were not their goal. These ancient Greek geometers were searching for a method that would in theory give them the exact solution—not a good approximation to the solution—to each of the three problems.

These three problems are probably more important in the history of algebra than in the history of geometry. In algebra the search for the solutions of these problems gave birth to a new concept of what algebra is. In the 19th century, after some extraordinary breakthroughs in algebraic thought, these problems were disposed of once and for all. Nineteenth-century mathematicians discovered that the reason these problems had remained unsolved for 2,000 years is that they are unsolvable. Remember: This was proved by using algebra, not geometry. The ideas required to prove that these problems are unsolvable represented a huge step forward in the history of algebra.

The geometric algebra described by Euclid was logically rigorous, but it was too simple to be very useful. Elementary results had been obtained by sophisticated techniques. The reliance on formal, very sophisticated geometric reasoning made it difficult to extend the ideas described by Euclid. A new approach to algebra was needed.

Diophantus of Alexandria

Diophantus is often described as the father of algebra. He was, perhaps, the only one of the great Greek mathematicians to devote himself fully to the study of algebra as a discipline separate from geometry. We know little of his life. The dates of his birth and death are unknown. We do know that he lived in Alexandria, Egypt. It is generally believed that he was alive during the third century C.E., but even this is not certain; some scholars believe that he was alive during the second century C.E., and some believe that he was alive during the fourth century C.E. What are thought to be the facts of his life are usually summed up in this ancient mathematics problem:

> God granted him to be a boy for the sixth part of his life, and adding a twelfth part to this, He clothed his cheeks with down; He lit him the light of wedlock after a seventh part, and five years after his marriage He granted him a son. Alas! Late-born child; after attaining the measure of half his father's life, chill Fate took him. After consoling his grief by this science of numbers for four years he ended his life.
>
> *(Reprinted by permission of the publishers and the Trustees of Loeb Classical Library from* Greek Anthology: *Volume V, Loeb Classical Library. Volume L 86, translated by W. R. Paton, Cambridge, Mass.: Harvard University Press, 1918)*

By solving the (linear) equation that is described in the problem, we learn that Diophantus lived to be 84 years old.

Diophantus's contribution to algebra consists of two works, the more famous of which is entitled *Arithmetica*. The other is *On Polygonal Numbers*. Neither work exists in its entirety. *Arithmetica* originally consisted of 13 volumes. Six volumes were preserved in the original Greek, and in the 1970s previously unknown Arabic translations of four more volumes were discovered. Even less of

On Polygonal Numbers has come down to us; it is known through a set of excerpts.

Arithmetica is arranged much as the *Nine Chapters* and the Mesopotamian problem texts are. It is essentially a long list of problems. The exception occurs at the beginning of the first volume, in which he attempts to give an account of the foundations of algebra. This is historic because it is the first time that anyone tried to do this.

With respect to the number system that he uses, he describes rational numbers—numbers that can be represented as fractions with whole numbers in the numerator and the denominator—and negative numbers. He gives rules for working with negative numbers, and he seems comfortable enough with this system. But when solving problems, he clearly prefers solutions that are nonnegative.

Unlike the *Nine Chapters* and the other books mentioned previously, *Arithmetica* is largely devoid of nonmathematical references. There are no references to the division of corn, the height of a tree, or the area of a field, and approximate solutions, no matter how accurate, are not acceptable to Diophantus. In this sense *Arithmetica* is more philosophical than practical. Although Diophantus certainly knows about incommensurable (irrational) numbers, he does not consider them to be acceptable solutions to any of his equations.

Another important contribution that Diophantus makes to algebra is his use of symbolism. All of the works that we have examined so far, whether written in Mesopotamia, Egypt, or China, were of a rhetorical character—that is, everything is expressed in words. This format tends to hinder progress in mathematics because it obscures the ideas and techniques involved. Diophantus introduced abbreviations and some symbols into his work. We call this mixture of abbreviations, words, and a few symbols syncopated algebra. Diophantus's syncopated algebra lacks the compact form of contemporary algebraic equations. It is not especially easy to read, but he went further toward developing a specialized system of symbols than any of his predecessors.

The problems that Diophantus studied often had multiple solutions. The existence of multiple solutions for a single problem would immediately catch the eye of any contemporary mathematician, but Diophantus usually seems not to care. If he can find even one solution he seems content. Did he know that in some cases other solutions exist? It is not always clear. On the other hand, Diophantus is very interested in *how* a solution is found, and he sometimes describes more than one method for solving the same problem. It is clear that algorithms are a primary focus for him.

It is tempting to see in Diophantus's exhausting list of problems and solutions the search for a rigorous theory of algebraic equations that is analogous to the highly developed system of geometry that the Greeks had developed centuries earlier. If that was his goal, he did not achieve it. There is no overarching concept to Diophantus's algebra. It is, instead, a collection of adroitly chosen problems, whose solutions more often than not depend on a clever trick rather than a deeper theoretical understanding. Nevertheless, *Arithmetica* served as a source of insight and inspiration for generations of Islamic and European mathematicians. And about 1,500 years after Diophantus wrote *Arithmetica*, his work inspired the French mathematician Pierre de Fermat to attempt to generalize one of the problems that he found in *Arithmetica* about representing one square as the sum of two squares. This gave rise to what is now called Fermat's last theorem, one of the most famous of all mathematical problems and one that was not solved until late in the 20th century.

Greek algebra—whether it is like that found in *Elements* or in *Arithmetica*—is characterized by a higher level of abstraction than that found in other ancient mathematically sophisticated cultures. Both the choice of problems and method of presentation were unique among the cultures of antiquity, and the Greek influence on future generations of Arab and European mathematicians was profound. New approaches to algebra that were eventually developed elsewhere, however, would prove to be equally important.

3

ALGEBRA FROM INDIA TO NORTHERN AFRICA

The tradition of Greek mathematical research ended in the third century C.E. with the death of Hypatia (ca. 370–415) in Alexandria. Hypatia was a prominent scholar and mathematician. She wrote commentaries on the works of Diophantus, Apollonius, and Ptolemy, but all of her work has been lost. We know of her through the works and letters of other scholars of the time. Hypatia was murdered in a religious dispute. Shortly thereafter many of the scholars in Alexandria left, and mathematical research at Alexandria, the last of the great Greek centers of learning, ended.

Mathematics, however, continued to develop in new ways and in new locations. In the Western Hemisphere the Mayan civilization was developing a unique and advanced form of mathematics. We know of some of their accomplishments, but most of their work was destroyed by Spanish conquerors in the 16th century. Another new and important center of mathematical research developed on the Indian subcontinent, but before examining the accomplishments of these mathematicians it is important to say a few words about terminology.

The mathematical tradition that developed on the Indian subcontinent during this time is sometimes called Indian mathematics. It was not created entirely in what is now India. Some of it arose in what is now Pakistan, and, in any case, India was not united under a central government during the period of interest to us. There was no India in the modern sense. There are some histories of "Indian" mathematics that use the term *Hindu mathematics*, but not all of the mathematicians who contributed to the development of this mathematical tradition were themselves Hindu. There are no other terms in general use. We use the terms *Indian mathematics* and *Hindu mathematics* interchangeably because they are the two common names for this mathematical tradition, but neither term is entirely satisfactory. We look forward to the time when better, more descriptive terminology is developed to describe the accomplishments of this creative and heterogeneous people.

There are widely varying claims made about the history of Indian mathematics. Some scholars think that a sophisticated Hindu mathematical tradition goes back several thousand years, but the evidence for this claim is indirect. Very few records from the more remote periods of Indian history have survived. Some of the earliest records of Indian mathematical accomplishments are the *Sulvasutras*, a collection of results in geometry and geometric algebra. The dating of these works is also a matter of dispute. Some scholars believe that they date to the time of Pythagoras, but others claim they were written several centuries after Pythagoras's death. Mathematically the *Sulvasutras* are, in any case, not especially sophisticated when compared with the Hindu works that are of most interest to us. In fact, it is their simplicity that is the best indicator that they preceded the works for which we do have reliable dates.

Despite their simplicity, the *Sulvasutras* contain many qualities that are characteristic of much of the Indian mathematical tradition. It is important to review these special characteristics, because Indian mathematics is quite distinct from that of the other mathematically sophisticated cultures that preceded it. Moreover, even when there is overlap between the mathematics of India and that of ancient Greece or Mesopotamia, it is clear that Indian mathematicians perceived mathematics differently. The mathematics of the Indians is often compared unfavorably to Greek mathematics, but such comparisons are not especially helpful. Hindu mathematics is better appreciated on its own terms. Mathematics occupied a different place in the culture of the Hindus than it did in the culture of the Greeks.

One characteristic of Hindu mathematics is that almost all of it—problems, rules, and definitions—is written in verse. This is true of the *Sulvasutras* and virtually all later works as well. Another characteristic property that we find in the *Sulvasutras* as well as later Hindu mathematics is that there are no proofs that the rules that one finds in the texts are correct. Ancient Indian texts contain almost no mathematical rigor, as we understand the term today. The rules that one finds in these texts were sometimes illustrated with one or more examples. The examples were sometimes followed with challenges directed to the reader, but there was little in the way of motivation or justification for the rules themselves. This was not simply a matter of presentation. The mathematicians who created this highly imaginative approach to mathematics must have had only a minimal interest in proving that the results they obtained were correct, because mistakes in the texts themselves often went unnoticed. Many of the best Hindu works contain a number of significant errors, but these works also contain important discoveries, some of which have had a profound effect on the entire history of mathematics.

Another important difference between Indian mathematics and the mathematics of other cultures with advanced mathematical traditions is that other cultures perceived mathematics as a separate field of study. In the Indian cultural tradition, mathematics was not usually treated as an independent branch of knowledge. There are very few ancient Sanskrit texts devoted solely to mathematics. Instead mathematical knowledge was usually conveyed in isolated chapters in larger works about astronomy. Astronomy and religion were very much intertwined in the classical culture of the Indian subcontinent. To many of the most important Hindu mathematicians, mathematics was a tool for better understanding the motions and relative locations of objects in the night sky. It was not a separate academic discipline.

Brahmagupta and the New Algebra

The astronomer and mathematician Brahmagupta (ca. 598–ca. 670) was one of the most important of all Indian mathematicians. Not

Sundial at the ancient observatory at Ujjain. For centuries, Ujjain was an important center for astronomical research. The unusual structures in the background were created to facilitate the making of accurate astronomical measurements.

much is known about his life. It is known that he lived in Ujjain, a town located in what is now central India. In Brahmagupta's time Ujjain was home to an important astronomical observatory, and Brahmagupta was head of the observatory. Brahmagupta's major work is a book on astronomy, *Brahma-sphuta-siddhānta* (The opening of the universe). Written entirely in verse, Brahmagupta's masterpiece is 25 chapters long. Most of the book contains information about astronomical phenomena: the prediction of eclipses, the determination of the positions of the planets, the phases of the Moon, and so on. Just two of the chapters are about mathematics, but those two chapters contain a great deal of important algebra.

Brahmagupta's work, like that of other Hindu mathematicians, contains plenty of rules. Most are stated without proof; nor does he provide information about how he arrived at these rules or why he believes them to be true. Many rules are, however, followed by problems to illustrate how the rules can be applied. Here, for example, is Brahmagupta's "rule of inverse operation":

> Multiplier must be made divisor; and divisor, multiplier; positive, negative and negative, positive; root [is to be put] for square; and square, for root; and first as converse for last.
>
> (*Brahmagupta and Bhaskara.* Algebra with Arithmetic and Mensuration. *Translated by Henry Colebrook. London: John Murray, 1819)*

By modern standards this is a fairly terse explanation, but by the standards of the day it was comparatively easy reading. To understand why, it helps to know that Brahmagupta, like many Indian mathematicians, probably grew up reading just this type of explanation. Indian astronomical and mathematical knowledge was generally passed from one generation to the next within the same family. Each generation studied astronomy, mathematics, and astrology and contributed to the family library. Brahmagupta's father, for example, was a well-known astrologer. Mathematical writing and astronomical writing were important parts of Brahmagupta's family tradition. He would have been accustomed to this kind of verse, but he advanced well beyond what he inherited from his forebears.

One of the most important characteristics of Brahmagupta's work is his style of algebraic notation. It is, like that of Diophantus, syncopated algebra. Syncopated algebra uses specialized symbols and abbreviations of words to convey the ideas involved. For instance, Brahmagupta used a dot above a number to indicate a negative number. When formulating an equation containing one or more unknowns, Brahmagupta called each unknown a different color. His use of colors is completely analogous to the way that we are taught to use the letters x, y, and z to represent variables when we first learn algebra. To simplify his notation he preferred to use an abbreviated form of each color word. One section of his book is even called Equations of Several Colors.

One consequence of his notation is that his mathematical prose is fairly abstract, and this characteristic is important for two reasons. First, a condensed, abstract algebraic notation often makes mathematical ideas more transparent and easy to express. Second, good algebraic notation makes adopting a very general and inclusive approach to problem solving easier, and generality is just what Brahmagupta achieved.

Brahmagupta considered the equation that we would write as $ax + by = c$, where a, b, and c are integers (whole numbers), called coefficients, that could be positive, negative, or zero. The letters x and y denote the *variables* that are meant to represent whole number solutions to the equation. Brahmagupta's goal was to locate whole numbers that, when substituted for x and y, made the equation a true statement about numbers.

Brahmagupta's very broad understanding of what a, b, and c represent stands in sharp contrast with the work of Diophantus. Diophantus preferred to consider only equations in which the coefficients are positive. This required Diophantus to break Brahmagupta's single equation into several special cases. If, for example, b was less than 0 in the preceding equation, Diophantus would add $-b$ to both sides of the equation to obtain $ax = -by + c$. (If b is negative, $-b$ is positive.) This equation, with the b transposed to the other side, was a distinct case to Diophantus, but Brahmagupta, because he did not distinguish between positive and negative coefficients, had to consider only the single equation $ax + by = c$. This allowed him to achieve a more general, more modern, and more powerful approach to the solution of algebraic equations. Furthermore, he accepted negative numbers as solutions, a concept with which his Greek predecessors had difficulty.

This highly abstract approach to the solution of algebraic equations is also characteristic of Brahmagupta's work with second-degree algebraic equations. When he solved second-degree algebraic equations, also called quadratic equations, he seemed to see all quadratic equations as instances of the single model equation $ax^2 + bx + c = 0$, where the coefficients a, b, and c could represent negative as well as nonnegative numbers. Brahmagupta was willing to accept negative solutions here as well. He also accepted rational and irrational numbers as solutions. (A *rational number* can be represented as the quotient of two whole numbers. An *irrational number* is a number that cannot be represented as the quotient of two whole numbers.) This willingness to expand the number system to fit the problem, rather than to restrict the problem to fit the number system, is characteristic of much of the best Indian mathematics.

Finally, Brahmagupta was interested in indeterminate equations. (An *indeterminate equation* is a single equation, or a system of equations, with many solutions.) When considering these types of problems he attempted to find all possible solutions.

Brahmagupta's work is algorithmic in nature. To Brahmagupta learning new math meant learning new techniques to solve equations. Today many of us think of mathematics as the search for solutions to difficult word problems, but mathematics has always been about more than finding the right solutions. The Greeks, for example, were often more concerned with discovering new properties of geometric figures than they were with performing difficult calculations. Brahmagupta was familiar with other approaches to mathematics, but he was motivated by problems that involved difficult calculations. He wanted to find calculating techniques that yielded answers, and he had a very broad idea of what constituted an answer. The *Brahma-sphuta-siddhānta* was quickly recognized by Brahmagupta's contemporaries as an important and imaginative work. It inspired numerous commentaries by many generations of mathematicians.

Mahavira

The mathematician Mahavira (ca. 800–ca. 870), also known as Mahaviracharya, was one of those inspired by *Brahma-sphuta-siddhānta* (Compendium of the essence of mathematics). Mahavira lived in southern India. He was an unusual figure in the history of Hindu mathematics. He was not, for example, a Hindu. He was a member of the Jain religion. (Jainism is a small but culturally important religious sect in present-day India.) He was not an astronomer. His book, called *Ganita Sara Samgraha*, is the first book in the Indian mathematical tradition that confines its attention to pure mathematics. It is sometimes described as a commentary on Brahmagupta's work, but it is more than that. Mahavira's book is an ambitious attempt to summarize, improve upon, and teach Indian mathematical knowledge as he understood it. Mahavira's book was very successful. It was widely circulated and used by students for several centuries.

There are traditional aspects of Mahavira's book. As with Brahmagupta's great work, *Brahma-sphuta-siddhānta*, Mahavira's book is written in verse and consists of rules and examples. The rules are stated without proof. Coupled with his very traditional presentation is a very modern approach to arithmetic. It is presented in a way that is similar to the way arithmetic is taught today.

In addition to his presentation of arithmetic, Mahavira demonstrated considerable skill manipulating the Hindu system of numeration: He constructed math problems whose answers read the same forward and backward. For example: $14287143 \times 7 = 100010001$. (Notice that the answer to the multiplication problem is a sort of numerical palindrome.) He was also interested in algebraic identities. (An identity is a mathematical statement that is true for all numbers.) An example of one of the identities that Mahavira discovered is $a^3 = a(a + b)(a - b) + b^2 (a - b) + b^3$. These kinds of identities sometimes facilitate calculation. They also demonstrate how various algebraic quantities relate one to another.

Word problems were also important to Mahavira. He included numerous carefully crafted problems in *Ganita Sara Samgraha*. Some of the problems are elementary, but some require a fair bit of algebra to solve.

Mahavira exercises his algebraic insights on two other classes of problems. In one section of the book he studies combinatorics. Combinatorics, which generally requires a fairly extensive knowledge of algebra, deals with the way different combinations of objects can be chosen from a fixed set. It is the kind of knowledge that is now widely used in the study of probability. He shows, for example, that the number of ways r objects can be chosen from a set containing n objects is

$$\frac{n\,(n-1)\,(n-2)\ldots(n-r+1)}{r\,(r-1)\,(r-2)\ldots 2\cdot 1}$$

where we have written his result in modern notation. This is an important formula that is widely used today.

The second class of algebra problems is geometric in origin. In Mahavira's hands even the geometry problems—and there are a number of them—are just another source of algebraic equations. For example, he attempts to find the dimensions of two triangles with the following properties: (1) the areas of the triangles are equal and (2) the perimeter of one is twice that of the other. This problem leads to some fairly sophisticated algebra and is a nice example of an indeterminate problem—it has many solutions.

More generally, there are several points worth noting about Mahavira's work. First, like Brahmagupta's work, Mahavira's writings are a highly syncopated approach to algebra. (Algebra is called syncopated when it is expressed in a combination of words, abbreviations, and a few specialized symbols.) Second, the emphasis in much of the book is on developing the techniques necessary to solve algebraic problems. It is a tour de force approach to solving various types of equations, but he provides no broader context into which we can place his results. Each problem stands on its own with no consideration given to a broader theory of equations. Third, there are no proofs or carefully developed logical arguments. He shows the reader results that he believes are important, but he often does not show the reader why he considers the results correct. His ideas are creative, but because of his lack of emphasis on mathematical proofs when he makes an error, even a glaring error, he sometimes fails to catch it. For example, when he tries to compute the area of an ellipse, he gets it wrong. Given the level of mathematics in Mahavira's time, this was admittedly a difficult problem. Perhaps he could not have solved the problem by using the mathematics available at the time, but with a more rigorous approach to the problem he might have been able to discover what the answer is *not.*

Bhaskara and the End of an Era

The discoveries of Brahmagupta, Mahavira, and many other mathematicians in the Indian tradition probably found their highest expression in the work of the mathematician and astronomer

Bhaskara (1114–ca. 1185). Bhaskara, also known as Bhaskaracharya and Bhaskara II, was the second prominent Indian mathematician of that name. (We will have no reason to refer to the first.) Bhaskara was born in southern India, in the city of Bijapur in the same general region in which Mahavira was born. Unlike Mahavira, but like Brahmagupta, Bhaskara was an astronomer. He eventually moved to Ujjain, where he became head of the astronomical observatory there. It was the same observatory that Brahmagupta had directed several centuries earlier.

Bhaskara's main work, *Siddhānta Siromani* (Head jewel of accuracy), is a book about astronomy and mathematics. It is divided into four sections, covering arithmetic, algebra, the celestial sphere, and various planetary calculations. Like the other texts we have considered, the *Siddhānta Siromani* is written in verse, although Bhaskara also provides an additional section written in prose that explains some of the mathematics found in the main body of the work. Sanskrit scholars have praised Bhaskara's work both for the quality of its poetry and for its mathematical content.

Bhaskara uses a highly syncopated algebraic notation. He solves a variety of determinate and indeterminate equations, and he is open to the possibility that the solutions to the equations that he solves may be negative as well as positive, and irrational as well as rational. He looks at very general first- and second-degree algebraic equations and seems comfortable with coefficients that are negative as well as positive. He even suggests special rules for doing arithmetic with certain irrational numbers. In many ways the work that Bhaskara did on second-degree algebraic equations is identical to work that high school students do today. Although this point may sound elementary, it was not. Mathematicians took millennia to extend their idea of number, their idea of solution, and their computational techniques to solve these types of equations. Furthermore, there are many aspects of Bhaskara's work with algebraic equations that were not surpassed anywhere in the world for several centuries.

The Leelavati and the Bijaganita, the two sections of his work that are mathematical in nature, are full of word problems to challenge the reader. One problem describes a bamboo plant, 32

A problem by Bhaskara: Before the plant broke it was 32 cubits tall. After it breaks the distance from the top of the plant (now on the ground) to the base is 16 cubits. At what height did the break occur?

cubits long, growing out of level ground. The wind springs up and breaks the plant. The top of the plant falls over, and the tip of the plant just touches the ground at a distance of 16 cubits from the base of the stalk. Bhaskara challenges the reader to compute the distance above the ground at which the stalk snapped. Interestingly, this same problem can also be found in ancient Chinese mathematical literature. (The answer is that the stalk snapped 12 cubits above the ground.)

Bhaskara's interest in the technical issues involved in solving particular equations allowed him to make great progress in special cases, and his work with the quadratic equation was very general, but in most cases, the progress that Bhaskara achieves is incremental progress. He absorbs the work of his predecessors and extends it. Most of what he did, from his use of verse, to his indifference to the concept of proof, to his choice of problems, and to his preference for algebraic as opposed to geometric methods, is reminiscent of the work of Indian mathematicians who preceded him. What distinguishes his work is that it is generally more advanced than that of his predecessors. He expresses his ideas with greater clarity. His approach is more general, that is, more abstract, and so he sees more deeply into each problem. Finally his work is more complete. The *Siddhānta Siromani* influenced many generations of mathematicians. It was a major achievement. It is sometimes described as the most important mathematical text to emerge from the classical Indian mathematical tradition.

POETRY AND ALGEBRA

It is an oft-repeated remark that in a poem, the poetry is the part that is lost in translation. If this is true for translations of verse between modern languages, the "loss of poetry" must be even more pronounced when ancient Sanskrit verse is translated to modern English. Nevertheless, skillful translations are the only means that most of us have of appreciating the poetry in which the mathematicians of ancient India expressed themselves. Here are two word problems, originally composed in verse in Sanskrit, by Bhaskara (followed by solutions using modern notation):

1.) One pair out of a flock of geese remained sporting in the water, and saw seven times the half of the square-root of the flock proceeding to the shore tired of the diversion. Tell me, dear girl, the number of the flock.

The algebraic equation to be solved is $(7/2)\sqrt{x} = x - 2$.
The solutions to the equation are $x = 16$ and $x = 1/4$.
The only reasonable solution to the word problem is $x = 16$.

2.) Out of a heap of pure lotus flowers, a third part, a fifth and a sixth, were offered respectively to the gods Siva, Vishńu and the Sun; and a quarter was presented to Bhavánĭi. The remaining six lotuses were given to the venerable preceptor. Tell quickly the whole number of flowers. (ibid.)

The algebraic equation to be solved is

$$x - \left(\frac{1}{3} + \frac{1}{5} + \frac{1}{6} + \frac{1}{4}\right)x = 6$$

The solution is $x = 120$.

(Brahmagupta and Bhaskara. Algebra with Arithmetic and Mensuration. Translated by Henry Colebrook. London: John Murray, 1819)

Islamic Mathematics

The origins of Indian mathematics, Egyptian mathematics, and Mesopotamian mathematics, to name three prominent examples, lie thousands of years in the past. Records that might help us understand how mathematics arose in these cultures are sometimes too sparse to provide much insight. This is not the case with Islamic mathematics.

Historically Islamic culture begins with the life of Muhammad (570–632). Historical records are reasonably good. We can refer to documents by Islamic historians as well as their non-Muslim neighbors. We know quite a bit about how mathematics in general, and algebra in particular, arose in the Islamic East, and this is important, because within 200 years of the death of the Prophet Muhammad great centers of learning had been established. A new and important mathematical tradition arose. This new tradition had a profound influence on the history of mathematics: Algebra was the great contribution of Islamic mathematicians. But the term *Islamic mathematics* must be used with care.

Islamic mathematics is the term traditionally given to the mathematics that arose in the area where Islam was the dominant religion, but just as the term *Hindu mathematics* is not entirely

The Great Mosque at Samarra was built about 60 miles (96 km) from Baghdad, which was where al-Khwārizmī made his home, at about the time that the mathematician lived.

satisfactory, neither is *Islamic mathematics* quite the right term. Although Islam was the dominant religion in the region around Baghdad in what is now Iraq when algebraic research flourished, Jews and Christians also lived in the area. For the most part, they were free to practice their religions unmolested. Although most of the prominent mathematical scholars of the time followed the Islamic faith, there was also room for others at even the most prominent institutions of higher learning. A number of Christian scholars, for example, helped to translate the ancient Greek mathematical texts that were stored at the House of Wisdom in Baghdad, one of the great centers of learning at the time. There was a notable 10th-century Jewish mathematician who published "Islamic" mathematics named Abu 'Otman Sahl ibn Bishr, ibn Habib ibn Hani; and one of the most prominent mathematicians of his day, Ali-sabi Thabit ibn Qurra al-Harrani, was a Sabean, a member of a sect that traced its roots to a religion of the ancient Mesopotamians. Despite this diversity, *Islamic mathematics* is the name often given to this mathematics because the Islamic faith had a strong cultural as well as religious influence.

Sometimes this mathematics is called Arabic, but not all the mathematicians involved were Arabic, either. Of the two choices, Arabic or Islamic mathematics, Islamic mathematics seems the more accurate description. Islam affected everything from governmental institutions to architectural practices. So we adopt the common practice of calling our subject Islamic mathematics, even though math, in the end, has no religious affiliation.

The history of Islamic mathematics begins in earnest with the life of al-Ma'mūn (786–833). Although al-Ma'mūn is an important figure in the history of algebra, he was no mathematician. He is best remembered for his accomplishments as a political leader. He was the son of the caliph Hārūn ar-Rashīd. (The caliphs were absolute rulers of their nations.) Ar-Rashīd had another son, al-Ma'mūn's half brother, named al-Amīn. After the father's death the two brothers, al-Ma'mūn and al-Amīn, led their respective factions in a brutal four-year civil war over succession rights to the caliphate. In the end al-Amīn lost both the war and his life.

As caliph, al-Ma'mūn proved to be a creative, if ruthless, political leader. He worked hard, though not entirely successfully, to heal the division that existed between the Shī'ite and Sunnite sects of Islam. In Baghdad he established the House of Wisdom, an important academic institution where Greek texts in mathematics, science, and philosophy were translated and disseminated. When these works could not be obtained within the caliphate, he obtained them from the libraries of Byzantium, a sometimes-hostile power. He established astronomical observatories, and he encouraged scholars to make their own original contributions. His work bore fruit. A new approach to algebra developed in Baghdad at this time.

Al-Khwārizmī and a New Concept of Algebra

A number of mathematicians responded to al-Ma'mūn's words of encouragement and contributed to the development of a new concept of algebra. Mathematically speaking, it was a very creative time. One of the first and most talented mathematicians was named Mohammed ibn-Mūsā al-Khwārizmī (ca. 780–ca. 850). Al-Khwārizmī described what happened in these words:

> [al-Ma'mūn] has encouraged me to compose a short work on Calculating by (the rules of) Completion and Reduction, confining it to what is easiest and most useful in arithmetic.
>
> *(Al-Khwārizmī, Mohammed ibn-Mūsā.* Robert of Chester's Latin Translation of the Algebra of al-Khwārizmī. *Translated by Karpinski, Louis C. New York: The Macmillan Company, 1915)*

Al-Khwārizmī's *approach* to algebra was new and significant, but many of the results that he obtained were not. Nor was he the only mathematician of his time to use the new approach. In recent historical times scholars have discovered the work of another Islamic mathematician, Abd-al-Hamid ibn-Turk, who wrote a book about algebra that was similar to al-Khwārizmī's. This second text was written at about the same time that al-Khwārizmī's work was published. The existence of Abd-al-Hamid's book indicates that some of the mathematical ideas described by al-Khwārizmī may not have originated with him. In that sense, al-Khwārizmī may, as Euclid was, have been more of a skilled expositor than an innovator. There is not enough information to know for sure. Nevertheless al-Khwārizmī's book had the greatest long-term influence. Even the author's name became part of the English language. Al-Khwārizmī's name was mispronounced often enough in Europe to take on the form *algorismi*, and this word was later shortened to the words *algorithm*, a specialized method for solving mathematical problems, and *algorism*, the so-called Arabic system of numerals. Furthermore, the second word in the title of one of al-Khwārizmī's books, *Hisāb al-jabr wa'l muqābala*, eventually found its way into English as the word *algebra*.

Al-Khwārizmī's book *Hisāb al-jabr wa'l muqābala* has little in common with those of Brahmagupta and Diophantus. For one thing, the problems that he solves are for the most part less advanced. Second, he avoids solutions that involve 0 or negative numbers. He avoids problems in indeterminate analysis—that is, problems for which multiple solutions exist—and he writes without any specialized algebraic notation. Not only does he avoid the use of letters or abbreviations for variables, he sometimes even avoids using numerals to represent numbers. He often prefers to write out the numbers in longhand. Even the motivation for Al-Khwārizmī's book was different from that of his predecessors. Diophantus seems to have had no motivation other than an interest in mathematics. Brahmagupta's motivation stemmed from his interest in mathematics and astronomy. But al-Khwārizmī wrote that al-Ma'mūn had encouraged him to develop a mathematics that would be of use in solving practical problems such as the "digging of canals" and the "division of estates."

Much of the first half of al-Khwārizmī's book *Hisāb al-jabr* is concerned with the solution of second-degree algebraic equations, but his method is not nearly as general as Brahmagupta's. Unlike Brahmagupta, he does not perceive all quadratic equations as instances of a single general type. Instead, what we would call "the" quadratic equation he perceived as a large number of separate cases. For example, he considers quadratic equations, such as $x^2 = 5x$, and he identifies the number 5 as a solution. (We would also recognize $x = 0$ as a solution, but al-Khwārizmī does not acknowledge 0 as a legitimate solution.) Because he uses rhetorical algebra, that is, an algebra devoid of specialized, algebraic symbols, his description of the equation $x^2 = 5x$ and its solution take some getting used to:

> A square is equal to 5 roots. The root of the square then is 5, and 25 forms its square which, of course, equals five of its roots.
>
> *(Al-Khwārizmī, Mohammed ibn-Mūsā.* Robert of Chester's Latin Translation of the Algebra of al-Khwārizmī. *Translated by Karpinski, Louis C. New York: The Macmillan Company, 1915)*

He plods his way from one special case to the next, and in this there is nothing new. At first it seems as if al-Khwārizmī, as his predecessor Brahmagupta and his far-away contemporary Mahavira did, sees algebra simply as a collection of problem-solving techniques. But this is not so. After establishing these results he shifts focus; it is this shift in focus that is so important to the history of algebra. After solving a number of elementary problems, he returns to the problems that he just solved and *proves* the correctness of his approach. In the field of algebra this is both new and very important.

Al-Khwārizmī's tool of choice for his proofs is geometry, but he is not interested in geometry as a branch of thought in the way that the ancient Greeks were. He is not interested in *studying* geometry; he wants to use it to provide a proof that his algebraic reasoning was without flaws. Recall that it was the lack of proofs in Hindu algebra that made it so difficult for those mathematicians to separate the true from the false. Al-Khwārizmī, by contrast, wanted to build his algebra on a solid logical foundation, and he was fortunate to have a ready-made model of deductive reasoning on hand: the classics of Greek geometry.

The geometry of the Greeks would certainly have been familiar to al-Khwārizmī. Throughout his life the translators associated with the House of Wisdom were busy translating ancient Greek works into Arabic, and there was no better example of careful mathematical reasoning available anywhere in the world at this time than in the works of the Greeks. Their works are filled with rigorous proofs. Al-Khwārizmī had the concept for a rigorous algebra and a model of mathematical rigor available. It was his great insight to combine the two into something new.

Al-Khwārizmī's interest in developing procedures for computing with square roots also bears mentioning. He begins with the very simplest examples, among them the problem of multiplying the square root of 9 by the number 2. Here is how he describes the procedure:

> Take the root of nine to be multiplied. If you wish to double the root of nine you proceed as follows: 2 by 2 gives 4, which you multiply by 9, giving 36. Take the root of this, i.e. 6, which is found to be two roots of nine, i.e. the double of three. For three, the root of nine, added to itself gives 6.
>
> (*Al-Khwārizmī, Mohammed ibn-Mūsā.* Robert of Chester's Latin Translation of the Algebra of al-Khwārizmī. *Translated by Karpinski, Louis C. New York: The Macmillan Company, 1915)*

In our notation we express this idea as $2\sqrt{9} = \sqrt{4 \cdot 9} = \sqrt{36} = 6$, which again emphasizes the importance and utility of our modern system of notation. He extends this simple numerical example into several more general algebraic formulas. For example, we would express one of his rhetorical equations as follows: $3\sqrt{x} = \sqrt{9x}$.

It is not clear why al-Khwārizmī avoided the use of any sort of algebraic symbolism. Without any specialized algebraic notation his work is not easy to read despite the fact that he is clearly a skilled expositor. Al-Khwārizmī's work had an important influence on the many generations of mathematicians living in the Near East, Northern Africa, and Europe. On the positive side, his concept of incorporating geometric reasoning to buttress his algebraic arguments was widely emulated. On the negative side, his highly rhetorical approach would prove a barrier to rapid progress. What is most important is that Al-Khwārizmī's work established a logical foundation for the subject he loved. His work set the standard for rigor in algebra for centuries.

A PROBLEM AND A SOLUTION

The following is a problem that was posed and solved by al-Khwārizmī in his algebra. It is a nice example of rhetorical algebra, that is, algebra expressed entirely in words and without the use of specialized algebraic symbols.

> If you are told, "ten for six, how much for four?" then *ten* is the measure; *six* is the price; the expression *how much* implies the unknown number of the quantity; and *four* is the number of the sum. The number of the measure, which is *ten,* is inversely proportionate to the number of the sum, namely, *four.* Multiply, therefore, ten by four, that is to say, the two known proportionate numbers by each other; the product is forty. Divide this by the other known number, which is that of the price, namely, six. The quotient is six and two-thirds; it is the unknown number, implied in the words of the question *how much?* it is the quantity, and inversely proportionate to the six, which is the price.

(Al-Khwārizmī, Mohammed ibn-Mūsā. Robert of Chester's Latin Translation of the Algebra of al-Khwārizmī. Translated by Louis C. Karpinski, New York: The Macmillan Company, 1915)

In our notation al-Khwārizmī solved the problem that we would express as $10/6 = x/4$.

Omar Khayyám, Islamic Algebra at Its Best

The astronomer, poet, mathematician, and philosopher Omar Khayyám (ca. 1050–1123) was perhaps the most important of all Islamic mathematicians after al-Khwārizmī. Omar was born in Neyshābūr (Nishāpūr) in what is now northeastern Iran. He also died in Neyshābūr, and between his birth and death he traveled a great deal. Political turbulence characterized Omar's times, and moving from place to place was sometimes a matter of necessity.

Omar was educated in Neyshābūr, where he studied mathematics and philosophy. As a young man he moved about 500 miles (800 km) to Samarqand, which at the time was a major city, located in what is now Uzbekistan. It was in Samarqand that he became well known as a mathematician. Later he accepted an invitation to work as an astronomer and director of the observatory at the city of Esfahan, which is located in central Iran. He remained there for about 18 years, until the political situation became unstable and dangerous. Funding for the observatory was withdrawn, and Omar moved to the city of Merv, now Mary, in present-day Turkmenistan. During much of his life Omar was treated with suspicion by many of his contemporaries for his freethinking and unorthodox ideas. He wrote angrily about the difficulty of doing scholarly work in the environments in which he found himself, but in retrospect he seems to have done well despite the difficulties.

Omar described algebra, a subject to which he devoted much of his life, in this way:

> By the help of God and with His precious assistance, I say that Algebra is a scientific art. The objects with which it deals are absolute numbers and measurable quantities which, though themselves unknown, are related to "things" which are known, whereby the determination of the unknown quantities is possible. Such a thing is either a quantity or a unique relation, which is only determined by careful examination. What one searches for in the algebraic art are the relations which lead from the known to the unknown, to discover which is the object of Algebra as stated above. The perfection of this art consists in knowledge of the scientific method by which one determined numerical and geometric quantities.
>
> *(Kasir, Daoud S.* The Algebra of Omar Khayyám. *New York: Columbia University Press, 1931. Used with permission)*

This is a good definition for certain kinds of algebra even today, almost a thousand years later. The care with which the ideas in the definition are expressed indicates that the author was a skilled writer in addition to being a skilled mathematician, but he is generally remembered as either one or the other. In the West, Omar Khayyám is best remembered as the author of *The Rubáiyát of Omar Khayyám*, a collection of poems. This collection of poems was organized, translated into English, and published in the 19th

No.	Month	Length
1	Farvardin	31
2	Ordibehesht	31
3	Khordad	31
4	Tir	31
5	Mordad	31
6	Shahrivar	31
7	Mehr	30
8	Aban	30
9	Azar	30
10	Dey	30
11	Bahman	30
12	Esfand	29 or 30

A small group of scientists, of which Omar Khayyám was the most prominent member, devised the Jalali calendar. With some modest modifications, this has become today's Persian calendar (pictured above).

century. It has been in print ever since and has now been translated into all the major languages of the world. The *Rubáiyát* is a beautiful work, but Omar's skill as a poet was not widely recognized in his own time, nor is it the trait for which he is best remembered in Islamic countries today.

Omar's contemporaries knew him as a man of extraordinarily broad interests. Astronomy, medicine, law, history, philosophy, and mathematics were areas in which he distinguished himself. He made especially important contributions to mathematics and to the revision of the calendar. His revision of the calendar earned him a certain amount of fame because the calendar in use at the time was inaccurate in the sense that the calendar year and the astronomical year were of different lengths. As a consequence over time the seasons shifted to different parts of the calendar year. This variability made using the calendar for practical, seasonal predictions difficult. Correcting the calendar involved collecting better astronomical data and then using this data to make the necessary computations. This is what Omar did. It was an important contribution because his calendar was extremely accurate, and its accuracy made it extremely useful.

In the history of algebra, Omar Khayyám is best remembered for his work *Al-jabr w'al muqābala* (Demonstration concerning the completion and reduction of problems; this work is also known as Treatise on demonstration of problems of algebra). The *Al-jabr w'al muqābala* is heavily influenced by the ideas and works of Al-Khwārizmī, who had died two centuries before Omar wrote his algebra. As with al-Khwārizmī, Omar does not see all quadratic equations as instances of the single equation $ax^2 + bx + c = 0$. Instead, he, too, divides quadratic equations into distinct types, for example, "a number equals a square," which we would write as $x^2 = c$; "a square and roots equal a number," which we would write as $x^2 + bx = c$; and "a square and a number equal a root," which we would write $x^2 + c = x$. (He made a distinction, for example, between $x^2 + bx = c$ and $x^2 + c = bx$ because Omar also prefers to work with positive coefficients only.)

Omar even borrows al-Khwārizmī's examples. He uses the same equation, $x^2 = 5x$, that al-Khwārizmī used in his book, and there are the by-now standard geometric demonstrations involving the proofs of his algebraic results. All of this is familiar territory and would have seemed familiar even to al-Khwārizmī. But then Omar goes on to consider equations of the third degree—that is, equations that we would write in the form $ax^3 + bx^2 + cx + d = 0$.

Omar classifies third-degree equations by using the same general scheme that he used to classify equations of the second degree, and then he begins to try to solve them. He is unsuccessful in finding an algebraic method of obtaining a solution. He even states that one does not exist. (A method was discovered several centuries later in Europe.) Omar does, however, find a way to represent the solutions by using geometry, but his geometry is no longer the geometry of the Greeks. He has moved past traditional Euclidean geometry. Omar uses numbers to describe the properties of the curves in which he is interested. As he does so he broadens the subject of algebra and expands the collection of ideas and techniques that can be brought to bear on any problem.

Omar's synthesis of geometric and algebraic ideas is in some ways modern. When he discusses third-degree algebraic equations, equations that we would write as $ax^3 + bx^2 + cx + d = 0$, he represents his ideas geometrically. (Here a, b, c and d represent numbers and x is the unknown.) For example, the term x^3, "x cubed," is interpreted as a three-dimensional cube. This gives him a useful conceptual tool for understanding third-degree algebraic equations, but it also proves to be a barrier to further progress. The problem arises when he tries to extend his analysis to fourth-degree equations, equations that we would write as $ax^4 + bx^3 + cx^2 + dx + e = 0$. Because he cannot imagine a four-dimensional figure, his method fails him, and he questions the reality of equations of degree higher than 3.

To his credit Omar was aware of the close relationship between algebraic equations and the number system. However, his narrow concept of number prevented him from identifying many solutions that Hindu mathematicians accepted without question. This may seem to be a step backward, but his heightened sense of rigor was an important step forward. There are important relationships between the degree of an algebraic equation and the properties of the numbers that can appear as solutions. (The *degree* of an equation is the largest exponent that appears in it. Fourth-degree equations, for example, contain a variable raised to degree 4, and no higher power appears in the equation.) In fact, throughout much of the history of mathematics it was the study of algebraic equations that required mathematicians to consider more carefully their concept of what a number is and to search for ways in which the number system could be expanded to take into account the types of solutions that were eventually discovered. Omar's work in algebra would not be surpassed anywhere in the world for the next several centuries.

The work of al-Khwārizmī and Omar exemplifies the best and most creative aspects of Islamic algebra. In particular, their synthesis of algebra and geometry allowed them to think about algebraic questions in a new way. Their work yielded new insights into the relations that exist between algebra and geometry. They provided their successors with new tools to investigate algebra, and they attained a higher standard of rigor in the study of algebra. Although Indian mathematicians sometimes achieved more advanced results than their Islamic counterparts, Indian mathematicians tended to develop their mathematics via analogy or metaphor. These literary devices can be useful for discovering new aspects of mathematics, but they are of no use in separating the mathematically right from the mathematically wrong. Islamic mathematicians emphasized strong logical arguments—in fact, they seemed to enjoy them—and logically rigorous arguments are the only tools available for distinguishing the mathematically true from the mathematically false. It is in this sense that the

LEONARDO OF PISA

There was one prominent European mathematician during the period in which Islamic mathematics flourished. He received his education in northern Africa from an Islamic teacher. As a consequence, he owed much of his insight to Islamic mathematics. He was the Italian mathematician Leonardo of Pisa, also known as Fibonacci (ca. 1170–after 1240). Leonardo's father, Guglielmo, was a government official in a Pisan community situated in what is now Algeria. During this time Leonardo studied mathematics with a Moor. (The Moors were an Islamic people who conquered Spain.) From his teacher he apparently learned both algebra and the Hindu base 10 place-value notation. He later wrote that he enjoyed the lessons. Those lessons also changed his life.

As a young man Leonardo traveled throughout North Africa and the Middle East. During his travels he learned about other systems of notation and other approaches to problem solving. He eventually settled down in Pisa, Italy, where he received a yearly income from the city.

Leonardo produced a number of works on mathematics. He described the place-value notation and advocated for its adoption. His efforts helped to spread news of the system throughout Europe. (Leonardo only used place-value notation to express whole numbers. He did not use the decimal notation to write fractions.) His description of the Indian system of notation is his most long-lasting contribution, but he also discovered what is now known as the Fibonacci series, and he was renowned for his skill in algebra as well. He studied, for example, the equation that we would write as $x^3 + 2x^2 + 10x = 20$. This equation was taken from the work of Omar Khayyám. In his analysis Leonardo apparently recognizes that the solution he sought was not a simple whole number or fraction. He responds by working out an approximation–and he recognizes that his answer *is* an approximation–that is accurate to the ninth decimal place. Leonardo, however, expressed his answer as a base 60 fraction. Unfortunately Leonardo does not explain how he found his answer, an approximation that would set the European standard for accuracy for the next several centuries.

algebra developed by the Islamic mathematicians is—especially during the period bracketed by the lives of al-Khwārizmī and Omar—much closer to a modern conception of algebra than is that of the Indians.

4

ALGEBRA AS A THEORY OF EQUATIONS

Art, music, literature, science, and mathematics flourished in Europe during the Renaissance, which had its origins in 14th-century Italy and spread throughout Europe over the succeeding three centuries. Just as art, music, and science changed radically during the Renaissance, all pre-Renaissance mathematics is profoundly different from the post-Renaissance mathematics of Europe. The new mathematics began with discoveries in algebra.

Many of the best European mathematicians of this period were still strongly influenced by the algebra of al-Khwārizmī, but in the space of a few years Italian mathematicians went far beyond all of the algorithms for solving equations that had been discovered anywhere since the days of the Mesopotamians. Mathematicians found solutions to whole classes of algebraic equations that had never been solved before. Their methods of solution were, by our standards, excessively complicated. The algorithms developed by Renaissance era mathematicians were also difficult and sometimes even counterintuitive. A lack of insight into effective notation, poor mathematical technique, and an inadequate understanding of what a number is sometimes made recognizing that they had found a solution difficult for them. Nevertheless, many problems were solved for the first time, and this was important, because these problems had resisted solution for thousands of years.

The new algorithms also exposed large gaps in the understanding of these mathematicians. To close those gaps they would have to expand their concept of number, their collection of problem solving techniques, and their algebraic notation. The algebraic solution of these new classes of problems was a major event in the history of mathematics. In fact, many historians believe that the modern era in mathematics begins with publication of the Renaissance era algebra book *Ars magna*, about which we will have more to say later.

To appreciate what these Renaissance era mathematicians accomplished, we begin by examining a simple example. The example is a quadratic equation, an algebraic equation of second degree. The remarks we make about quadratic equations guide our discussion of the more complicated equations and formulas used by the mathematicians of the Renaissance. Our example is taken from the work of al-Khwārizmī. He was an expert at this type of problem, but because his description is a little old-fashioned, and because we also want to discuss his problem in modern notation, we introduce a little terminology first. A quadratic, or second-degree, equation is any equation that we can write in the form $ax^2 + bx + c = 0$. In this equation, the letter x is the unknown. The number or numbers that, when substituted for x, make the equation a true statement are called the *roots* of the equation, and the equation is solved when we find the root or roots. The letters a, b, and c are the *coefficients*. They represent numbers that we assume are known. In the following excerpt, al-Khwārizmī is describing his method of solving the equation $x^2 + 21 = 10x$. In this example the coefficient a equals 1. The coefficient b is -10. (Al-Khwārizmī prefers to transpose the term $-10x$ to the right side of the equation because he does not work with negative coefficients.) Finally, the c coefficient equals 21. Here is al-Khwārizmī's method for solving the equation $x^2 + 21 = 10x$:

> A square and 21 units equal 10 roots. . . . The solution of this type of problem is obtained in the following manner. You take first one-half of the roots, giving in this instance 5, which multiplied by itself gives 25. From 25 subtract the 21 units to which we have just referred in connection with the squares. This gives 4, of which you extract the square root, which is 2. From the half

of the roots, or 5, you take 2 away, and 3 remains, constituting one root of this square which itself is, of course, 9.

(*Al-Khwārizmī, Mohammed ibn-Mūsā.* Robert of Chester's Latin Translation of the Algebra of al-Khwārizmī. *Translated by Karpinski, Louis C. New York: The Macmillan Company, 1915)*

Al-Khwārizmī has given a rhetorical description of an application of the algorithm called the quadratic formula. Notice that what al-Khwārizmī is doing is "constructing" the root, or solution of the equation, from a formula that uses the coefficients of the equation as input. Once he has identified the coefficients he can, with the help of his formula, compute the root. We do the same thing when we use the quadratic formula, although both our formula and our concept of solution are more general than those of al-Khwārizmī. In fact, we learn two formulas when we learn to solve equations of the form $ax^2 + bx + c = 0$. The first is $x = \frac{-b}{2a} + \frac{\sqrt{b^2 - 4ac}}{2a}$, and the second is $x = \frac{-b}{2a} - \frac{\sqrt{b^2 - 4ac}}{2a}$. These are the formulas that allow us to identify the roots of a quadratic equation provided we know the coefficients.

Various rhetorical forms of these formulas were known to al-Khwārizmī and even to Mesopotamian mathematicians. They are useful for finding roots of second-degree equations, but they are useless for computing the roots of an equation whose degree is not 2. Until the Renaissance, *no one in the history of humankind* had found corresponding formulas for equations of degree higher than 2. No one had found a formula comparable to the quadratic formula for a third-degree equation, that is, an equation of the form $ax^3 + bx^2 + cx + d = 0$, where a, b, c, and d are the coefficients. This was one of the great achievements of the Renaissance.

There is one more point to notice about the preceding formulas for determining the solutions to the second-degree equations: They are exact. These formulas leave no uncertainty at all about the true value for x. We can compare these formulas with the solution that Leonardo of Pisa obtained for the third-degree equation given in the preceding chapter. His approximation was accurate to the billionth place. This is far more accurate than he (or we) would need for any practical application, but there is still some uncertainty about the true value for x.

From a practical point of view, Leonardo completely solved his problem, but from a theoretical point of view, there is an important distinction between his answer and the exact answer. His approximation is a rational number. It can be expressed as a quotient of two whole numbers. The exact answer, the number that he was searching for, is an irrational number. It *cannot* be expressed as a quotient of two whole numbers. Leonardo's solution was, for the time, a prodigious feat of calculation, but it fails to communicate some of the mathematically interesting features of the exact solution. Leonardo's work shows us that even during the Middle Ages there were algorithms that enabled one to compute highly accurate *approximations* to at least some equations of the third degree, but there was no general algorithm for obtaining exact solutions to equations of the third degree.

The New Algorithms

The breakthrough that occurred in Renaissance Italy was unrelated to finding useful approximations to algebraic equations. It involved the discovery of an algorithm for obtaining exact solutions of algebraic equations.

The discovery of exact algorithms for equations of degree higher than 2 begins with an obscure Italian academic named Scipione del Ferro (1465–1526). Little is known of del Ferro, nor are scholars sure about precisely what he discovered. Some historians believe that he was educated at the University of Bologna, but there are no records that indicate that he was. What is certain is that in 1496 he joined the faculty at the University of Bologna as a lecturer in arithmetic and geometry and that he remained at the university for the rest of his life.

Uncertainty about del Ferro's precise contribution to the history of algebra arises from the fact that he did not publish his ideas and discoveries about mathematics. He was not secretive. He apparently shared his discoveries with friends. Evidently he learned

Ancient buildings in Bologna, Italy. During the Renaissance, Bologna was an important center for mathematical research. (EdLab at Teachers College, Columbia University)

how to solve certain types of cubic equations. These equations had resisted exact solution for thousands of years, so del Ferro's discovery was a momentous one. Del Ferro did not learn how to solve every cubic equation, however.

To appreciate the difficulties that they faced, it is important to keep in mind that European mathematicians of del Ferro's time did not use negative coefficients, so they did not perceive a cubic equation as a single case as we do today. Today we say that a cubic equation is *any* equation that can be written in the form $ax^3 + bx^2 + cx + d = 0$. But where we see unity, they saw a diversity of types of cubic equations. They classified equations by the side of the equals sign where each coefficient was written. Where we would write a negative coefficient they carefully transposed the term containing the negative coefficient to the other side of the equation so that the only coefficients they considered were positive. For example, they looked at the equations $x^3 + 2x = 1$ and $x^3 = 2x + 1$ as separate cases. Furthermore, they would also consider any cubic equation with an x^2 term, such as $x^3 + 3x^2 = 1$, as a case separate from, say, $x^3 + 2x = 1$, because the former has an x^2 term and no x term, whereas the latter equation has an x term but no x^2 term. The number of such separate cases for a third-degree equation is quite large.

Although we cannot be sure exactly what types of cubic equations del Ferro solved, many scholars believe that he learned to solve one or both of the following types of third-degree algebraic equations: (1) $x^3 + cx = d$ and/or (2) $x^3 = cx + d$, where in each equation the letters c and d represent positive numbers. Whatever del Ferro learned, he passed it on to one of his students, Antonio Maria Fior.

News of del Ferro's discovery eventually reached the ears of a young, creative, and ambitious mathematician and scientist named Niccolò Fontana (1499–1557), better known as Tartaglia. Tartaglia was born in the city of Brescia, which is located in what is now northern Italy. It was a place of great wealth when Tartaglia was a boy, but Tartaglia did not share in that wealth. His father, a postal courier, died when Tartaglia was young, and the family was left in poverty. It is often said that Tartaglia was self-taught. In one story the 14-year-old Tartaglia hires a tutor to help him learn to read but has only enough money to reach the letter *k*. In 1512, when Tartaglia was barely a teenager, the city was sacked by the French. There were widespread looting and violence. Tartaglia suffered severe saber wounds to his face, wounds

Niccolò Fontana, also known as Tartaglia. His discovery of a method for solving arbitrary third-degree equations had a profound effect on the history of mathematics. (Smithsonian Institution)

that left him with a permanent speech impediment. (Tartaglia, a name which he took as his own, began as a nickname. It means "stammerer.")

When Tartaglia heard the news that del Ferro had discovered a method of solving certain third-degree equations, he began the search for his own method of solving those equations. What he discovered was a method for solving equations of the form $x^3 + px^2 = q$. Notice that this is a different type of equation from those that had been solved by using del Ferro's method, but both algorithms have something important in common: They enable the user to construct solution(s) using only the coefficients that appear in the equation itself. Tartaglia and del Ferro had found formulas for third-degree equations that were similar in concept to the quadratic formula.

When Tartaglia announced his discovery, a contest was arranged between him and del Ferro's student, Antonio Maria Fior. Each mathematician provided the other with a list of problems, and each was required to solve the other's equations within a specified time. Although he initially encountered some difficulty, Tartaglia soon discovered how to extend his algorithm to solve those types of problems proposed by Fior, but Fior did not discover how to solve the types of problems proposed by Tartaglia. It was a great triumph for Tartaglia.

Tartaglia did not stop with his discoveries in algebra. He also wrote a physics book, *Nova Scientia* (A new science), in which he tried to establish the physical laws governing bodies in free fall, a subject that would soon play an important role in the history of science and mathematics. Tartaglia had established himself as an important mathematician and scientist. He was on his way up.

It is at this point that the exploits of the Italian gambler, physician, mathematician, philosopher, and astrologer Girolamo Cardano (1501–76) become important to Tartaglia and the history of science. Unlike del Ferro, who published nothing, Cardano published numerous books describing his ideas, his philosophies, and his insights on every subject that aroused his curiosity, and he was a very curious man. He published the first book on probability. As a physician he published the first clinical description of typhus, a serious disease that is transmitted through the bite of certain insects. He also wrote about philosophy, and he seemed to enjoy writing about himself as well. His autobiography is entitled *De propria vita* (Book of my life). In the field of algebra, Cardano did two things of great importance: He wrote the book *Ars magna* (Great art), the book that many historians believe marks the start of the modern era in mathematics, and he helped an impoverished boy named Lodovico Ferrari (1522–65).

At the age of 14 Ferrari applied to work for Cardano as a servant, but unlike most servants of the time, Ferrari could read and write. Impressed, Cardano hired him as his personal secretary instead. It soon became apparent to Cardano that his young secretary had great potential, so Cardano made sure that Ferrari received an excellent university education. Ferrari learned Greek, Latin, and mathematics at the university where Cardano lectured, and when Ferrari was 18, Cardano resigned his post at the university in favor of his former secretary. At the age of 18 Ferrari was lecturing in mathematics at the University of Milan.

Meanwhile Tartaglia's success had attracted Cardano's attention. Although Tartaglia had discovered how to solve cubic equations, he had not made his algorithm public. He preferred to keep it secret. Cardano wanted to know the secret. Initially he sent a letter requesting information about the algorithm, but Tartaglia refused the request. Cardano, a capable mathematician in his own right and a very persistent person, did not give up. He continued to write to Tartaglia. They argued. Still Tartaglia would not tell, and still Cardano persisted. Their positions, however, were not equal. Tartaglia, though well known, was not well off. By contrast, Cardano was wealthy and well connected. He indicated that he could help Tartaglia find a prestigious position, which Tartaglia very much wanted. Cardano invited Tartaglia to his home, and, in exchange for a promise that Cardano would tell no one, Tartaglia shared his famous algorithm with his host.

It was a mistake, of course. Tartaglia is said to have recognized his error almost as soon as he made it. Cardano was of no help in finding Tartaglia a position, but with the solution to the third-degree equation firmly in hand, Cardano asked his former servant, secretary, and pupil, Ferrari, to solve the general fourth-degree equation, and Ferrari, full of energy and insight, did as he was asked. He discovered a formula that enabled the user to construct the root(s) to a fourth-degree equation using only the coefficients that appeared in the equation itself. Now Cardano knew how to solve both third- and fourth-degree equations, and that is the information Cardano published in *Ars magna.*

Tartaglia was furious. He and Cardano exchanged accusations and insults. The whole fight was very public, and much of the public was fascinated. Eventually a debate was arranged between Tartaglia and Ferrari, who was an intensely loyal man who never forgot who gave him help when he needed it. It was a long debate, and it did not go well for Tartaglia. The debate was not finished when Tartaglia left. He did not return. Tartaglia felt betrayed and remained angry about the affair for the rest of his life.

In some ways Cardano's *Ars magna* is an old-fashioned book. It is written very much in the style of al-Khwārizmī: It is a purely rhetorical work, long on prose and bereft of algebraic notation. That is one reason that it is both tedious and difficult for a modern reader to follow. In the manner of al-Khwārizmī, Cardano avoids negative coefficients by transposing terms to one side of the equation or another until all the numbers appearing in the equation are nonnegative. In this sense, *Ars magna* belongs to an earlier age.

The significance of *Ars magna* lies in three areas. First, the solutions that arose in the course of applying the new algorithms were often of a very complicated nature. For example, numbers such as $\sqrt[3]{287\tfrac{1}{2}+\sqrt{80449\tfrac{1}{4}}}+\sqrt[3]{287\tfrac{1}{2}-\sqrt{80449\tfrac{1}{4}}}-5$, a solution that Cardano derives in his book for a fourth-degree equation, inspired many mathematicians to reconsider their ideas of what a number is. This turned out to be a very difficult problem to resolve, but with the new algorithms, it was no longer possible to avoid asking the question.

Second, Cardano's book marks the first time since the Mesopotamians began pressing their ideas about quadratic equations into clay slabs that anyone had published general methods for obtaining exact solutions to equations higher than second degree. Algebra had always seemed to hold a lot of promise, but its

ALGEBRA AS A TOOL IN SCIENCE

During the Renaissance great progress was made in obtaining exact solutions to algebraic equations. There was a certain excitement associated with the work of Tartaglia, Ferrari, and others because these mathematicians were solving problems that had resisted solution for millennia. There was also a highly abstract quality to their work: It was now possible to solve fourth-degree equations, for example, but opportunities to *use* fourth-degree equations to solve practical problems were not especially numerous.

There was, however, another trend that was occurring during the Renaissance, the application of algebra to the solution of problems in science. There is no better example of a scientist's reliance on algebra as a language in which to express ideas than in the work of the Italian scientist, mathematician, and inventor Galileo Galilei (1564–1642).

One of Galileo's best-known books, *Dialogues Concerning Two New Sciences,* is filled with algebra. It is not a book about algebra. It is a book about science, in which Galileo discusses the great scientific topics of his time: motion, strength of materials, levers, and other topics that lie at the heart of classical mechanics. To express his scientific ideas he uses a rhetorical version of an algebraic function.

In the following quotation, taken from *Dialogues,* Galileo is describing discoveries he had made about the ability of objects to resist fracture:

> Prisms and cylinders which differ in both length and thickness offer resistances to fracture . . . which are directly proportional to the cubes of the diameters of their bases and inversely proportional to their lengths.

(Galileo Galilei. Dialogues Concerning Two New Sciences. Translated by Henry Crew and Alfonso de Salvio. New York: Dover Publications, 1954)

Galileo is describing the physical characteristics of real objects with algebraic functions. Unfortunately he lacks a convenient algebraic notation to express these ideas.

In his use of algebra, Galileo was not alone. During the Renaissance scientists discovered that algebra was often the most convenient way that they had to express their ideas. The synthesis that occurred between algebra and science during the Renaissance accelerated interest in algebra. It probably also accelerated progress in science

(continues)

ALGEBRA AS A TOOL IN SCIENCE
(continued)

because it made new abstract relations between different properties more transparent and easier to manipulate. Algebra *as a symbolic language* was gaining prominence in mathematics. As notation improved and insight deepened into how algebra could be used, algebraic notation became the standard way that mathematicians expressed their ideas in many branches of mathematics. Today algebra has so thoroughly permeated the language of mathematics and the physical sciences that it is doubtful that the subject matter of these important disciplines could be expressed independently of the algebraic notation in which they are written.

actual utility had been limited because mathematicians knew only enough algebra to solve relatively simple problems. The problems that were solved by the Mesopotamian, Chinese, Indian, and Islamic mathematicians were by and large simple variations on a very small group of very similar problems. This changed with the publication of *Ars Magna.*

Finally, Cardano's book made it seem at least possible that similar formulas might exist for algebraic equations of fifth degree and higher. This possibility inspired many mathematicians to begin searching for algorithms that would enable them to find exact solutions for equations of degree higher than 4.

François Viète, Algebra as a Symbolic Language

Inspired by the very public success of Tartaglia and Ferrari and the book of Cardano, the study of algebra spread throughout much of Europe. One of the first and most obvious barriers to further progress was the lack of a convenient symbolism for expressing the new ideas, but this condition was changing, albeit in a haphazard way. Throughout Europe various algebraic symbols were introduced. Mathematicians in different geographical or linguistic regions employed different notation. There were several symbols pro-

Ancient Poitiers, France. Viète studied law at the University of Poitiers, but he soon turned his attention to mathematical research. (Middlebury College, Language Schools, Graduate Program)

posed for what we now know as an equals sign (=). There were also alternatives for +, -, ×, and so on. It took time for the notation to become standardized, but all of these notational innovations were important in the sense that they made algebra easier. Rhetorical algebra can be slow to read and unnecessarily difficult to follow. Ordinary everyday language, the kind of language that we use in conversation, is not the right language in which to express algebra, and the higher the level of abstraction becomes, the more difficult the rhetorical expression of algebra is to read. Nor was the lack of a suitable notation the only barrier to progress.

Algebra is about more than symbols. Algebra is about ideas, and despite the creativity of del Ferro, Tartaglia, Ferrari, and others, the algebra of much of the 16th century was similar in concept to what Islamic mathematicians had developed centuries earlier. For most mathematicians of the time, algebra was still about finding roots of equations. It was a very concrete subject. The equation was like a question; the numbers that satisfied the equation were the answers. A successful mathematician could solve several different types of equations; an unsuccessful mathematician could not. At the time algebra was a collection of problem-solving techniques. It was the search for formulas. The formulas might well be complicated, of course, but the goal was not. One of the first mathematicians to understand that algebra is about more than developing techniques to solve equations was the French mathematician François Viète (1540–1603).

Viète was born into a comfortable family in Fontenay-le-Comte, a small town located in the west of France not far from the Bay of Biscay. He studied law at the University of Poitiers. Perhaps his initial interest in law was due to his father, who was also a lawyer, but the legal profession was not for Viète. Within a few years of graduation he had given up on law and was working as a tutor for a wealthy family. His work as a tutor was a quiet beginning to an eventful life. As with many French citizens, Viète's life was profoundly influenced by the political instability that long plagued France. For Viète, the cause of the turmoil was religious tension between the Roman Catholic majority and the Protestant minority, called Huguenots. Viète's sympathies lay with the Huguenots.

While he was working as a tutor, Viète began his research into mathematics. He left his job as a tutor in 1573, when he was appointed to a government position. Fortunately for mathematics, in 1584 he was banished from government for his Huguenot sympathies. Viète moved to a small town and for five years devoted himself to the study of mathematics. It was, mathematically speaking, the most productive time of his life.

Viète understood that the unknowns in an algebraic equation could represent *types* of objects. His was a much broader view of an equation than simply as an opportunity to find "the answer." If the unknown could represent a type or "species" of object, then algebra was about relationships between types. Viète's higher level of abstract thought led to an important notational breakthrough. It is to Viète that we owe the idea of representing the unknown in an equation with a letter. In fact, in his search for more general patterns Viète also used letters to represent known quantities. (The "known quantities" are what we have been calling coefficients.)

Although we have been using this notation since the beginning of this volume—it is hard to talk about algebra without it—historically speaking, this method of notation did not begin until Viète invented it.

Viète's method was to use vowels to represent unknown quantities and consonants to represent known quantities. (This is not quite what we use today; today we let letters toward the end of the alphabet represent unknowns and letters toward the beginning of the alphabet represent known quantities). Notice that by employing letters for the coefficients Viète deprives himself of any hope of finding numerical solutions. The compensation for this loss of specificity is that the letters made it easier for Viète to see broader patterns. The letters helped him identify relationships between the various symbols and the classes of objects that they represented.

Though some of Viète's ideas were important and innovative, others were old-fashioned or just plain awkward. Viète was old-fashioned in that he still had a fairly restricted idea of what constituted an acceptable solution. As had his predecessors, Viète accepted only positive numbers as legitimate solutions.

Viète had an unusual and, in retrospect, awkward idea for how unknowns and coefficients should be combined. He interpreted his unknowns as if there were units attached to them. We have already encountered a similar sort of interpretation. Recall that Omar Khayyám had conceptual difficulties in dealing with fourth-degree equations *because* he interpreted an unknown as a length. For Omar an unknown length squared represented a (geometric) square, an unknown length cubed was a (three-dimensional) cube, and as a consequence there was no immediate way of interpreting an unknown raised to the fourth power. In a similar vein, Viète required all terms in an equation to be "homogeneous" in the sense that they all had to have the same units. The equation that we would write as $x^2 + x = 1$, an equation without dimensions, would have made little sense to Viète since it involved adding, for example, a line segment, x, to a square, x^2. Instead Viète insisted on assigning dimensions to his coefficients so that all terms had the same dimensions. For example, he preferred to work with equations like $A^3 - 3B^2A = B^2D$. This was very important to him, although in retrospect it is hard to see why. Succeeding generations of mathematicians perceived his requirement of homogeneity as a hindrance and abandoned it.

Viète's work was a remarkable mixture of the old and the new, and with these conceptual tools he began to develop a theory of equations. Although he knew how to solve all algebraic equations up to and including those of the fourth degree, he went further than simply identifying the roots. He was, for example, able to identify certain cases in which the coefficients that appeared in the equations were functions of the equation's solutions. This is, in a sense, the reverse of the problem considered by Tartaglia and Ferrari, who found formulas that gave the solutions as functions of the coefficients. This observation allowed Viète to begin making new connections between the coefficients that appeared in the equation and the roots of the equation.

Viète also began to notice relationships between the degree of the equation and the number of roots of the equation. He demonstrated that at least in certain cases, the number of roots was the same as the degree of the equation. (He was prevented from drawing more general conclusions by his narrow conception of what a number, and hence a solution, is.)

All of these observations are important because there are many connections between the solutions of an algebraic equation and the form of the equation. It turns out that if one knows the coefficients and the degree of the equation then one also knows a great deal about the roots, and vice versa. Then as now, the exact solutions (roots) of an equation were sometimes less important to mathematicians than other, more abstract properties of the equation itself. Viète may well have been the first mathematician to think along these lines.

Viète eventually returned to government service, but he did not abandon his mathematical studies. He was a successful cryptographer during a war between France and Spain—so successful that the Spanish king complained to the pope that the French were using sorcery to break the Spanish codes. Viète also wrote books about astronomy and trigonometry, and he wrote about the three classical, unsolved problems of ancient Greece: trisection of an angle, squaring of a circle, and doubling of a cube. (During Viète's life, these problems had become fashionable again, and claims were made that all three of the problems had been solved. Viète rightly showed that all of the new proofs were faulty and that the problems remained unsolved.) Viète, a lawyer by training, was one of the most forward-thinking mathematicians of his time. His insights into algebra permanently and profoundly changed the subject.

Thomas Harriot

Very little is known of Thomas Harriot (1560–1621) before he enrolled in Oxford University. After he graduated, however, he became a public figure because of his association with Sir Walter Raleigh (ca. 1554–1618). Today, Raleigh is best remembered as an adventurer and writer. He sailed the Atlantic Ocean, freely confusing the national good with his own personal profit, and he attempted to establish a colony on Roanoke Island in present-day North Carolina. He sailed to present-day Guyana in search of gold to loot. He wrote about his adventures, and the stories of his exploits made him a popular figure with the general public and with Queen Elizabeth I, who became his patron and protector.

Walter Raleigh was also a serious student of mathematics. He was not, apparently, an insightful mathematician, but he knew enough to hire Thomas Harriot, who was probably the best mathematician in England at the time. Raleigh was interested in the mathematical problems associated with navigation, and Harriot, who was active in a number of scientific and technical areas, spent a good deal of his time researching how best to use observations of the Sun and stars to determine one's latitude as accurately as possible.

In addition to his work for Raleigh, Harriot was a creative scientist with an intense interest in the progress of science and mathematics. In some ways, Harriot was a more modern mathematician than Viète. His algebraic notation was simpler than Viète's and his concept of number broader than that of his French contemporary.

In particular, Harriot accepted positive, negative, and imaginary roots as solutions to algebraic equations, although he had only a vague idea of what an imaginary number is. His broader concept of solution enabled him to make an interesting and very important observation about polynomials and their roots.

Since antiquity, mathematicians interested in algebra had sought to solve the following problem: Given a polynomial, find the roots. Harriot introduced what might be called an inverse to the ancient problem: Given a collection of numbers, find a polynomial with those numbers as roots.

To see how this works, suppose that we are given the numbers a, b, and c. Harriot discovered that he could write a third degree polynomial with these numbers as roots. Here, in modern notation, is his solution: $(x - a)(x - b)(x - c)$. Multiplied out, this expression can be written as $x^3 - (a + b + c)x^2 + (ab + bc + ac)x - abc$. Harriot's was an important observation for two reasons.

First, recall that the product of several numbers can equal zero only if at least one of the numbers is zero. Consequently, the polynomial $(x - a)(x - b)(x - c)$ equals zero only when x equals a, or b, or c. Second, the expression $x^3 - (a + b + c)x^2 + (ab + bc + ac)x - abc$ shows how each coefficient can be written using only the roots of the polynomial. Once Harriot's observations about the inverse problem were generalized they would prove very useful to future generations of mathematicians.

Harriot might have exercised as much influence over the development of algebra as Viète if he had taken time to publish his results, but he did not. His early years were spent working for Raleigh, and his later years were in continual turmoil because of his former association with Raleigh. King James I, who succeeded Elizabeth I upon her death, disapproved of Raleigh's adventurism. Soon Raleigh was accused of conspiring to overthrow James and imprisoned. Harriot, tarnished by his association with Raleigh, spent a brief time in jail and encountered difficulty in finding further support for his work. For 40 years, Harriot had made important discoveries in mathematics and science, but he had published nothing, and so his discoveries were not widely known. Shortly before he died of cancer he asked a friend to sort through his work and arrange it for publication. Eventually a single book entitled *Artis analyticae praxis* was published describing some of Harriot's work. It is through this one slim volume that we know of Harriot's accomplishments in algebra.

Albert Girard

Near the end of Harriot's life, the Flemish mathematician and engineer Albert Girard (1590–1633) was active in mathematics. Girard was educated at the University of Leiden, in Holland, and for a time he worked with the Flemish scientist, mathematician, and engineer Simon Stevin. During his own time, Girard was known as a military architect, designing fortifications and the like, but today he is better remembered for an interesting observation—but not a proof—about the nature of algebraic equations. Girard correctly hypothesized that every polynomial of degree n has n roots.

To appreciate what was new about Girard's hypothesis, recall that al-Khwārizmī and Viète only accepted positive roots as valid roots. Consequently, for them it was false that every polynomial of degree n has n roots. In order for Girard's conjecture to be true, one must be willing to accept negative numbers and even complex numbers as roots. This was part of the problem: Even if one accepted that Girard's conjecture was correct, its meaning was not entirely clear because no one had a very detailed understanding of the number system.

Another way of understanding Girard's guess is by analogy with the set of natural numbers. Each natural number belongs to exactly one of the following three sets: the set of all prime numbers, the set of all composite numbers, or the set consisting of the number 1. (The number 1 constitutes its own class; it is neither prime nor composite.) A prime number is only divisible by itself and 1. The first five prime numbers, for example, are 2, 3, 5, 7, and 11. Any natural number larger than 1 that is not prime is called a composite number. Every composite number can always be written as a product of prime numbers. The number 462, for example, is composite, and it can be written as the product of four primes: $462 = 2 \times 3 \times 7 \times 11$. Consequently, prime numbers are

$$x^n + a_{n-1}x^{n-1} + a_{n-2}x^{n-2} + \ldots + a_1x + a_0 = (x - r_1)(x - r_2)(x - r_3)\ldots(x - r_n)$$

Girard hypothesized that every polynomial of degree n *has* n *roots. This is equivalent to the hypothesis that every polynomial of degree* n *can be written as a product of* n *linear factors.*

like building blocks. From the set of prime numbers, every natural number greater than 1 can be "constructed." Or to put it another way: Any natural number greater than 1 can be written as a product of primes.

Essentially, Girard speculated that linear factors—that is, expressions of the form $(x - a)$—have the same function in the set of all polynomials that prime numbers have in the set of all natural numbers. To appreciate Girard's conjecture in more modern notation, suppose that we are given a polynomial of the form $x^n + a_{n-1}x^{n-1} + a_{n-2}x^{n-2} + \ldots + a_2x^2 + a_1x + a_0$. (Recall that the a_js are real numbers, and the x^js represent the variable x raised to the power j, where j is always a natural number.) Girard asserted that there exist n linear factors $(x - r_1)(x - r_2), \ldots (x - r_n)$ such that when they are multiplied together one obtains the given polynomial. The resulting equation, which is a generalization of Harriot's observation, looks like this:

$$x^n + a_{n-1}x^{n-1} + a_{n-2}x^{n-2} + \ldots + a_2x^2 + a_1x + a_0$$
$$= (x - r_1)(x - r_2)(x - r_3) \ldots (x - r_{n-1})(x - r_n)$$

At the time, this was a very hard statement to prove. Given an arbitrary polynomial, no one knew how to prove that the necessary linear factors existed, and no one knew the nature of the roots that appear on the right side of the equation. Today, it is relatively easy to show that the linear factors exist; we have a much keener appreciation of the difficulties involved in finding the linear factors, and we have a much clearer understanding of the nature of the roots that appear in the factors.

To better appreciate how an incomplete understanding of the number system proved to be a barrier to progress in the study of algebraic equations, recall that complex numbers are defined as numbers of the form $a + bi$, where a and b are real numbers, and the letter i represents a number with the property that $i^2 = -1$. But this modern definition of a complex number, a definition that most students now encounter in their high school math class, was of little help to mathematicians of Girard's time. They examined equations such as $x^4 + 1 = 0$, and they could see that $\sqrt{i}$ was a root of this polynomial, but could $\sqrt{i}$ be written in the form $a + bi$? Is $\sqrt{i}$ a complex number? Today, we know the answer is yes, but in Girard's time no one knew the answer. No one knew how to expand the number system until it contained all solutions to all algebraic equations.

With the work of del Ferro, Tartaglia, and Ferrari on the one hand, and Viète, Harriot, and Girard on the other, mathematicians were faced with two very different paths for research. Mathematicians could search (and some did search!) for algorithms that would enable them to write the roots of polynomials of degree higher than four using only the coefficients that appear in the polynomials. This type of research was an extension of the discoveries of del Ferro, Tartaglia, and Ferrari. After two centuries of additional effort, the efforts of these mathematicians culminated in the surprising discovery that such algorithms do not exist. Alternatively, mathematicians could develop the necessary concepts to prove Girard's conjecture that every polynomial of degree n can be written as a product of linear factors. Some of the best mathematical minds of the next two centuries would work on this problem. It was solved almost as soon as a modern representation of the complex number system was developed.

The Fundamental Theorem of Algebra

Approximately two centuries separated Girard's inspired guess that every polynomial can be written as a product of linear factors and the first proof that he had guessed correctly. Many prolific and successful mathematicians tried their hand at producing a proof, and a modest amount of progress was made. Typical of the times were the efforts of the Swiss mathematician and scientist Leonhard Euler (1707–93), who proved that every polynomial of degree less than or equal to six could be written as a product of linear factors. Euler believed that the result was generally true—that is, that every polynomial could be written as a product of linear factors—but he was unable to prove it.

By the end of the 18th century, few mathematicians had any doubt that every polynomial could be written as a product of linear factors, but they were unable to show it. But mathematics is a deductive science. No result can be accepted as true unless it is proved true. The statement that every polynomial can be written as a product of linear factors cannot be accepted simply because it is true for every polynomial that has been examined. There are infinitely many polynomials. Most are of a degree so high that it is impossible to write the largest exponent appearing in the polynomial. It is just too large. Consequently, the situation cannot be resolved by examining individual cases. The statement must be shown to be a logical consequence of the axioms that form the basis of the subject. This was accomplished early in the 19th century—most famously by the German mathematician and scientist Carl Friedrich Gauss (1777–1855).

Gauss showed mathematical promise at a young age and was awarded a stipend by the duke of Brunswick. The stipend made it possible for Gauss to go through high school and university and to earn a Ph.D. His proof of what is now called the fundamental theorem of algebra was part of his Ph.D. thesis. It was his first proof, and there were some gaps in it, but it was a great step forward.

Gauss's proof of this important result did not lead him to a job immediately after graduation. However, he did not want a job. He preferred to study, and he was able to act independently because the stipend that he received from the duke continued for several years after he was awarded the Ph.D. During this time he devoted himself to his own mathematical research. He obtained a job as director of the astronomical observatory at Göttingen University only after the duke died and his stipend was discontinued. Gauss remained in his position as director of the observatory for his entire working life.

Despite the many discoveries Gauss made throughout his life, the fundamental theorem of algebra was an especially important idea for him. As we mentioned, his first attempt at proving the fundamental theorem of algebra, the attempt that earned him a Ph.D., had several gaps in it. He later revised the proof to correct its initial deficiencies. In fact Gauss never stopped tinkering with the fundamental theorem of algebra. He later published a third proof, and when he wanted to celebrate the 50th anniversary of receiving his Ph.D., he published a fourth proof. Each proof approached the problem from a slightly different perspective. He died not long after publishing his fourth proof, his mathematical career bracketed by the fundamental theorem of algebra.

Gauss's eventual success in developing a completely rigorous proof of the fundamental theorem was due, in part, to his firm grasp of the nature of complex numbers. It is no coincidence that he was also one of the first mathematicians to develop a clear, geometrical interpretation of the complex number system. Gauss represented the complex number system as points on a plane. Each complex number $a + bi$ is interpreted as a point (a, b) in the so-called complex plane (see the accompanying figure, on page 80). This geometric representation of the complex number system is the one that is in common use today. Gauss was not the only person to have this particular insight. Though the idea seems simple enough, it was a very important innovation. This is demonstrated by the fact that after centuries of work, two of the earliest individuals to discover this clear and unambiguous interpretation of the complex number system also discovered proofs of the fundamental theorem of algebra.

Another early proof of the fundamental theorem of algebra was given by Jean Robert Argand (1768–1822). Argand was Swiss-born. He was a quiet, unassuming man. Little is known of his early background or even of his education. We do know that he lived in Paris and worked as a bookkeeper and accountant, that he was married and had children, and that as much as he enjoyed the study of mathematics, it was for him just a hobby.

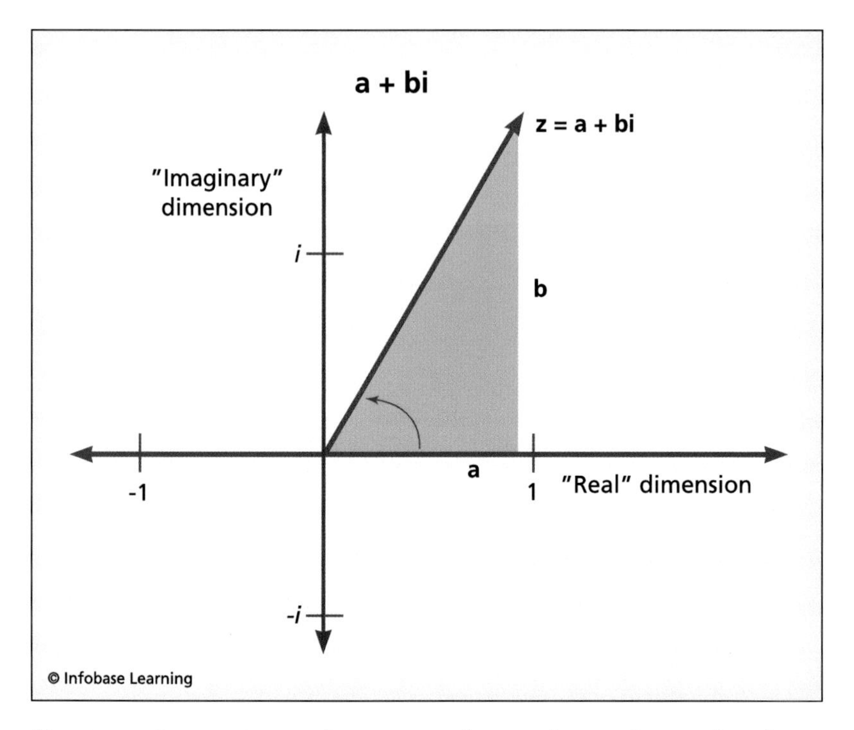

Every complex number can be represented as a point on the complex plane.

Argand was not especially aggressive in making himself known to other mathematicians. In 1806 he published a thin mathematics book at his own expense. In it one finds two of the most important ideas of the time: a geometrical representation of the complex numbers and a proof of the fundamental theorem of algebra. Because the book was published anonymously, years later, when the work had finally attracted the attention of some of the best mathematicians of the day, a call went out for the unknown author to identify himself and claim credit for the ideas contained in the work. It was only then that Argand stepped forward to identify himself as the author. He later published a small number of additional papers that commented on the work of other authors or elaborated on the ideas contained in his original, anonymously published book.

Argand's geometric representation of the complex numbers is today known as the Argand diagram. It is the interpretation of the complex numbers that students learn first when introduced

THE IMPORTANCE OF POLYNOMIALS

For much of the history of the human race, mathematicians have worked to understand the mathematical structure of polynomials. They have developed algorithms to find the roots of polynomials, they have developed algorithms that allow them to approximate the roots of polynomials, and they have studied the mathematical relationships that exist between algebraic (polynomial) equations and their roots. The study of polynomials has motivated the development of many important mathematical ideas.

Polynomials also play an important role in scientific and engineering computations. In many of the mathematical equations that arise in these disciplines, the unknown is not a number but a function. That function may represent the path of a rocket through space or, in meteorology, the position of a high-pressure front as it moves across Earth's surface. Equations that have functions, instead of numbers, for solutions are often exceedingly difficult to solve. In fact, as a general rule, the precise solutions are often impossible to calculate. The strategy that applied scientists adopt, therefore, is to construct a function that approximates the exact solution. A polynomial, or a set of polynomials, is often the ideal choice for an approximating solution. There are two main reasons polynomials are so widely used.

First, polynomials are well-understood mathematical functions. In addition to the fundamental theorem of algebra there are a host of other theorems that describe their mathematical properties. These theorems enable scientists and engineers to calculate with polynomials with relative ease.

Second, there are many polynomials to choose from. This means that in many problems of practical importance there are sufficiently many polynomials to enable the scientist or engineer to calculate a very accurate approximation to the solution by using only polynomials. This method is often used despite the fact that the exact solution to the equation in which they have an interest is not a polynomial at all.

These two facts have been known to mathematicians since the 19th century, but the computational difficulties involved in calculating the desired polynomials often made applying these ideas too difficult. With the advent of computers, however, many of the computational difficulties disappeared, so that from a practical point of view polynomials are now more important than ever.

to the subject. His proof that every algebraic equation of degree n with complex coefficients has n roots would also seem familiar to students interested in more advanced algebra. His approach to proving the fundamental theorem is similar to a common modern proof.

The fundamental theorem of algebra is the culmination of a theory of equations that began with the work of ancient Mesopotamian scribes pressing triangular shapes into slabs of wet clay on the hot plains of Mesopotamia thousands of years ago. It illuminates basic connections between polynomials and the complex number system. It does not solve every problem associated with polynomials, of course. There were still questions, for example, about the computational techniques needed to approximate the roots of polynomials and about the role of polynomials in broader classes of functions. Furthermore special classes of polynomials would eventually be identified for their utility in solving practical computational problems in science and engineering. (This research would be further accelerated by the invention of computers.) And, finally, the fundamental theorem itself sheds no light on why the methods that had proved so useful for finding the roots of algebraic equations of degree less than 5 would, in general, prove ineffective for finding the roots of equations of degree 5 or more. Nevertheless the fundamental theorem shows how several properties of polynomials that had been of interest to mathematicians for the last 4,000 years are related.

- It relates the degree of an equation to the number of its solutions.
- It demonstrates that, in theory, any polynomial can be factored.
- It shows that the complex number system contains all solutions to the set of all algebraic equations.

Research into algebra did not end with the fundamental theorem of algebra, of course; it shifted focus from the study of the solutions of polynomials toward a more general study of the logical structure of mathematical systems.

5

ALGEBRA IN GEOMETRY AND ANALYSIS

In the 17th century mathematicians began to express geometric relationships algebraically. Algebraic descriptions are in many ways preferable to the geometric descriptions favored by the ancient Greeks. They are often more concise and usually easier to manipulate. But to obtain these descriptions, mathematicians needed a way of connecting the geometric ideas of lines, curves, and surfaces with algebraic symbols. The discovery of a method for effecting this connection—now called analytic geometry—had a profound impact on the history of mathematics and on the history of science in general.

In order to appreciate the importance of applying algebraic methods to the study of geometry, it helps to know a little about how the ancient Greeks studied curves. The Greek mathematician Apollonius of Perga (ca. 262 B.C.E.–ca. 190 B.C.E.), for example, was one of the most successful of all Greek mathematicians, and his major work, called *Conics*, was entirely devoted to the study of four curves: the ellipse, the circle, which is closely related to the ellipse, the parabola, and the hyperbola. These are the so-called conic sections.

Apollonius described each of the conic sections as the intersection of a plane with a double cone. (A double cone looks like two ice cream–like cones joined together at their respective vertices. Here is how Apollonius described a double cone: Begin with a circle and a point directly above the center of the circle. Call the point *P*. Next imagine a line passing through *P* and some point on the circle. Now pivot the line through the point *P* in such a way that it remains in contact with the circle. The line will trace out a double cone in space. The exact shape of the cone will depend on the distance between the point P and the center of the circle.) If the plane cuts all of the way across a single cone, then the result is a circle or an ellipse. A circle is produced when the cone's line of symmetry makes a right angle with the plane; change the angle slightly and the result is an ellipse. If the plane cuts across the cone in such a way that it is parallel with a line that passes through P and remains in contact with the cone, then the result is a parabola. Finally, if the plane cuts both the upper and lower cones, the result is a hyperbola.

DISCOURS
DE LA METHODE
Pour bien conduire ſa raiſon, & chercher
la verité dans les ſciences.
PLUS
LA DIOPTRIQVE.
LES METEORES.
ET
LA GEOMETRIE.
Qui ſont des eſſais de cete METHODE.

A LEYDE
De l'Imprimerie de IAN MAIRE.
CIↃ IↃ C XXXVII.
Avec Privilege.

A philosopher and scientist as well as a mathematician, Descartes revealed connections between algebra and geometry that contributed to progress in both fields in his Discours de la Methode. *Its publication marks the beginning of a new era in mathematics.*

The preceding description of the conic sections is very wordy because it uses no algebra. Instead, it describes these curves in the same way that Apollonius understood them, which, when compared with modern methods, is a very awkward way indeed. But Apollonius's geometric construction was typical of the way that the Greeks described all the curves that they studied—a method of description so awkward that although the Greek mathematical tradition spanned six centuries, they studied only about one dozen curves. Their dependence on diagrams and complicated sentences proved a barrier to progress, and that barrier remained in place for more than 1,000 years.

Coordinate systems were what enabled mathematicians

to surmount what the 17th-century mathematician René Descartes called "the tyranny of the diagram." By establishing a correspondence between points on a plane and ordered pairs of numbers, mathematicians were able to express planar curves (geometry) as relations between variables (algebra). Most of us become so familiar with coordinate systems at such an early age that we fail to appreciate their importance. But coordinate systems enabled mathematicians to interpret algebraic statements in the language of geometry and sometimes solve algebraic problems using geometric methods. Coordinate systems also enabled mathematicians to express geometric problems in the language of algebra, and when they did so they discovered that difficult geometry problems sometimes had simple algebraic solutions. Progress in one branch of mathematics led to progress in the other. The result of this effort was what we now call analytic geometry. The foundational concepts of analytic geometry were proposed almost simultaneously by two 17th-century mathematicians, René Descartes and Pierre de Fermat.

René Descartes

The French mathematician, scientist, and philosopher René Descartes (1596–1650) was one of the more colorful characters in the history of mathematics. Although we will concentrate on his ideas about mathematics, his contributions to several branches of science are just as important as his mathematical innovations, and today he is perhaps best remembered as a philosopher.

Descartes's mother died when he was an infant. His father, a lawyer, ensured that Descartes received an excellent education. As a youth Descartes displayed a quick intellect, and he was described by those who knew him at the time as a boy with an endless series of questions. He attended the Royal College at La Flèche and the University of Poitiers, but the more education he received, the less pleasure he seemed to derive from it. Given his academic record this is a little surprising. He was a talented writer who demonstrated a real gift for learning languages. He also displayed an early interest in science and math. The Royal College, where he received his early education, accommodated his idiosyncrasies: The rector at the school allowed Descartes to spend his mornings in bed. Descartes enjoyed lying in bed thinking, and he apparently maintained this habit for most of his life. Nevertheless by the time he graduated from college he was confused and disappointed. He felt that he had learned little of which he could be sure. It was a deficiency that he spent a lifetime correcting.

After college Descartes wandered across Europe for a number of years. On occasion he enlisted in an army. This was not an uncommon way for a young gentleman to pass the time. He claimed that as a young man he enjoyed war, though there are conflicting opinions about how much time he spent fighting and how much time he spent "lying in" each morning. (Ideas about military discipline have changed in the intervening centuries.) In addition to his military adventures, Descartes took the time to meet intellectuals and to exchange ideas. This went on for about a decade. Eventually, however, he settled in Holland, where he remained for almost two decades, writing and thinking.

Holland was a good place for Descartes. His ideas were new and radical, and like most radical ideas, good and bad, Descartes's ideas were not especially popular. In a less tolerant country, he would have been in great danger, but because he was under the protection of the Dutch leader, the prince of Orange, he was safe from physical harm. Though he was not physically attacked for his ideas, there was a period when his books were banned.

During his stay in Holland Descartes applied himself to exploring and describing his ideas in mathematics, science, and philosophy. In mathematics his major discoveries can be found in the book *Discours de la méthode* (Discourse on method), especially in an appendix to this work that describes his ideas on geometry. It is in the *Discours* that Descartes makes the necessary connections between geometry and algebra that resulted in a new branch of mathematics. It is also in this book that he developed most of the algebraic symbolism that we use today. With very few exceptions, Descartes's algebra resembles our algebra. (The reason is that our algebra is modeled on Descartes's.) Most modern readers can understand Descartes's own equations without difficulty.

One of Descartes's simpler and yet very important contributions was to reinterpret ideas that were already known. Since the days of the ancient Greeks, an unknown was associated with a line segment. If we call the unknown x, the product of x with itself was interpreted as a square. This is why we call the symbol x^2 "x squared." This geometric interpretation had been a great conceptual aid to the Greeks, but over the intervening centuries it had become a barrier to progress. The difficulty was not with x^2 or even $x \times x \times x$, written x^3 and called "x cubed." The symbol x^3 was interpreted as a three-dimensional cube. The problem with this geometric interpretation was that it required one to imagine a four-dimensional "cube" for the product of x with itself four times, a five-dimensional cube for the product of x with itself five times, and so forth. This impeded understanding. The great mathematician Omar Khayyám, for example, was unable to assign a meaning to a polynomial of degree 4, because he was not able to see past this type of geometric interpretation of the symbol x^4.

Descartes still imagined the variable x as representing a line segment of indeterminate length. His innovation was the way he imagined higher powers of x and, more generally, the geometric interpretation he gave products. Descartes, for example, simply imagined x, x^2, x^3, x^4, and so forth, as representing lengths of line segments, and products of two different variables, x and y, as representing the length of a third line segment of length xy instead of a rectangle of area xy as the Greeks and their successors had imagined. To make the idea palatable, he described it geometrically (see the sidebar Descartes on Multiplication).

Today, Descartes's name is used to describe a particular type of coordinate system, the Cartesian coordinate system. As with all coordinate systems—and over the intervening centuries scientists, engineers, and mathematicians have created many different coordinate systems—the Cartesian system is a method for establishing a correspondence between points and numbers. A two-dimensional Cartesian system is formed by identifying a special point, which is called the origin, and a line passing through the origin, which we will call the x-axis. A second point on the x-axis is used to establish a direction and a distance. The distance from the origin to this second point is taken as one unit, and the direction one travels from the origin to the second point identifies the direction of increasing x. The line that passes through the origin and is perpendicular to the x-axis is the y-axis. By convention, the positive direction on the y-axis is determined by that of the x-axis: Rotate the x-axis 90 degrees about the origin in a counterclockwise direction, and the rotated x-axis will be pointing in the direction of increasing y. Often distances along the y-axis are measured according to the same scale as distances along the x-axis, but this is not necessary, and sometimes it is advantageous to use a different scaling along the y-axis. Once the origin, the axes, and

DESCARTES ON MULTIPLICATION

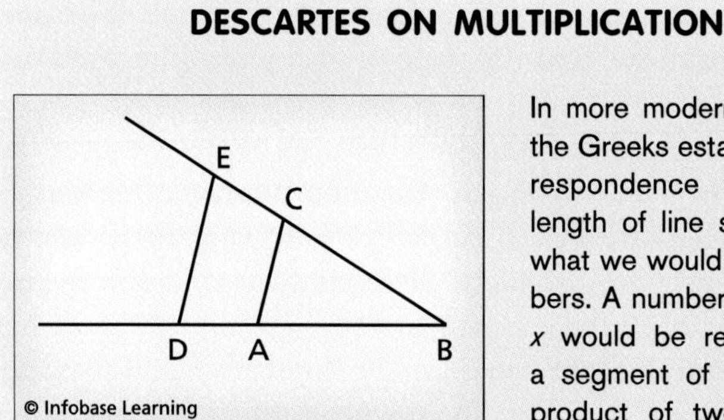

Descartes's geometric interpretation of the operation of multiplication

In more modern terminology, the Greeks established a correspondence between the length of line segments and what we would call real numbers. A number of magnitude x would be represented by a segment of length x. The product of two numbers x and y was represented as a rectangle with the segment of length x forming one side of the rectangle and the segment of length y forming the other. This works well until one wants to consider products of numbers u, v, x, and y. Most of us have a difficult time picturing a way of orienting line segments of length u, v, x, and y so as to form a four-dimensional rectangular solid.

Descartes's innovation was to use triangles rather than rectangles and imagine all products as simply line segments of the appropriate length. We use the accompanying diagram to paraphrase Descartes's ideas on multiplication. Suppose

- the distance from A to B is one unit long

the scaling have been determined, a unique pair of numbers—the x and y coordinates—can be placed in correspondence with each point on the plane, and every point on the plane can be placed in correspondence with a unique pair of numbers. Questions about points can now be rephrased as questions about ordered pairs of numbers and vice versa. Conceptually similar Cartesian coordinate systems can be used to describe three-dimensional space, which requires an ordered triplet of numbers, four-dimensional space, which requires an ordered "four-tuple" of numbers, and so forth. Cartesian coordinate systems are often described as one of Descartes's important contributions, but the situation is more complicated.

- the distance from A to C is x units long and
- the distance from B to D is y units long

and construct a segment passing through D that is parallel to the line AC. Segment DE is this parallel line segment. The triangles ABC and DBE are similar, so the ratios of their corresponding sides are equal. In symbols this is written as

$$AC/1 = DE/BD$$

or using x and y in place of AC and BD, respectively,

$$x/1 = DE/y$$

or, finally, solving for DE, we get

$$xy = DE$$

With this diagram Descartes provided a new and more productive geometric interpretation of arithmetic.

Descartes's innovation freed mathematicians from the limiting ideas of Greek and Islamic mathematicians about the meaning of multiplication and other arithmetic operations. He also showed that the requirement of homogeneity that had made using Viète's algebra so awkward was unnecessary.

Descartes's understanding of analytic geometry was profoundly different from ours. We acknowledge his contribution by calling the most common of all coordinate systems in use today the Cartesian coordinate system, but Descartes made little use of Cartesian coordinates. To be sure, he recognized the value of coordinates as a tool in bridging the subjects of algebra and geometry, but he generally used oblique coordinates. (In an oblique coordinate system, the coordinate axes do not meet at right angles.) Furthermore because he questioned the reality of negative numbers, he refrained from using negative coordinates. As a consequence Descartes restricted himself to what we would call the first quadrant, that part of the coordinate plane where both coordinates are positive. He did, however, recognize and exploit the connection between equations and geometric curves, and this was extremely important.

In Descartes's time conic sections—ellipses, parabolas, and hyperbolas—were still generally described as Apollonius had described them. Descartes explored the connections between the geometric descriptions of conic sections and algebraic equations. He did this by examining the connections between geometry and the *algebraic* equation $y^2 = ay - bxy + cx - dx^2 + e$. (In this equation x and y are the variables and a, b, c, d, and e are the coefficients.) Depending on how one chooses the coefficients one can obtain an algebraic description of any of the conic sections. For example, if a, b, and c are chosen to be 0 and d and e are positive, then the equation describes an ellipse. If, on the other hand, a, b, d, and e are taken as 0 and c is not 0 then the graph is a parabola. Descartes went much further in exploring the connections between algebra and geometry than any of his predecessors, and in doing so he demonstrated how mathematically powerful these ideas are.

What, in retrospect, may have been Descartes's most important discoveries received much less attention from their discoverer than they deserved. Descartes recognized that when one equation contains two unknowns, which we call x and y, there is generally more than one solution to the equation. In other words, given a value for x, we can, under fairly general conditions, find a value for y such that together the two numbers satisfy the one equation. The set of all such solutions as x varies over some interval forms a curve. These observations can be made mathematically precise, and the

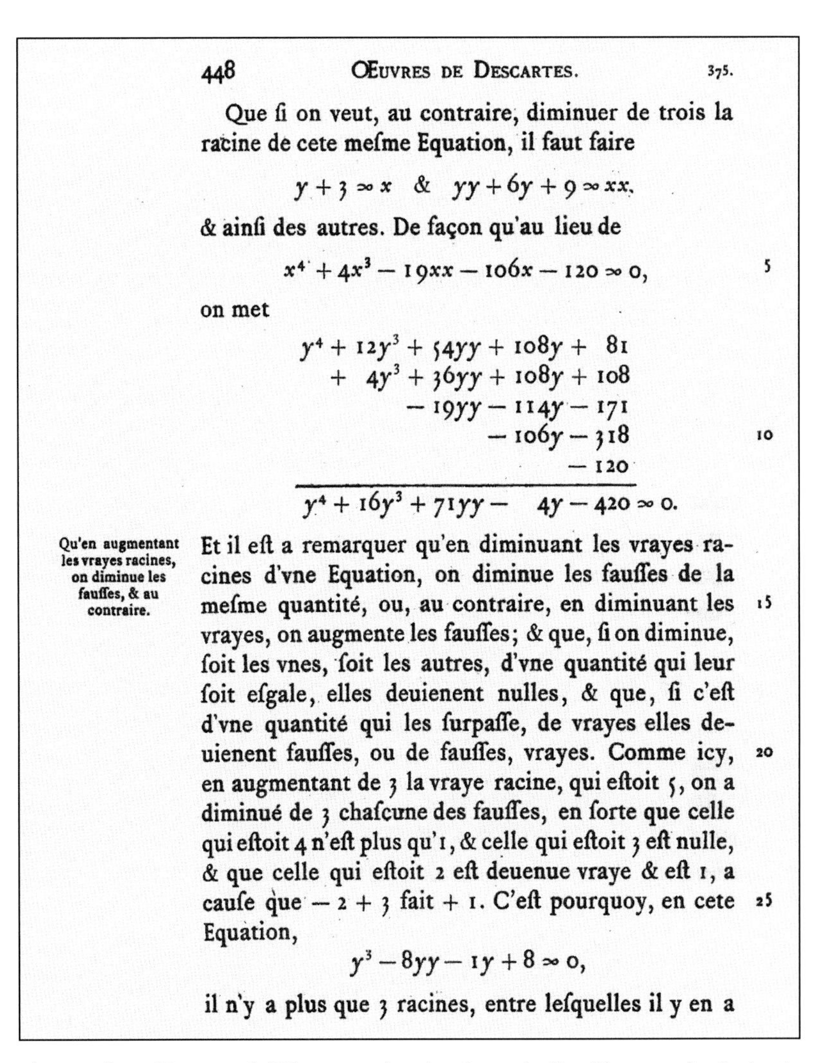

448 ŒUVRES DE DESCARTES. 375.

Que ſi on veut, au contraire, diminuer de trois la racine de cete meſme Equation, il faut faire

$$y + 3 \infty x \quad \& \quad yy + 6y + 9 \infty xx.$$

& ainſi des autres. De façon qu'au lieu de

$$x^4 + 4x^3 - 19xx - 106x - 120 \infty 0,$$ 5

on met

$$\begin{array}{r} y^4 + 12y^3 + 54yy + 108y + 81 \\ + 4y^3 + 36yy + 108y + 108 \\ - 19yy - 114y - 171 \\ - 106y - 318 \\ - 120 \\ \hline y^4 + 16y^3 + 71yy - 4y - 420 \infty 0. \end{array}$$ 10

Qu'en augmentant les vrayes racines, on diminue les fauſſes, & au contraire.

Et il eſt a remarquer qu'en diminuant les vrayes racines d'vne Equation, on diminue les fauſſes de la meſme quantité, ou, au contraire, en diminuant les vrayes, on augmente les fauſſes; & que, ſi on diminue, ſoit les vnes, ſoit les autres, d'vne quantité qui leur ſoit eſgale, elles deuienent nulles, & que, ſi c'eſt d'vne quantité qui les ſurpaſſe, de vrayes elles deuienent fauſſes, ou de fauſſes, vrayes. Comme icy, en augmentant de 3 la vraye racine, qui eſtoit 5, on a diminué de 3 chaſcune des fauſſes, en ſorte que celle qui eſtoit 4 n'eſt plus qu'1, & celle qui eſtoit 3 eſt nulle, & que celle qui eſtoit 2 eſt deuenue vraye & eſt 1, a cauſe que $-2+3$ fait $+1$. C'eſt pourquoy, en cete Equation,

$$y^3 - 8yy - 1y + 8 \infty 0,$$

il n'y a plus que 3 racines, entre leſquelles il y en a

A page from Descartes's Discours *showing how similar Descartes's algebraic notation is to modern notation* (Courtesy of University of Vermont)

precise expression of these ideas is often called the fundamental principle of analytic geometry. Descartes knew the fundamental principle of analytic geometry, but he seems to have considered it less important than some of the other ideas contained in his work.

The fundamental principle of analytic geometry is important because it freed mathematicians from the paucity of curves that had been familiar to the ancient Greeks. Descartes had discovered a method for generating infinitely many new curves: Simply write one equation in two variables; the result, subject to a few not-very-demanding conditions, is another new curve. Descartes went even further. A single equation that involves exactly three variables—x, y, and z, for example—in general, describes a surface. This is called the fundamental principle of solid analytic geometry, and it, too, was known to Descartes. Today this is recognized as a very important idea, but its importance does not seem to have been recognized by Descartes. He gives a clear statement of the principle but does not follow it with either examples or further discussion. He understood the idea, but he did not use it.

The principles of analytic and solid geometry, so clearly enunciated by Descartes, were important because they pointed to a way of greatly enriching the vocabulary of mathematics. Conic sections and a handful of other curves, as well as cylinders, spheres, and some other surfaces, had been studied intensively for millennia, in part because few other curves and solids had mathematical descriptions. Descartes's insights had made these restrictions a thing of the past. His algebra and the fundamental principles of analytic geometry and solid geometry changed the development of mathematics in a fundamental way.

Descartes's discoveries in science, mathematics, and philosophy eventually attracted the attention of the queen of Sweden. Queen Christina invited him to become a member of her court, and Descartes accepted. Descartes was not a man who liked the cold. Nor did he like to get up early in the morning. (He had maintained his habit of spending his mornings in bed throughout his life.) On his arrival in September, Descartes must have been dismayed to learn that Queen Christina liked to receive her instruction from her new philosopher at five in the morning. Descartes died in the cold of a Swedish December, less than five months after arriving at the queen's court.

Pierre de Fermat

The French mathematician and lawyer Pierre de Fermat (1601–65) also discovered analytic geometry, and he did so independently of René Descartes. Little is known of Fermat's early life. He was educated as a lawyer, and it was in the field of law that he spent his working life. He worked in the local parliament in Toulouse, France, and later he worked in the criminal court. We also know that he had an unusual facility with languages. He spoke several languages and enjoyed reading classical literature. He is best remembered for his contributions to mathematics, which were profound.

Today much of what we know of Fermat is derived from the numerous letters that he wrote. He maintained an active correspondence with many of the leading mathematicians of his time. His letters show him to be humble, polite, and extremely curious. He made important contributions to the development of probability theory, the theory of numbers, and some aspects of calculus, as well as analytic geometry. Mathematics was, however, only one of his hobbies.

One activity that Fermat shared with many of the mathematicians of his time was the "reconstruction" of lost ancient texts. By the early 17th century some of the works of the ancient Greek mathematicians had again become available. These were, for the

Ancient (and modern) Beaumont de Lomagne, France, birthplace of Pierre de Fermat (Office of Tourism, Beaumont de Lomagne)

most part, the same texts with which we are familiar today. Most of the ancient texts, however, had been lost in the intervening centuries. Although the works had been lost, they had not been forgotten. The lost works were often known through commentaries written by other mathematicians. The ancient commentaries described the work of other mathematicians, but often they were much more than simple descriptions. Sometimes a commentary contained corrections, suggestions, or alternative proofs of known results; on occasion, the commentaries even contained entirely new theorems that extended those appearing in the work that was the subject of the commentary. Other times, however, a commentary simply mentioned the title of a work in passing. In any case, much of what we know about Greek mathematics and Greek mathematicians we know through the commentaries. It had become fashionable among mathematicians of the 17th century to try to reconstruct lost works on the basis of information gained from these secondary sources. Fermat attempted to reconstruct the book *Plane Loci* by Apollonius on the basis of information contained in a commentary written by the Greek geometer Pappus of Alexandria.

We will never know how close Fermat was in his reconstruction of the original, but the effort was not wasted. While attempting the reconstruction Fermat discovered the fundamental principle of analytic geometry: Under very general conditions, a single equation in two variables describes a curve in the plane.

As Descartes did, Fermat worked hard to establish algebraic descriptions of conic sections. Hyperbolas, ellipses, and parabolas were, after all, the classic curves of antiquity, and any attempt to express geometry in the language of algebra had at least to take these curves into account in order to be successful. Fermat was extremely thorough in his analysis. Again as Descartes did, he analyzed a very general second-degree equation in the variables x and y. Fermat's method was to manipulate the equation until he had reduced it to one of several standard forms. Each standard form represented a class of equations that were similar in the sense that each equation in the class could be transformed into a standard form via one or more elementary operations. (The standard

FERMAT'S LAST THEOREM

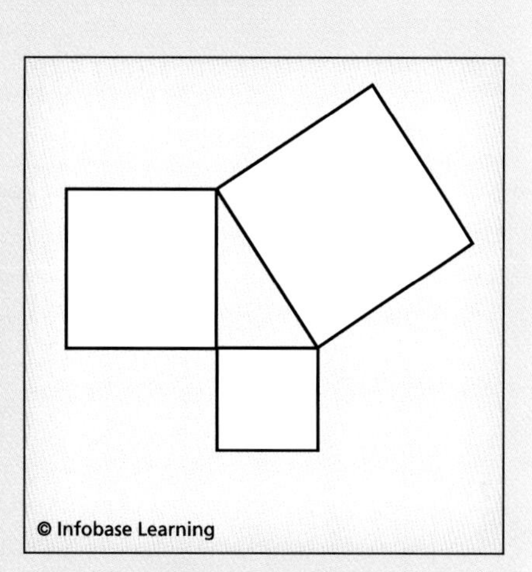

The sum of the areas of the two smaller squares equals the area of the largest square. The lengths of the sides of the triangle on which the squares are constructed are in the ratio 3:4:5.

One of Fermat's most famous insights is his so-called last theorem. This problem, which was finally solved late in the 20th century, is one of the most famous problems in the history of mathematics. It can, however, be understood as a generalization of a much older problem, the problem of finding Pythagorean triples. A Pythagorean triple is a set of three natural numbers with the property that if each number of the triple is squared then the sum of the two smaller squares equals the largest square. For example, the set (3, 4, 5) is a Pythagorean triple because, first, each number in the triple is a natural number, and, second, the three numbers satisfy the equation $x^2 + y^2 = z^2$, where we can let x, y, and z represent 3, 4, and 5, respectively. Another way of understanding the same problem is that we have represented the number 25, which is a perfect square, as the sum of two smaller perfect squares, 9 and 16. There are infinitely many Pythagorean triples, a fact of which the Mesopotamians seemed fully aware. (The Mesopotamians' work on Pythagorean triples is discussed in chapter 1.)

The generalization of the idea of Pythagorean triples of interest to Fermat involved writing natural numbers greater than 2 in place of the exponent in the equation $x^2 + y^2 = z^2$. The resulting equation is $x^n + y^n = z^n$, where n belongs to the set {3, 4, 5, . . .}. When n is equal to 3 we can interpret the problem geometrically: We are searching for three cubes, each of which has an edge that is an integral number of units long, such that the volume of the largest cube equals the sum of the two smaller volumes. When n is greater than 3 we can describe the problem in terms of hyper-cubes, but in higher dimensions there is

(continues)

FERMAT'S LAST THEOREM
(continued)

no easy-to-visualize generalization of the two- and three-dimensional interpretations.

Fermat's goal, then, was to find a triplet of natural numbers that satisfies any one of the following equations $x^3 + y^3 = z^3$, $x^4 + y^4 = z^4$, $x^5 + y^5 = z^5$, . . . He was unable to find a single solution for any exponent larger than 2. In fact, he wrote that he had found a wonderful proof that *there are no solutions* for any n larger than 2, but the margin of the book in which he was writing was too small to contain the proof of this discovery. No trace of Fermat's proof has ever been discovered, but his cryptic note inspired generations of mathematicians, amateur and professional, to try to develop their own proofs. Before World War I a large monetary prize was offered, and this inspired many more faulty proofs.

Throughout most of the 20th century mathematicians proved that solutions did not exist for various special cases. For example, it was eventually proved that if a solution did exist for a particular value of n, then n had to be larger than 25,000. *Most* natural numbers, however, are larger than 25,000, so this type of result hardly scratches the surface. In the late 20th century, Fermat's theorem was finally proved by using mathematics that would have been entirely unfamiliar to Fermat. The British-born mathematician Andrew Wiles devised the proof, which is about 150 pages long.

Wiles and many others do not believe that Fermat actually had a proof. They think that the proof that Fermat thought he had discovered actually had an error in it. This kind of thing is not uncommon in a difficult logical argument; Wiles himself initially published an incorrect proof of Fermat's theorem. Nevertheless, unlike most of his successors, Fermat was an epoch-making mathematician. As a consequence it would be wrong to discount completely the possibility that he had found a valid proof using only mathematics from the 17th century, but it does not seem likely.

form of the equation depended on the initial values of the coefficients.) Finally, Fermat showed that each of these standard forms described the intersection of a plane with a double cone. He had found a correspondence between a class of curves and a class of equations. This analysis was an important illustration of the utility of the new methods.

As Descartes did, Fermat used coordinates as a way of bridging the separate disciplines of algebra and geometry. Fermat, too, was comfortable using oblique coordinates as well as what we now call Cartesian coordinates.

Fermat and Descartes were each well aware of the work of the other. They even corresponded with each other through the French priest and mathematician Marin Mersenne (1588–1648). Mersenne was a friend of both men and a talented mathematician in his own right. In addition he opened his home to weekly meetings of mathematicians in the Paris area and worked hard to spread the news about discoveries in mathematics and the sciences throughout Europe.

Despite the many similarities in their work on analytic geometry and the fact that they both made their discoveries known to Mersenne, Descartes had much more influence on the development of the subject than did Fermat. One reason was that Fermat did not publish very much. In fact, Fermat only published a single paper during his lifetime. It was only later that his writings were collected and made generally available. Moreover, unlike Descartes, who had a flair for good algebraic notation, Fermat used the older, more awkward notation of François Viète.

Fermat's principal mathematical interest, however, was number theory, not analytic geometry. Although he tried to interest others in problems in the theory of numbers, Fermat was largely unsuccessful. For the most part, he worked on his favorite subject alone. His isolation, however, seemed to pose no barrier to creative thinking. He discovered a number of important results as well as a famous conjecture called Fermat's last theorem (see the sidebar).

The New Approach

Descartes and Fermat developed a new symbolic language that enabled them to bridge the gap that had separated algebra from geometry. This language contributed to progress in both fields.

Algebraic operations represented the manipulation of geometric objects, and geometric manipulations could now be expressed in a compact, algebraic form. Descartes and Fermat had made an important conceptual breakthrough, and unlike many other new mathematical ideas, these ideas were immediately recognized by their contemporaries as valuable.

Mathematicians interested in geometry exploited the fundamental principles of analytic and solid geometry to develop new ways of describing old curves and surfaces. They also explored the properties of entirely new curves and surfaces. The exploration

POLAR COORDINATES

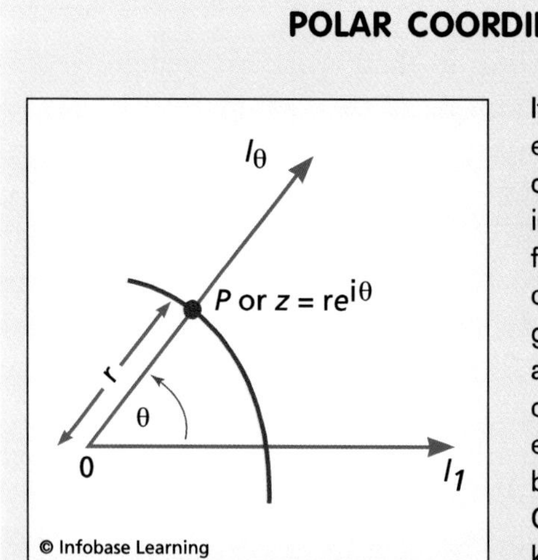

Polar coordinates provide still another way to visualize complex numbers. Each method of displaying complex numbers reveals something new about the numbers themselves.

It would be difficult to overestimate the effect of introducing coordinate systems into mathematics. For the first time, algebraic equations could be associated with graphs and graphs could be analyzed algebraically, and often the transition from geometry to algebra and from algebra to geometry was easy. Coordinate systems were a key factor in rapid progress in both algebra and geometry because a new insight into one discipline often translated into a new insight into the other.

Since Descartes introduced his coordinates, researchers have invented many other coordinate systems. Each system is a different way of "organizing" space. The choice of coordinate system depends on the needs of the researcher. One of the most widely used non-Cartesian coordinate systems is called the polar coordinate system. It is an alternative way of placing coordinates on the plane.

of geometry from an algebraic point of view and the application of geometry to algebra challenged many fine mathematicians. Descartes and Fermat had opened up a new mathematical landscape, and for several generations thereafter, mathematicians worked to extend the ideas and techniques that Descartes and Fermat had pioneered.

The geometric interpretation of algebraic quantities also influenced other branches of mathematics and science. Perhaps most importantly, the language of analytic geometry, somewhat modified and augmented, became the language of analysis, that branch

A polar coordinate system requires a point, O, which corresponds to the origin, and a single ray emanating out from O, which we will call l_1. To identify a point P that is different from O, first specify the distance from O to P. If r represents the distance from O to P, then the set of all points that are r units away from O is a circle of radius r with O as its center. Consequently, knowing r only allows us to conclude that P is one of the points on the circle. Next extend a ray from O to P. Call this ray l_θ. The Greek letter θ represents the angle (measured counterclockwise) that is formed by l_1 and l_θ. The unique point where l_θ intersects the circle of radius r center O is the point P. Each pair of coordinates (r, θ) determines a unique point on the plane. (See the diagram.)

Polar coordinates also enable us to represent complex numbers in a way different from that of the Argand diagram. Recall from page 80, every complex number z can be represented on the Argand diagram in the form $a + bi$, where a and b are real numbers. To convert this representation to a polar representation, notice that the distance from the origin to $a + bi$ is $\sqrt{a^2 + b^2}$. (This is just the Pythagorean theorem.) Call this distance r. To complete the representation, we use the fact that $e^{i\theta}$ identifies a point on the unit circle. (The "unit circle" is the circle of radius 1 centered at the origin.) If a ray is drawn from O through $e^{i\theta}$, that ray, which we will again call l_θ, makes an angle of θ *radians* with l_1. (Radians are just another set of units for measuring angles. The conversion factor is determined by the equation $360° = 2\pi$ *radians*.) The expression $re^{i\theta}$ is, therefore, that point on the complex plane that lies r units away from O along the ray l_θ. It is called the polar representation of a complex number. (The angle θ is called the argument of the complex number. Refer, for example, to the interview with Dr. Bonita Saunders in the Afterword on page 187.)

of mathematics that arose out of calculus. Calculus was discovered twice, once by the British physicist and mathematician Sir Isaac Newton (1643–1727), and again independently by the German philosopher, mathematician, and diplomat Gottfried Wilhelm Leibniz (1646–1716).

The new analysis enabled the user to solve problems in geometry and physics that had previously been too difficult. In fact, early in the development of analysis certain problems in geometry that Descartes himself had believed to be unsolvable were solved. The techniques the analysts used often required a great deal of analytic geometry. Newton, for example, invented and employed a number of coordinate systems to facilitate his study of both physics and geometry. Some of these coordinate systems have proved to be more important than others, but in every case they were extensions of the concepts of Descartes and Fermat: Each coordinate system established a correspondence between ordered sets of real numbers and (geometric) points. Each coordinate system served as a bridge between the magnitudes of geometry—those continually varying quantities, such as length, area, and volume—and the numbers and symbols of algebra.

Newton often interpreted the variables that arose in his studies as representing geometric magnitudes. In his studies of physics, however, Newton sometimes interpreted variables as quantities of another sort: forces, accelerations, and velocities. We take symbolic notation for granted today, but the symbolic language developed by some of these mathematicians contributed substantially to progress in the mathematical and physical sciences. Newton's notation was, however, only a modest extension of the notation used in the analytic geometry of his time. Newton absorbed the ideas of Descartes and Fermat and used these ideas throughout his work. He managed to develop a new branch of mathematics that used their notation, but he did not contribute much new notation himself.

Leibniz, who was much more gifted in languages than was Newton, greatly extended the notation of Descartes and Fermat to create a highly expressive symbolic language that was ideally suited to the new mathematics. He used this notation to express the ideas of analysis in a much more sophisticated way than that of Newton. He, too, generally interpreted the symbols that arose in his study of calculus as geometric or physical quantities. It was one of Leibniz's great accomplishments to extend the language of analytic geometry until it fit the problems in which he had an interest.

The role of good notation is sometimes expressed by saying that with good notation the pencil becomes as smart as the holder. To see the difference that good algebraic notation makes, knowing something about the early history of calculus is helpful. British mathematicians were more heavily influenced by Newton than they were by Leibniz. They considered it a matter of national honor to use the notation of their countryman. Unfortunately for them, Newton's notation was not expressive enough to be especially useful. In continental Europe, however, mathematicians wholeheartedly adopted Leibniz's notation, which was far superior to that of Newton. Leibniz devised his symbols to embody several basic concepts of calculus in order to communicate his ideas more effectively. This system facilitated discovery both for him and for those who followed. As a consequence calculus initially evolved much more slowly on the British Isles than it did on the Continent.

Today Leibniz's notation is still used in analysis, and the interpretation of algebraic symbols as geometric magnitudes or as physical quantities is still one of the basic conceptual tools of the geometer and the analyst. So thoroughly have algebraic notation and language pervaded geometry and analysis that it is doubtful whether mathematicians who specialize in these subjects could express their discoveries without the use of algebraic notation. But this was just the beginning. Algebra changed radically more than once in the years following the revolution of Descartes and Fermat.

6

THE SEARCH FOR NEW STRUCTURES

Early in the 19th century the nature of algebra changed again. Extraordinary new ideas were introduced. They changed the nature of every branch of mathematics that depends on algebra—and today *every* branch of mathematics depends on algebra. They caused mathematicians to perceive their subject in new ways, and this new perspective enabled them to imagine and solve entirely new kinds of problems.

When the new algebra was first introduced, its importance was not generally recognized. Some of the first groundbreaking papers were dismissed because the reviewers, who were among the best mathematicians of their day, did not understand the ideas involved. To those responsible for the innovations, however, the power of the new ideas and techniques was apparent. Some of the first applications of the new algebra involved solving some of the oldest, most intractable problems in the history of mathematics. For example, the new algebra enabled mathematicians to prove that the three classic problems of antiquity, the squaring of the circle, the trisection of the angle, and the doubling of the cube (all performed with a straightedge and compass) are unsolvable. In addition, they showed that the problem of finding an algorithm for factoring any fifth-degree polynomial—an algorithm similar in spirit to the one that Tartaglia discovered in the 16th century for factoring a third-degree polynomial—could not be solved because the algorithm does not exist.

These very important discoveries were made under very difficult conditions. We often forget how important disease and violence were in shaping much of the history of Europe. Their role is revealed in their effects on the lives of these highly creative mathematicians. These young people lived short, hard lives. They faced one difficulty after another as best they could, and they never stopped creating mathematics. On the night before he expected to die, the central figure in this mathematical drama, a young mathematician named Évariste Galois, spent his time hurriedly writing down as much of what he had learned about mathematics as possible so that his insights, which were wholly unrecognized during his brief life, would not be lost.

Beginning in the 19th century, mathematicians became increasingly preoccupied with the identification and study of algebraic "structures," a term used to denote abstract algebraic concepts shared by very different-looking mathematical systems. (ImageF1)

Broadly speaking the mathematical revolution that occurred in algebra early in the 19th century was a move away from computation and toward the identification and exploitation of the structural underpinnings of mathematics. Underlying any mathematical system is a kind of logical structure. Often the structure is not immediately apparent, but research into these structures has generally proved to be the most direct way of understanding the mathematical system itself. About 200 years ago mathematicians began to identify and use some of these structures, and they have been busy extending their insights ever since.

Niels Henrik Abel

The Norwegian mathematician Niels Henrik Abel (1802–29) was one of the first and most important of the new mathematicians. As with many 17th-, 18th-, and 19th-century mathematicians, he was the son of a minister. The elder Abel was also a political activist, and he tutored Niels at home until the boy was 13 years old. Niels Abel attended secondary school in Christiania, now called Oslo. While there, he had the good fortune to have a mathematics teacher named Bernt Holmboe, who recognized his talent and worked with him to develop it. Under Holmboe's guidance Abel studied the works of earlier generations of mathematicians, such as Leonhard Euler, as well as the mathematical discoveries of his contemporaries, such as Carl Friedrich Gauss. In addition to exposing Abel to some of the most important works in mathematics, Holmboe also suggested original problems for Abel to solve. Abel's ability to do mathematics even at this young age was stunning.

Abel's father died shortly before his son was to enroll in university. The family, not rich to begin with, was left impoverished. Once again, Holmboe helped. He contributed money and helped raise additional funds to pay for Abel's education at the University of Christiania. Still under the tutelage of Holmboe, Abel began to do research in advanced mathematics. During his last year at the university, Abel searched for an algorithm that would enable him to solve all algebraic equations of fifth degree. (Recall that an algebraic equation is any equation of the form $a_n x^n + a_{n-1} x^{n-1} + \ldots + a_1 x + a_0 = 0$, where the a_j are numbers, called coefficients, and the x^j is the variable x raised to the jth power. (Throughout the rest of this chapter, we will always suppose that all coefficients appearing in all polynomials that we consider are rational numbers.) The degree of the equation is defined as the highest exponent appearing in the equation. A second-degree, or quadratic, equation, for example, is any equation of the form $a_2 x^2 + a_1 x + a_0 = 0$.) Abel thought that he had found a general solution for all such equations, but he was quickly corrected. Far from being discouraged, he continued to study algebraic equations of degree greater than 4.

After graduation Abel wanted to meet and trade ideas with the best mathematicians in Europe, but there were two problems to overcome. First, he did not speak their languages; second, he had no money. With the help of a small grant he undertook the study of French and German so that he could become fluent enough to engage these mathematicians in conversation. During this time he also proved that there was no general algebraic formula for solving equations of the fifth degree.

Recall that centuries earlier Niccolò Fontana, also known as Tartaglia, had found an algorithm that enabled him to express solutions of any third-degree algebraic equation as a function of the coefficients appearing in the equations. Shortly thereafter Lodovico Ferrari had discovered an algorithm that enabled him to express solutions of any fourth-degree equation as functions of the coefficients. Similar methods for identifying the solutions to all second-degree equations had been discovered even earlier.

What had never been discovered—despite much hard work by many mathematicians—were similar methods that could enable one to express the roots of arbitrary equations of degree higher than 4 as functions of the coefficients. Abel showed that, at least in the case of fifth-degree equations, the long-sought-after formula did not exist. This, he believed, was a demonstration of his talent that would surely attract the attention of the mathematicians he wanted to meet. In 1824 he had the result published in pamphlet form at his own expense, and in 1825 he left Norway with a small sum given him by the Norwegian government to help him in his studies.

He was wrong about the pamphlet. He sent his pamphlet to Carl Friedrich Gauss, but Gauss showed no interest. This is puzzling since Abel had just solved one of the most intractable problems in mathematics. Although Gauss was no help, during the winter of 1825–26, while in Berlin, Abel made the acquaintance of the German mathematician August Leopold Crelle, the publisher of a mathematics journal. Abel and Crelle became friends, and subsequently Crelle published a number of Abel's papers on mathematics, including his work on the insolubility of fifth-degree equations. Abel also traveled to Paris and submitted a paper to the Academy of Sciences. He hoped that this would gain him the recognition that he believed he deserved, but again nothing happened. Throughout much of his travel Abel had found it necessary to borrow money to survive. He eventually found himself deeply in debt, and then he was diagnosed with tuberculosis.

Abel returned to Norway in 1827. Still heavily in debt and without a steady source of income, he began to work as a tutor. Meanwhile news of his discoveries in algebra and other areas of mathematics had spread throughout the major centers of mathematics in Germany and France. Several mathematicians, including Crelle, sought a teaching position for him in the hope of providing Abel with a better environment to study and a more comfortable lifestyle. Meanwhile Abel continued to study mathematics in the relative isolation of his home in Norway. He died before he was offered the job that he so much wanted.

Évariste Galois

Today the French mathematician Évariste Galois (1811–32) is described as a central figure in the history of mathematics, but during his life he had little contact with other mathematicians. This, however, was not for lack of trying. Galois very much wanted to be noticed.

Évariste Galois was born into a well-to-do family. Nicolas-Gabriel Galois, his father, was active in politics; Adelaide-Marie Demante, his mother, taught Galois at home until he was 12 years old. Because Évariste Galois was dead before his 21st birthday—and because the last several years of his life were extremely turbulent—it is safe to say that he received much of his formal education from his mother. In 1823 Galois enrolled in the Collège Royal de Louis-le-Grand. Initially he gave no evidence of a particular talent for mathematics. Soon, however, he began to do advanced work in mathematics with little apparent preparation. By the time he was 16 he had begun to examine the problem of finding roots to algebraic equations. This problem had already been solved by Abel, but Galois was not aware of this at the time.

Évariste Galois. Before his death at the age of 20, he completely changed the nature of algebraic research and the history of mathematics. (Michael Avandam)

Galois was off to a good start, but his luck soon took a turn for the worse. He submitted two formal papers describing his discoveries to the Academy of Sciences in Paris. These papers were sent to the French mathematician Augustin-Louis Cauchy (1789–1857) for review. Cauchy was one of the most prominent mathematicians of his era. He certainly had the imagination and the mathematical skill required to understand Galois's ideas, and a positive review or recommendation from Cauchy would have meant a lot to Galois. Cauchy lost both papers. This occurred in 1829, the same year that Galois's father committed suicide. Eight months later, in 1830, Galois tried again. He submitted another paper on the solution of algebraic equations to the Academy of Sciences. This time the paper was forwarded to the secretary of the academy, the French mathematician and Egyptologist Joseph Fourier (1768–1830). Fourier died before any action was taken on Galois's paper. The paper that was in Fourier's possession was lost as well. Meanwhile Galois had twice applied for admission to the École Polytechnique, which had one of the best departments of mathematics in France. It was certainly the school to attend if one wanted to work as a mathematician. Both times Galois failed to gain admission.

Galois shifted his emphasis and enrolled in the École Normale Supérieure. He hoped to become a teacher of mathematics, but as his father had, Galois became involved in politics. Politics was important to Galois, and he was not shy about making his ideas known. At the time this activity involved considerable personal risk.

France had been embroiled in political instability and violence since before Galois was born: The French Revolution began in 1789. It was followed by a period of political terror, during which thousands of people were executed. The military leader and later emperor Napoléon Bonaparte eventually seized power and led French forces on several ultimately unsuccessful campaigns of conquest. The results were the defeat of the French military and Napoléon's imprisonment in 1815. Napoléon's defeat did nothing to resolve the conflict between those who favored monarchy and those who favored democracy. Galois was one of the latter. In 1830 the reigning French monarch, Charles X, was exiled, but he was replaced with still another monarch. The republicans—Galois among them—were disappointed and angry. Galois wrote an article expressing his ideas and was expelled from the École Normale Supérieure. He continued his activism. He was arrested twice for his views. The second arrest resulted in a six-month jail sentence.

Despite these difficulties Galois did not stop learning about mathematics. In 1831 he tried again. He rewrote his paper and resubmitted it to the academy. This time the paper fell into the hands of the French mathematician Siméon-Denis Poisson (1781–1840). In the history of mathematics, Poisson, like Cauchy and Fourier, is an important figure, but with respect to his handling of Galois's paper, the best that can be said is that he did not lose it. Poisson's review of Galois's paper was brief and to the point: He (Poisson) did not understand it. Because he did not understand it, he could not recommend it for publication. He suggested that the paper be expanded and clarified.

This was the last opportunity Galois had to see his ideas in print. In 1832 at the age of 20 years and seven months, Galois was challenged to a duel. The circumstances of the duel are not entirely clear. Romance and politics are two common, and presumably mutually exclusive, explanations. In any case Galois, although he was sure he would not survive the duel, accepted the challenge. He wrote down his ideas about algebra in a letter to a friend. The contents of the letter were published four months after Galois died in the duel. This was the first publication in the branch of mathematics today known as Galois theory.

Galois Theory and the Doubling of the Cube

To convey some idea of how Galois theory led to a resolution of the three classical unsolved problems in Greek geometry we examine the problem of doubling the cube. Originally the problem was stated as follows: Given a cube, find the dimensions of a second cube whose volume is precisely twice as large as the volume of the first. If we suppose that the length of an edge on the first cube is one unit long, then the volume of the first cube is one cubic unit: Volume = length × width × height. The unit might be a meter, an inch, or a mile; these details have no effect on the problem. If the volume of the original cube is one cubic unit then the problem reduces to finding the dimensions of a cube whose volume is two cubic units. If we suppose that the letter x represents the length of one edge of the larger cube, then the volume of this new cube is x^3, where x satisfies the equation $x^3 = 2$. In other words, $x = \sqrt[3]{2}$, where the notation $\sqrt[3]{2}$ (called the cube root of 2) represents the number that, when cubed, equals 2. The reason that the problem was so difficult is that it called for the construction of a segment of length $\sqrt[3]{2}$ unit *using nothing but a straightedge and compass.* It turns out that this is impossible.

To show that it is not possible to construct a segment of length $\sqrt[3]{2}$ we need two ideas. The first idea is the geometric notion of a constructible number. The second is the algebraic notion of a field. We begin with an explanation of a constructible number.

We say that a number x is constructible if given a line segment one unit long, we can construct a line segment x units long using only a straightedge and compass. (From now on when we use the word *construct*, we mean "construct using only a straightedge and compass.") A straightedge and compass are very simple implements. There is not much that can be done with them. We can, for example, use the compass to measure the distance between two points by placing the point of the compass on one geometric point and adjusting the compass so that the other point of the compass is on the second geometric point. This creates a "record" of the distance between the points. Also if we are given a line, we can use the compass to construct a second line perpendicular to the first. Besides these there are a few other basic techniques with which every geometry student is familiar. All other geometric constructions are some combination of this handful of basic techniques.

Some numbers are easy to construct. For example, given a segment one unit long, it is easy to construct a segment two units long. One way to accomplish this is to extend the unit line segment, and then use the compass to measure off a second line segment that is one unit long and placed so that it is end to end with the original unit segment. This construction proves that the number 2 is constructible. In a similar way, we can construct a segment that is n units long where n is any natural number. Our first conclusion is that all natural numbers are constructible.

We can also use our straightedge and compass to represent the addition, subtraction, multiplication, and division of natural numbers. To add two natural numbers—which we call m and n—we just construct the two corresponding line segments—one of length m and one of length n—and place them end to end. The result is a line segment of length $m + n$. In a similar way we can represent the difference of the numbers $n—m$: To accomplish this we just measure "out" n units, and "back" m units. It is also true, although we do not show it, that given any two natural numbers m and n, we can construct a line segment of length mn and a line segment of length m/n, provided, of course, that n is not 0. What this indicates is that every rational number is constructible.

Some irrational numbers are also constructible. We can, for example, use a straightedge and compass to construct a square each of whose sides is one unit long. The diagonal of the square is of length $\sqrt{2}$ units long, as an application of the Pythagorean theorem demonstrates. This shows that $\sqrt{2}$ is also a constructible number. We can even construct more complicated-looking numbers. For example, because $\sqrt{2}$ is constructible, we can also construct a line segment of length $1 + \sqrt{2}$. We can use this segment to construct a square with sides of length $1 + \sqrt{2}$. The diagonal of this square is of length $\sqrt{6 + 4\sqrt{2}}$, as another application of the Pythagorean theorem shows. This proves that this more complicated-looking number is constructible as well. These processes can be repeated as many times as desired. The result can be some very complicated-looking numbers. The question then is, Can $\sqrt[3]{2}$ be constructed by some similar sequence of steps?

If we can show that $\sqrt[3]{2}$ is not constructible then we will have demonstrated that it is impossible to double the cube by using a straightedge and compass as our only tools. To do this we need the algebraic concept of a field.

We define a *field* as any set of numbers that is *closed* under addition, subtraction, multiplication, and division except we must explicitly exclude division by zero. (Division by zero has no meaning.) By closed we mean that if we combine any two numbers in the set through the use of one of the four arithmetic operations, the result is another number in the set. For example, the rational numbers form a field because no matter how we add, subtract, or multiply any pair of rational numbers the result is another rational number, and if we divide any rational number by a nonzero rational number the result is another rational number. Similarly, the set of real numbers forms a field. There are also many different fields that contain all of the rational numbers but do not contain all of the real numbers. These "intermediary" fields are the ones that are important to proving the impossibility of doubling the cube.

To see an example of one of these intermediary fields, consider the set of all numbers of the form $a + b\sqrt{2}$, where a and b are chosen from the set of rational numbers. No matter how we add, subtract, multiply, or divide two numbers of the form $a + b\sqrt{2}$ the result is always another number of the same form. This field is called an extension of the rational numbers. We say that we have adjoined $\sqrt{2}$ to the rational numbers to obtain this extension. Every number in the field consisting of $\sqrt{2}$ adjoined to the rational numbers, which we represent with the symbol $Q(\sqrt{2})$, is constructible. (Notice that when $b = 0$ the resulting number is rational. This shows that the field of rational numbers is a *subfield* of $Q(\sqrt{2})$—that is, the field of rational numbers is a proper subset of $Q(\sqrt{2})$.)

Having created the extension $Q(\sqrt{2})$ we can use it to make an even larger field by adjoining the square root of some element

(text continues on page 114)

DOUBLING THE CUBE WITH A STRAIGHTEDGE AND COMPASS IS IMPOSSIBLE

Using the information in the text, we can show how the "new algebra" can be used to complete the proof that it is impossible to construct $\sqrt[3]{2}$ with a straightedge and compass. To appreciate the proof one needs to keep in mind two facts:

1. The number $\sqrt[3]{2}$ is irrational.
2. The graph of the polynomial $y = x^3 - 2$ crosses the x-axis only once.

Here is the proof: Suppose that we adjoin $\sqrt{k_1}$, $\sqrt{k_2}$, $\sqrt{k_3}$, . . ., $\sqrt{k_n}$ to the field of rational numbers, one after another, in the same way that $\sqrt{2}$ and then $\sqrt{a + b\sqrt{2}}$ are adjoined to the rational numbers in the main body of the text. Our hypothesis is that if we adjoin enough of these numbers to the field of rational numbers, we eventually create a field that contains $\sqrt[3]{2}$. (We can use this hypothesis to create two contradictions that prove that doubling the cube with a straightedge and compass is impossible.)

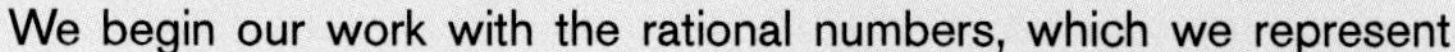

We begin our work with the rational numbers, which we represent with the letter F_0. By fact 1, $\sqrt[3]{2}$ does not belong to F_0 so we have to adjoin at least one number to F_0 in order that our new field will contain $\sqrt[3]{2}$. We adjoin $\sqrt{k_1}$ to the rational numbers where k_1 belongs to F_0 but $\sqrt{k_1}$ does not. We get a new field that we call F_1. (Every number in F_1 is of the form $a + b\sqrt{k_1}$, where a, b, and k_1 are chosen from F_0, the field of rational numbers.) Next we choose k_2 from F_1 and then adjoin $\sqrt{k_2}$ to F_1 to create a new field, which we call F_2. (The numbers in F_2 are of the form $c + d\sqrt{k_2}$, where c and d represent numbers taken from F_1.) We continue the process until we reach F_n, which is obtained by adjoining $\sqrt{k_n}$ to the field F_{n-1}. The ele-

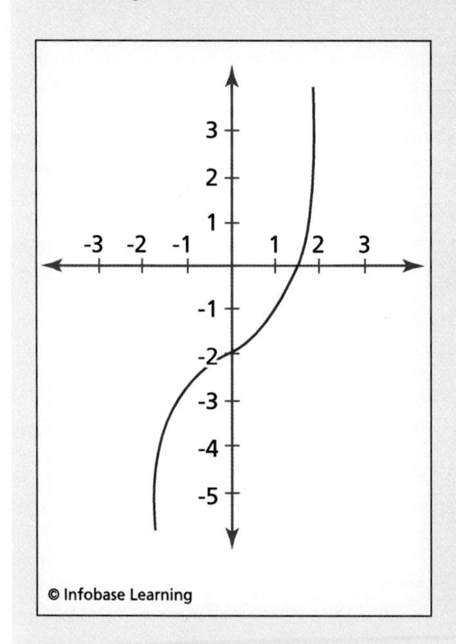

The graph of the equation $y = x^3 - 2$. *Notice that the graph crosses the* x*-axis only once.*

ments in F_n are of the form $p + q\sqrt{k_n}$, where p, q, and k_n belong to the field F_{n-1}. (The fields are like traditional Russian matryoshka dolls, each one fitting inside a slightly larger one, beginning with the smallest, the field of rational numbers, and ending with the largest, F_n.) We assume that each time we adjoin some $\sqrt{k_j}$ it makes the field to which we adjoin it bigger. In other words, we suppose that it is never the case that $\sqrt{k_j}$ belongs to F_{j-1}, the field from which k_j was drawn. (Otherwise every number of the form $e + f\sqrt{k_j}$ would be in F_{j-1} and we would not have made F_{j-1} bigger by adjoining $\sqrt{k_j}$ to it.) Finally we assume that we stop as soon as we have a field that contains $\sqrt[3]{2}$. This means that F_n contains $\sqrt[3]{2}$ but F_{n-1} does not.

To prove that the cube cannot be doubled by using a straightedge and compass, we work with the equation $\sqrt[3]{2} = p + q\sqrt{k_n}$. This equation must be true for some numbers p, q, and k_n in the field F_{n-1} because we have assumed that $\sqrt[3]{2}$ lies in F_n and every number in F_n can be written in this form. We use this equation for $\sqrt[3]{2}$ to obtain two contradictions. The contradictions show that the hypothesis that $\sqrt[3]{2}$ belongs to F_n is impossible.

The computations go like this: Cube both sides of the equation $\sqrt[3]{2} = p + q\sqrt{k_n}$–that is, multiply each side by itself three times–to get $2 = (p^3 + 3q^2k_n) + (3p^2q + b^3k_n)\sqrt{k_n}$. Now consider $(3p^2q + b^3k_n)$, the coefficient of $\sqrt{k_n}$.

- Contradiction 1: If $(3p^2q + b^3k_n)$ is not equal to 0, then we can solve for $\sqrt{k_n}$ in terms of numbers that all belong to the field F_{n-1}. Since F_{n-1} is a field we conclude that $\sqrt{k_n}$ belongs to F_{n-1} and our assumption that F_n is bigger than F_{n-1} was in error. This is the first contradiction.

- Contradiction 2: If the number $(3p^2q + b^3k_n)$ equals 0, then cube the number $p - q\sqrt{k_n}$ to get $(p^3 + 3q^2k_n) - (3p^2q + b^3k_n)\sqrt{k_n}$. Since $(3p^2q + b^3k)$ is 0 it must be the case that $p - q\sqrt{k_n}$ is also a cube root of 2. [Because if $(3p^2q + b^3k_n)$ is 0 then both the cube of $p - q\sqrt{k_n}$ and the cube of $p + q\sqrt{k_n}$ are equal, and we have already assumed that $p + q\sqrt{k_n}$ is the cube root of 2.] Therefore, the graph of $y = x^3 - 2$ must cross the x axis at $p - q\sqrt{k_n}$ and at $p + q\sqrt{k_n}$. This contradicts fact number 2.

The situation is hopeless. If we assume that $(3p^2q + b^3k_n)$ is not 0 we get a contradiction. If we assume that $(3p^2q + b^3k_n)$ is 0 we get a contradiction. This shows that our assumption that we could construct $\sqrt[3]{2}$ was in error, and we have to conclude that $\sqrt[3]{2}$ is not constructible with a straightedge and compass. This is one of the more famous proofs in the history of mathematics.

(text continued from page 111)
of $Q(\sqrt{2})$. The element we adjoin is of the form $\sqrt{a + b\sqrt{2}}$. Every number in this field has the form $c + d\sqrt{a + b\sqrt{2}}$, where c and d are chosen from $Q(\sqrt{2})$ and $a + b\sqrt{2}$ is positive. We can do this as often as we want. Each new field can be chosen so that it is larger than the previous one. Every number in each such extension is constructible, and conversely, every constructible number belongs to a field that is formed in this way.

To complete the proof we need only show that no matter how many times we extend the rational numbers in the manner just described, the resulting field never contains the number $\sqrt[3]{2}$. The proof uses the concept of field and requires us to complete a few complicated-looking multiplication problems and recall a bit of analytic geometry (see the sidebar for details).

Algebraic Structures

Some fields are smaller than others. To repeat an example already given, the field of rational numbers is "smaller" than the field defined as $Q(\sqrt{2})$, because every number in $Q(\sqrt{2})$ is of the form $a + b\sqrt{2}$ where a and b are rational numbers; if we consider the case where b is 0 and a is any rational number, then it is apparent that $Q(\sqrt{2})$ contains every rational number. However, when $b = 1$ and $a = 0$, we can see that $Q(\sqrt{2})$ also contains $\sqrt{2}$, which is not rational. Because the rational numbers are a proper subset of $Q(\sqrt{2})$, we can say that the field of rational numbers is a subfield of $Q(\sqrt{2})$.

For each algebraic equation there is always a smallest field that contains all the roots of the equation. This is the field we obtain by adjoining the smallest possible set of numbers to the set of rational numbers. This field, which is determined by the roots of the polynomial of interest, is important enough to have its own name. It is called the splitting field. Depending on the polynomial, the splitting field can have a fairly complicated structure. The numbers that make up the field can sometimes be difficult to write down; they are usually not constructible; and as with all the fields that we consider, the splitting field contains infinitely many numbers. Furthermore, it must be closed under the four arithmetic operations: addition, subtraction, multiplication, and division. Fields are complicated objects. It was one of Galois's great insights that he was able to rephrase the problem of solving algebraic equations so that it was simple enough to solve. His solution involved another type of algebraic structure called a group.

The idea of a group is one of the most important ideas in mathematics. There are many kinds of groups. We can create an example of a symmetry group by cutting a square out of the center of a piece of paper. Suppose that, by moving clockwise about the square, we number each of the exterior corners as shown. Suppose, too, that we number the corresponding interior corners so that when we replace the square inside the square hole, each number on the square matches up with its mate (see Figure A).

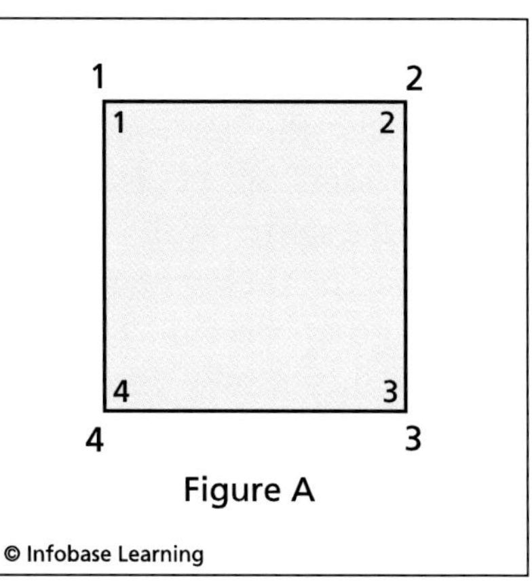

Figure A, the identity permutation

If we now rotate the square 90° clockwise about its center, the number 1 on the square matches up with the number 2 on the hole. The number 2 on the square matches the number 3 on the hole, and so on (see Figure B).

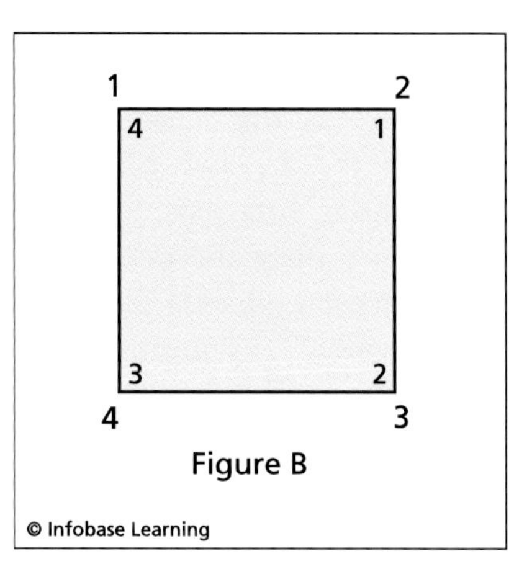

Figure B, a 90° clockwise rotation

If we rotate the square 180° clockwise out of its original

position, we get a new configuration (see Figure C).

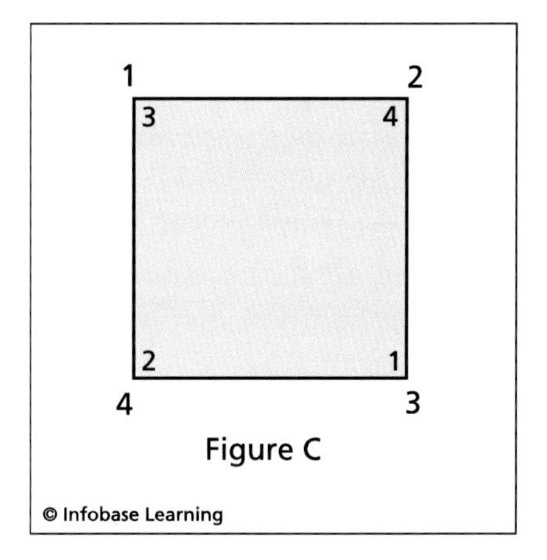

Figure C, a 180° clockwise rotation

There are two other rotations of interest. One entails rotating the square 270° clockwise—this yields a fourth configuration (see Figure D)—and the last rotation entails rotating the square 360° clockwise (see Figure A again). Notice that making the last rotation has the same effect as not moving the square at all.

All four of these rotations form a group because taken together they exhibit the following four properties, which are the defining properties of a mathematical group:

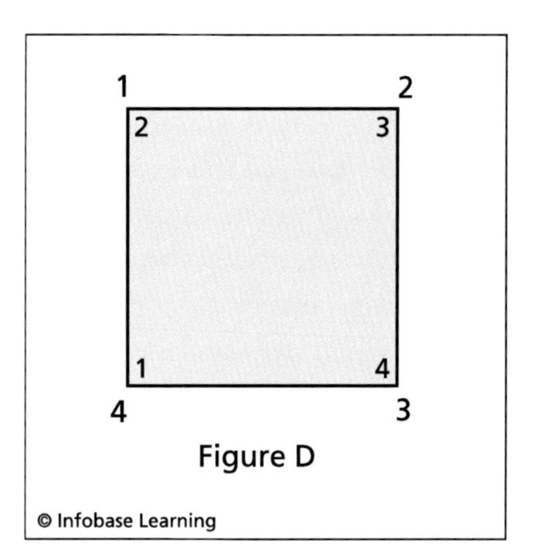

Figure D, a 270° clockwise rotation

1. One rotation followed by another yields a third.

2. Rotations are associative in the sense that if π_1, π_2, and π_3 are three rotations, then they obey the associative law—namely, $\pi_1 \circ (\pi_2 \circ \pi_3) = (\pi_1 \circ \pi_2) \circ \pi_3$, where the symbol "$\circ$" means that two rotations are combined by first performing the rotation on the right of the pair and then performing the rotation on the left member of the pair. The associative law means that it does not matter whether we first combine π_2 and π_3 and then combine π_1 with $\pi_2 \circ \pi_3$—this is the left side of the equation—or whether we combine π_1 with π_2 and then combine $\pi_1 \circ \pi_2$ with π_3 as is done on the right side of the equation.

3. The set of rotations associated with the square has an "identity rotation," which is usually represented with the letter e. When combined with any other rotation, the identity rotation leaves the other rotation unchanged. In symbols: $e \circ \pi = \pi \circ e = \pi$. (The identity rotation, which means rotating the square 0°, is analogous to the number zero in the group of real numbers under the operation of addition.)

4. Finally, for each rotation, another rotation exists that undoes the work of the first. This second rotation is called the inverse of the first. For example, rotation by 270° is the inverse of rotation by 90°.

The four-element group described in the previous paragraph is also a *subgroup*—that is, it is a group that is part of a larger group of symmetries of the square. We can get more symmetries in our group by "reflecting" the square about a line of symmetry. Physically this can be accomplished by flipping the square over along one of its lines of symmetry. For example, we can flip the square along a line connecting two opposite corners. Under these circumstances two corners of the square remain motionless while the other two corners swap places. If, for example, we reflect the square about the line connecting the corners 1 and 3, then corners 2 and 4 of the square change places while 1 and 3 remain motionless (see Figure E). This configuration (corners 1 and 3 fixed and corners 2 and 4 exchanged) is new; we cannot obtain this reflection through any sequence of rotations, but it is not the only reflection that we can generate. We can obtain still another configuration of the square within its

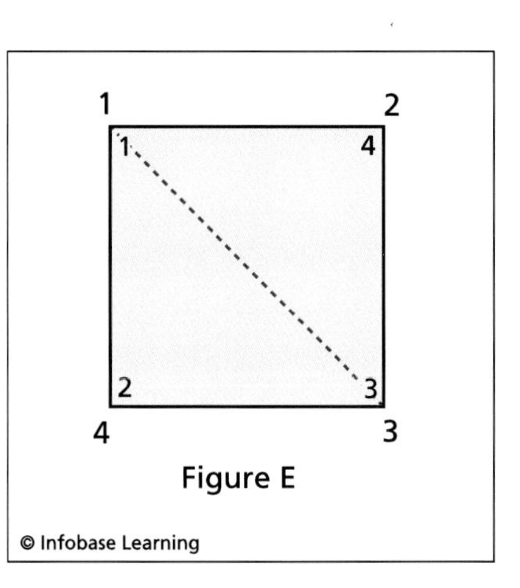

Figure E, a reflection of the square about the line connecting corners 1 and 3

hole by reflecting the square about its other diagonal (see Figure F). There are two other reflections—each is obtained by reflecting the square about lines passing through the center of the square and the midpoint of a side. We omit the details. Allowing all possible combinations of rotations and reflections gives us a larger group than the group of rotations. (Of course, there are mathematical formulas that do the same thing that we are doing with paper, but the mathematical methods are simply a symbolic substitute for rotating the square through multiples of 90° and reflecting it about its axes of symmetry.)

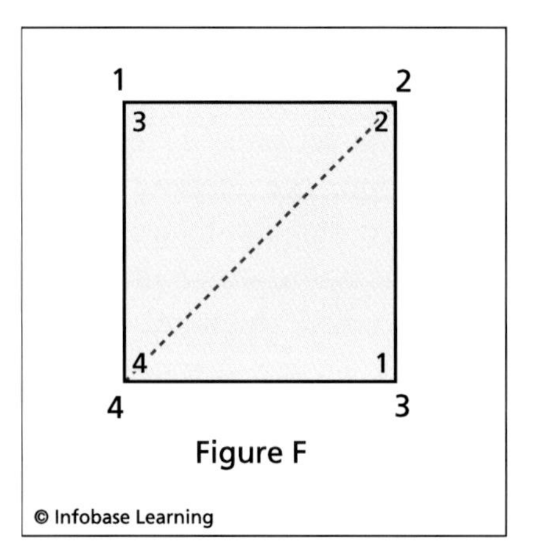

Figure F, a reflection of the square about the line connecting corners 2 and 4

The group of symmetry transformations of the square can also be interpreted as a group of permutations of the integers 1, 2, 3, and 4. A permutation group acts on a set of objects—in this case numbers—and changes their order. Rotation of the square by 90 degrees, for example, when applied to the set (1, 2, 3, 4) yields (4, 1, 2, 3). The outside numbers on the diagram show the original order, and the inside numbers show the new (or permuted) order. (See Figure B, page 115.)

Similarly, the reflection shown in Figure F permutes (1, 2, 3, 4) so that we obtain (3, 2, 1, 4). Mathematically, it does not matter whether we represent what we are doing as a group of symmetry transformations of the square or as a group of permutations of the numbers 1, 2, 3, and 4. The group of symmetry transformations and the corresponding group of permutations are two instances of the same abstract group. In what follows, we use permutation groups, but we could just as easily talk about symmetry groups of various geometric figures.

So far we have examined only the permutations obtainable by rotating and reflecting a square, but we can generate other, very different, permutation groups by using other geometric figures. Depending on the figures we choose to study, the permutation groups we generate may have more or fewer elements than the permutation group associated with the square. The subgroups associated with each permutation group also depend on the group that we study. Galois did not invent permutation groups, but he did find an extraordinarily creative use for them.

Galois noticed that to each (infinite) splitting field there corresponds a unique (finite) permutation group. The algebraic structure of two splitting fields is "the same" if they have the same permutation group. Even better, the permutation group contains important information about the splitting field and the algebraic equation from which the field is obtained. In particular, an algebraic equation can be solved if the permutation group has a certain structure. If the permutation group lacks this structure then there is no algorithm analogous to those discovered by Tartaglia and Ferrari that would enable one to solve the equation.

It might seem as if Galois simply swapped the difficulties of working with algebraic equations for the difficulties of working with splitting fields, then swapped the problems he encountered with splitting fields for a new set of problems associated with permutation groups. There is, however, a real advantage to studying the permutation group instead of the splitting field or the algebraic equation: The group problem is simple enough to solve. Unlike the splitting fields, which have four operations and infinitely many numbers, each permutation group has only one operation and finitely many elements. Galois swapped a harder problem for an easier problem. The group problem was manageable; the field problem was not.

This is an example of what is meant by *structure* in mathematics. Each splitting field has many properties in common with other fields—that is why they are all called fields—but there are differences between the fields as well. These finer points of structure are determined by the nature of the roots that are adjoined to the rational numbers in order to get the splitting field. The finer points of structure in the field determine the properties of the permutation group. In this sense the structure of the group reflects

THE IMPORTANCE OF GROUP THEORY

Group theory occupies a central place in contemporary mathematics. The four axioms that define a group have been supplemented in various ways to produce many different classes of groups. Some groups consist of finitely many elements; some have infinitely many elements; some are "finitely generated," which means that there is a finite subset of the group, which we can call $g_1, g_2, \ldots, g_n$, such that every element in the group can be represented as a product of powers of these n elements; and other groups are not finitely generated. Mathematicians have studied many special classes of groups, and many types of groups have not yet been thoroughly studied.

But the importance of group theory extends far beyond algebra because the group structure can be found imbedded in many different "systems," both in and outside mathematics. Sometimes the existence of the group structure is evident–the set of all integers under the operation of addition, for example, is a system with a group structure–but other times the group structure is less obvious. In geometry, for example, Euclid asserted that two triangles are congruent when one triangle can be moved so that it coincides with the other. This statement is not quite precise enough for modern mathematicians, but once it is made precise it can be proved that the set of all such motions forms a group, which is now called the group of Euclidean transformations. Euclidean geometry can, in fact, be characterized as the study of exactly those properties of a figure that remain unchanged by the group of Euclidean transformations. Other geometries have their own somewhat different transformation groups associated with them, and each transformation group has its own peculiar algebraic properties. By comparing transformation groups, one can determine how various geometries are related to each other. Group theory is an important tool in the study of geometry.

the structure of the field, but, because the group is easier to understand, solving problems associated with the field by studying its associated permutation group becomes possible.

Our description of Galois's work is a modern one. It describes how we see what Galois did, but it almost certainly does not describe Galois's own view of his work. Galois used permutation groups as a tool in order to better understand algebraic equations.

Group theory is useful in science as well. The special theory of relativity, for example, makes a number of exotic predictions about the nature of space and time and the ways that space and time are related, but the physical theory can be described in terms of a particular (mathematical) group called the group of Lorentz transformations. Mathematically, the special theory of relativity, which is most closely associated with the work of the German-born American physicist Albert Einstein (1879–1955), can be summarized by saying that the laws of nature are invariant with respect to the group of Lorentz transformations. A physical interpretation of the group of Lorentz transformations is that they reveal how measurements made in one inertial reference frame apply to a different inertial reference frame. (An observer's *reference frame* is the coordinate system most natural to the observer. A reference frame is *inertial* when it is moving at constant velocity.) Loosely speaking, the Lorentz transformations, which are named after the Dutch physicist Hendrik Antoon Lorentz (1853–1928), enable observers in different reference frames to view the world as others see it.

In chemistry, symmetry groups can be associated with many molecules. The symmetry group of a particular molecule is the largest group of spatial transformations that leaves the molecule in the same volume of space after the transformation as it was before. The idea is similar to the symmetry group associated with the square. (That group is described in figures A through F in this chapter.) The symmetry group carries information about the shape of the molecule, and it also carries information about the chemical properties of the molecule. The study of these groups is an important part of theoretical chemistry.

It would be hard to overstate the importance of group theory to the development of contemporary mathematics and science. Groups are some of the most fundamental logical structures in mathematics and nature, and they provide a framework in which many fundamental questions about mathematics and science can be asked and answered.

The object of his study was algebraic equations, not groups. It must have been more or less obvious to him as it is obvious to most people who work with permutation groups that all permutation groups share the four simple properties that define a group. (These properties are listed on page 116 in the discussion of the group of symmetry transformations of the square.) And Galois, who was one of the most inventive mathematicians of his era, must have recognized that all four of these very general statements are true, but he regarded these statements in a much narrower way than contemporary mathematicians. Today, mathematicians consider abstract groups, sets composed of elements represented by letters, and they suppose that there exists an operation on this set of letters, and they suppose that under this operation the letters combine according to the four group axioms. (They might also impose some other conditions depending on the problem they are considering.)

Groups can be used as tools to understand specific systems. This is what Galois did, but today they are also used to model classes of systems. Research into group theory can reveal facts about all of the systems that share a particular group structure—no matter how different-looking the systems are. Or to say the same thing in a slightly different way: Any facts that one can deduce from the study of an abstract group will apply to every system with the same group structure.

The mental leap that mathematicians made in passing from individual instances of groups to the study of abstract groups is similar to what an archaeologist does when studying the architecture of a vanished civilization for the first time. The first building that the archaeologist studies is always described in particulars such as, "This building used structural elements made from a certain material. The elements in the building were of a certain thickness. They were placed at specific points within the structure. They were joined in a certain way." But after the archaeologist has studied the ruins of many such buildings, it becomes possible to make very broad statements about the building practices of the civilization such as, "The buildings of this time period utilized the following materials. Structural elements were joined using the following techniques, etc." The general statements summarize the knowledge obtained from many particular observations.

Throughout the 19th century, mathematicians operated like our archaeologist encountering the first few buildings. They studied many different mathematical systems, and they discovered that groups were present in many of these systems. Originally, they concentrated on the specifics of the system and used groups as tools to further their understanding. They did not see the "big picture." They did not usually see the group structure as something that was worth studying in its own right. But early in the 20th century, mathematicians began to formulate more general statements about the structure of mathematical systems. The systems might be sets of numbers or polynomials or some other set of objects, and the operation might be addition or multiplication or some other operation entirely. After they had accumulated enough examples, they sought to isolate the properties shared by all of these systems and study only those shared properties. The result was a series of statements that were true for all such systems. Much of modern algebra now focuses on uncovering and understanding the logical architecture of mathematics, and groups are a big part of that architecture.

The discovery of these group methods—even in a limited way—required an especially creative mathematical mind. Galois's ideas represented a huge leap forward in mathematical thinking, and it would be some time before other mathematicians caught up. Today groups are one of the central concepts in all of mathematics. They play a prominent role in geometry, analysis, algebra, probability, and many branches of applied science as well. The search for the structures that underlie mathematics, and the search for criteria—analogous to Galois's permutation groups—that enable mathematicians to determine when two structures are really "the same" are now central themes of algebraic research. In many ways these ideas are responsible for the ever-increasing pace of mathematical progress. What we now call modern, or abstract, algebra begins with the work of a French teenager almost 200 years ago.

7

THE LAWS OF THOUGHT

Algebra changed radically more than once during the 19th century. Previously Descartes had interpreted his variables as magnitudes, that is, lengths of line segments. He used algebra as a tool in his study of geometry. Leibniz and Newton had interpreted the variables that arose in their computations as geometric magnitudes or as forces or accelerations. On the one hand, these interpretations helped them state their mathematical questions in a familiar context. They enabled Newton and Leibniz to discover new relationships among the symbols in their equations, so in this sense these interpretations were useful. On the other hand, these interpretations were not necessary. One can study the equations of interest to Descartes, Newton, and Leibniz without imposing any extramathematical interpretation on the symbols employed. At the time no one thought to do this.

In the 19th century mathematicians began to look increasingly inward. They began to inquire about the true subject matter of mathematics. The answer for many of them was that mathematics is solely concerned with relationships among symbols. They were not interested in what the symbols represented, only in the rules that governed the ways symbols are combined. To many people, even today, this sounds sterile. What is surprising is that their inquiries about relationships among symbols resulted in some very important, practical applications, the most notable of which is the digital computer.

Aristotle

This new and more abstract concept of mathematics began in the branch of knowledge called logic. Early in the 19th century, logic was synonymous with the works of the Greek philosopher Aristotle (384 B.C.E.–322 B.C.E.). Aristotle was educated at the academy of the philosopher Plato, which was situated in Athens. He arrived at the academy at the age of 17 and remained until Plato's death 20 years later. When Plato died, Aristotle left Athens and traveled for the next 12 years. He taught in different places and established two schools. Finally he returned to Athens, and at the age of 50 he established the school for which he is best remembered, the Lyceum. Aristotle taught there for the next 12 years. The Lyceum was a place that encouraged free inquiry and research. Aristotle himself taught numerous subjects and wrote about what he discovered. For Aristotle all of this abruptly ended in 323 B.C.E.,when Alexander the Great died. There was widespread resentment of Alexander in Athens, and Aristotle, who had been Alexander's tutor, felt the wrath of the public directed at him after the death of his former student. Aristotle left Athens under threat of violence. He died one year later.

This statue of Aristotle is a Roman copy of a Greek original. Aristotle's ideas about logic were central to Western thinking for 2,000 years. (Ludovisi Collection)

Part of Aristotle's contribution to logic was his study of something called the syllogism. This is a very formal, carefully defined type of reasoning. It begins with categorical statements, usually called categorical propositions. A proposition is a simple statement. "The car is black" is an example of a proposition. Many other expressions are not categorical propositions. "Do you wish you had a black car?" and "Buy the black car" are examples of statements that are not propositions. These types of sentences are not part of Aristotle's inquiry. Instead his syllogisms are defined only for the categorical proposition.

A categorical proposition is a statement of relationship between two classes. "All dogs are mammals" is an example of a categorical proposition. It states that every creature in the class of dogs also belongs to the class of mammals. We can form other categorical propositions about the class of dogs and the class of mammals. Some are more sensible than others:

- "Some dogs are mammals."
- "No dogs are mammals."
- "Some dogs are not mammals."

are all examples of categorical propositions. We can strip away the content of these four categorical propositions about the class of dogs and the class of mammals and consider the four general *types* of categorical expressions in a more abstract way:

- All x's are y's.
- Some x's are y's.
- No x's are y's.
- Some x's are not y's.

Here we can either let the xs represent dogs and the ys represent mammals or let the letters represent some other classes. We can even refrain from assigning any extramathematical meaning at all to the letters.

We can use the four types of categorical propositions to form one or more syllogisms. A syllogism consists of three categorical propositions. The first two propositions are premises. The third proposition is the conclusion. Here is an example of a syllogism:

- Premise 1: All dogs are mammals.
- Premise 2: All poodles are dogs.
- Conclusion: All poodles are mammals.

We can form similar sorts of syllogisms by using the other three types of categorical propositions.

Aristotle's writings were collected and edited by Andronicus of Rhodes, the last head of Aristotle's Lyceum. This occurred about three centuries after Aristotle's death. The *Organon*, as Andronicus named it, is the collection of Aristotle's writings on logic. It became one of the most influential books in the history of Western thought.

Aristotle's ideas on logic were studied, copied, and codified by medieval scholars. They formed an important part of the educational curriculum in Renaissance Europe. In fact, the *Organon* formed a core part of many students' education into the 20th century. But the syllogism tells us little about the current state of logic. Its importance is primarily historical: For about 2,000 years the syllogism was the principal object of study for those interested in logic. Many scholars thought that, at least in the area of logic, Aristotle had done all that could be done. They believed that in the area of logic no new discoveries were possible.

There is no doubt that Aristotle made an important contribution to understanding logic, because his was the first contribution. In retrospect, however, Aristotle's insights were very limited. Logic is more than the syllogism, because language is more than a set of syllogisms. Logic and language are closely related. We can express ourselves logically in a variety of ways, and not every set of logical statements can be reduced to a collection of syllogisms. Aristotle had found a way of expressing certain logical arguments, but his insights are too simple to be generally useful.

George Boole and the Laws of Thought

The 20th-century British philosopher and mathematician Bertrand Russell wrote that modern, "pure" mathematics began with the work of the British mathematician George Boole (1815–64). Not everyone agrees with Russell's assessment, but there can be little doubt that Boole, a highly original thinker, contributed many insights that have proved to be extremely important in ways both theoretical and practical. The following famous quotation, taken from his article "Mathematical Analysis of Logic," strikes many contemporary readers as radical in that he insists that mathematics is about nothing more than relationships among symbols:

> They who are acquainted with the present state of the theory of Symbolical Algebra, are aware, that the validity of the processes of analysis does not depend upon the interpretation of the symbols which are employed, but solely upon the laws of their combination. Every system of interpretation which does not affect the truth of the relations supposed, is equally admissible, and it is thus that the same process may, under one scheme of interpretation, represent the solution of a question on the properties of numbers, under another, that of a geometrical problem, and under a third, that of a problem of dynamics or optics. This principle is indeed of fundamental importance; and it may with safety be affirmed, that the recent advances of pure analysis have been much assisted by the influence which it has exerted in directing the current of investigation.
>
> (*Boole, George.* Mathematical Analysis of Logic: being an essay towards a calculus of deductive reasoning. *Oxford: B. Blackwell, 1965)*

Despite Boole's assertion that mathematics is about nothing more than symbolic relationships, Boole's insights have since found important applications, especially in the area of computer chip design.

Boole was born into a poor family. His father was a cobbler, who was interested in science, mathematics, and languages. His interest in these subjects was purely intellectual. He enjoyed learning, and he put his discoveries to use by designing and then creating various optical instruments; telescopes, microscopes, and cameras were all produced in the elder Boole's workshop. As a youth George Boole helped his father in the workshop, and it was from these experiences presumably that he developed an interest in the science of optics, a subject about which he wrote as an adult.

Despite their poverty, the Booles sent their son to various schools. These schools were not especially distinguished, but he learned a great deal from his father, and he supplemented all of this with a lot of independent study. He read about history and science; he enjoyed biographies, poetry, and fiction; and as many of the mathematicians described in this history did, Boole displayed an unusual facility with languages. While a teenager, and despite a good deal of adversity, he learned Latin, Greek, French, Italian, and German. His interest in learning languages began early. At the age of 14 he translated a poem from Latin to English and had the result published in a local newspaper. The publication of the translation set off a minor controversy when one reader wrote to the newspaper to question whether anyone so young could have produced such a skillful translation. Boole was clearly an outstanding student, but his formal education was cut short.

George Boole, founder of Boolean algebra. When he first proposed his ideas they were interesting but not especially practical, but 80 years later they were essential to the development of the computer. (MAA Mathematical Sciences Digital Library)

In 1831 when Boole was 16, his father's business became bankrupt. The penalties for bankruptcy were more serious then than they are now, and George Boole left school to help his family. He got a job, first as an assistant teacher, and later as a teacher. It was at this time that he began to concentrate his energy on learning mathematics. He later explained that he turned toward mathematics because at the time he could not afford to buy many books, and mathematics books, which required more time to be read, offered better value. Throughout this period of independent study Boole went through several teaching jobs. At the age of 20 when he had saved enough money, he opened his own boarding school in his hometown of Lincoln.

It is a tribute to his intellectual ambition and his love of mathematics that despite moving from job to job and later establishing and operating his own school, Boole found enough time to learn higher mathematics and to publish his ideas. He also began to make contacts with university professors so that he could discuss mathematics with other experts. Eventually Boole was awarded the Royal Society of London's first Gold Medal for one of his mathematics papers.

Boole never did attend college. His formal education ended permanently when he left school at the age of 16. Not everyone has the drive to overcome this kind of educational isolation, but it seemed to suit Boole. Boole's language skills enabled him to read important mathematical works in their original languages. He developed unique insights in both mathematics and philosophy. Soon Boole turned away from the type of mathematics that would have been familiar to every mathematician of Boole's time and directed his energy toward discovering what he called the "laws of thought."

Boole's inquiry into the laws of thought is a mathematical and philosophical analysis of formal logic, often called symbolic logic, logic, or, sometimes, logistic. The field of logic deals with the principles of reasoning. It contains Aristotle's syllogisms as a very special case, but Boole's inquiry extended far beyond anything that Aristotle envisioned. Having developed what was essentially a new branch of mathematics, Boole longed for more time to explore these ideas further. His duties at his own school as well as other duties at other local schools were enough to keep him very busy but not very well off. When Queen's College (now University College) was established in Cork, Ireland, Boole applied for a teaching position at the new institution. Between the time that he applied and the time that he was hired, three years passed. He had despaired of ever being offered the position, but in 1849 when he finally was, he accepted it and made the most of it.

Boole remained at Queen's College for the rest of his life. He married, and by all accounts he was extremely devoted to his wife. He was apparently regarded with both affection and curiosity by his neighbors: When Boole met anyone who piqued his curiosity, he immediately invited that person to his home for supper and conversation. He is often described as someone who was kind, generous, and extremely inquisitive. Boole died after a brief illness at the age of 49.

Boolean Algebra

Boole's great contribution to mathematics is symbolic logic. Boole sought a way of applying algebra to express and greatly extend classical logic. Recall that Aristotle's syllogisms were a way of making explicit certain simple logical relations between classes of objects. Boole's symbolic logic can do the same thing, but his concept of logic greatly extended Aristotle's ideas by using certain logical *operators* to explore and express the relations between various classes. His approach to logic depended on three "simple" operators, which we (for now) denote by AND, OR, and NOR. Over the intervening years, mathematicians have found it convenient to modify some of Boole's own ideas about these operators, but initially we restrict our attention to Boole's definitions. For purposes of this exposition it is sufficient to restrict our attention to careful definitions of the operators AND and OR.

Definition of AND: Given two classes, which we call x and y, the expression xANDy denotes the set of all elements that are common to both the classes x and y. For example, if x represents the class of all cars and y represents the set of all objects that are red, then xANDy represents the class of all red cars. This notation is hardly satisfactory, however, because Boole was interested in developing an *algebra* of thought. As a consequence Boole represented what we have written as xANDy as the "logical product" of x and y, namely, xy. This leads to the first unique aspect of Boole's algebra. Because the set of all elements common to the class x is the class x itself—that is, xANDx *is* x—Boole's algebra has the property that $xx = x$ or, using exponents to express this idea, $x^2 = x$. By repeating this argument multiple times we arrive at the statement $x^n = x$, where n represents any natural number. This equation does not hold true in the algebra that we first learn in junior and senior high school, but that does not make it wrong. It is, in fact, one of the defining properties of Boolean algebra. Finally, notice that with this definition of a logical product, the following statement is true: $xy = yx$. That is, the elements that belong to the class xANDy are identical to the elements that belong to the class yANDx.

Definition of OR: Boole's definition of the operator OR is different from that used in Boolean algebra today, but it coincides with one common usage of the word *or.* To appreciate Boole's definition of the operator OR, imagine that we are traveling through the country and arrive at a fork in the road. We have a decision to make: We can turn left or we can turn right. We cannot, however, simultaneously turn both left and right. In this sense, the word *or* is used in a way that is exclusive. We can take one action or the other but not both. Boole defined the OR operator in this exclusive sense. Given two classes, which we call x and y, the expression xORy means the set of elements that are in x but not in y together with the set of elements that are in y but not in x. In particular, the class xORy does not contain any elements that are in both x and y. If we again let x represent the class of all cars, and y represent the class of all objects that are red, then the class xORy contains all cars that are not red and every object that is red provided it is not a car. As we have written xORy, Boole used the expression $x + y$. Notice that with this interpretation of the symbol + it is still true that $x + y = y + x$. (We emphasize that Boole's definition of the OR operator is different from the definition in common use today. See the section Refining and Extending Boolean Algebra later in this chapter for a discussion of the difference.)

The axioms for Boole's algebra can now be expressed as follows. A Boolean algebra is, according to Boole, any theory that satisfies the following three equations:

1. $xy = yx$
2. $x(y + z) = xy + xz$
3. $x^n = x$, *where n is any natural number*

The first and third axioms have already been discussed. The meaning of the second axiom can best be explained via a Venn diagram (see the accompanying illustration).

The roles of the numbers 0 and 1 are especially important in Boole's algebra. The number 0 represents the empty set. If, for example, x represents the set of diamonds and y represents the set of emeralds, then xy, the set of objects that are both diamonds and emeralds, is empty—that is, with this interpretation of x and y, $xy = 0$. For Boole, the number 1 represents the *universe* under consideration, that is, the entire class of objects being considered. If we continue to let x represent the set of diamonds and y the set of emeralds, and if, in addition, we let 1 represent the set of all gemstones then we obtain the following three additional equations: (1) $1x = x$, (2) $1y = y$, and (3) $x(1 - y) = x$, where the expression $(1 - y)$ represents all the objects in the universal set that are not emeralds, so that the class of diamonds AND the class of gems that are not emeralds is simply the class of diamonds.

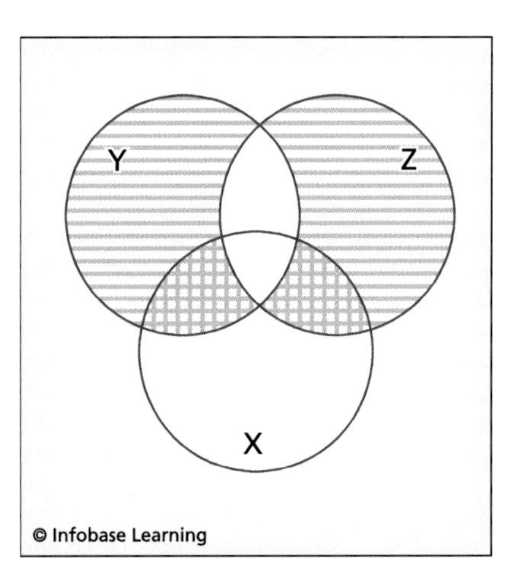

Venn diagram demonstrating that in Boolean algebra x(y + z) = xy + xz. *Note that* y + z *is the horizontally shaded area, and* x(y + z) *is the intersection of the horizontally shaded area with* x. *Alternatively,* xy *is the area common to the disc* x *and the disc* y, xz *is the area common to the disc* x *and the disc* z, *and* xy + xz *is the union of* xy *and* xz *minus their common area. Therefore,* x(y + z) = xy + xz.

Recall that in the excerpt from Boole quoted at the beginning of this chapter, he explicitly states, ". . . the validity of the processes of analysis does not depend upon the interpretation of the symbols which are employed." Having established the basic properties of his algebra, he is free to interpret the symbols in any manner convenient. This is important because there was another interpretation that Boole had in mind, and this interpretation has since become very important as well. To appreciate Boole's second interpretation, we imagine some proposition, which we call X, and we let the letter x represent the times when proposition X is true. Because a proposition is either true or false—it cannot be both true and false—the expression $1 - x$ must represent those times when X is false. For example, suppose we let X represent the statement "It is raining." Because we let x represent the times when statement X is true, x represents those times when it actually is raining. The expression $1 - x$ holds when the statement X is false: That is, $1 - x$ represents those times when it is not raining.

In a similar way we can interpret the logical product and the logical sum. If we have two propositions—we represent them with X and Y—then we can let the letter x represent those times when proposition X is true and y represent those times when proposition Y is true. The logical product xy now represents those times when propositions X and Y are *simultaneously* true. For example, let X represent the proposition "It is raining" and let Y represent the statement "It is windy." The expression xy represents those times when it is simultaneously rainy and windy.

In a similar way, the logical sum $x + y$ represents those times when, according to Boole, either proposition X is true or proposition Y is true, but *not* when they are simultaneously true. In our weather example, $x + y$ represents those times when it is either windy or rainy but not both.

Keep in mind that this in no way changes Boole's algebra. According to Boole, mathematics is only about the relationships among symbols, so from a mathematical perspective the interpretation that we place on the symbols is irrelevant. From the point of view of applications, however, the interpretation that we place on the symbols means everything, because it determines how we *use* the algebra. Boole's alternative interpretation of his algebraic symbols as representing true and false values has important applications. It enables one to calculate the true or false values associated

AND	T	F
T	T	F
F	F	F

© Infobase Learning

Truth table for the operation AND

ARISTOTLE AND BOOLE

For more than 2,000 years Aristotle's treatment of logic was not modified in any significant way. For Western scholars Aristotle's logic was the only type of logic. When propositions are written in ordinary language, understanding why they arrived at this conclusion is not difficult. Written rhetorically, Aristotle's syllogisms have no obvious generalizations. This changed entirely with the work of Boole. Boole's insights enabled him to show that Aristotle's conception was not only limited but also easily extended. Once Aristotle's treatment of the syllogism was expressed by using Boole's algebra, it was seen to be a particularly simple set of computations, and Aristotle's insights were seen to be a very small part of a much larger algebraic landscape.

In the following we list Aristotle's four categorical propositions. Each proposition is followed, in parentheses, by the same expression using Boole's algebra. Following Boole, we use the letter v to represent a nonempty class of objects:

- All x's are y's. ($xy = x$)
- Some x's are y's. ($v = xy$)
- No x's are y's. ($xy = 0$)
- Some x's are not y's. ($v = x(1 - y)$)

Notice that each proposition has been expressed as a simple algebraic equation.

Using Boole's algebra, we can express any syllogism via a set of algebraic equations. To illustrate, we rewrite the syllogism given near the beginning of this chapter involving mammals, dogs, and poodles. This requires three equations. Each line of the syllogism is written in words followed, in parentheses, by an algebraic equation that expresses the same idea. We let the letters m, d, and p represent mammals, dogs, and poodles, respectively.

- Premise 1: All dogs are mammals. ($dm = d$)
- Premise 2: All poodles are dogs. ($pd = p$)
- All poodles are mammals. ($pm = p$)

Notice that if we multiply the first equation by p we get $pdm = pd$, but pd equals p by the second equation. All that remains for us to do is write p in place of pd on the left and right in the equation $pdm = pd$. This yields the conclusion of the syllogism, namely, $pm = p$. This example shows how a syllogism can be reduced to an algebraic computation.

with a question or even a chain of equations. Engineers use these ideas to design logic circuits for computers.

Boole's applications of his algebra centered on the theory of probability and the philosophy of mind. Both applications are of a philosophical nature, and they are not well remembered now, principally because Boole's work in these two areas, although intellectually interesting, did not uncover much that was new even at the time. The principal application of Boolean algebra, which involves the design of computer hardware and software, would not be discovered until the 20th century.

Refining and Extending Boolean Algebra

Boole's conception was an important step forward, but it contained some logical problems of its own. The first difficulty was identified and corrected by the British logician and economist William Stanley Jevons (1835–82). The difficulty arises in the course of computations.

In mathematics—especially pure mathematics—the method by which we arrive at a solution is at least as important as the solution itself. When solving a math problem, any sequence of steps should have the property that *each step can be logically justified.* To express the same idea in a different way: There should be a mathematical reason for every step in the solution. This was not the case in Boole's own version of Boolean algebra, and much of the difficulty centered around Boole's definition of the OR operator.

Recall that given two classes of objects, which we call x and y, Boole defined $x + y$ to mean that class of objects that belongs to x or to y but not to both. The problem with this definition is that as one uses Boole's algebra to solve problems, one sometimes encounters expressions such as $x + x$. To obtain this expression, we just substitute the letter x for the letter y in the first sentence of this paragraph. We get that $x + x$ is *that class of objects that belongs to x or to x but not to both.* This is meaningless. Although Boole found ways to manipulate the expressions to obtain valid results, from a logical point of view his definition is not entirely satisfactory. Jevons proposed a new definition for the OR operator, and it is his definition that is in general use today.

Jevons defined the OR operator inclusively: Given two classes, which we call x and y, the expression xORy means the set consisting of all objects in x together with all objects in y. In particular, Jevons defined OR so that if an object is both in x and in y it also belongs to xORy. An object fails to belong to xORy only when it fails to belong to x, *and* it fails to belong to y. The main advantage of Jevons's definition is that it allows us to attach a reasonable definition to the expression $x + x$: The set of objects belonging to the class xORx equals the class x itself. Admittedly this sounds stilted, but it allows us to attach a meaning to the expression $x + x$ that is logically satisfactory. In particular, this definition enables us to write the equation $x + x = x$. This is a different sort of equation than the one we encounter in our first algebra courses, but it parallels Boole's own equation for logical multiplication, namely, $x^2 = x$. This new definition of the OR operator straightened out many of the logical difficulties that had arisen in computing with Boole's algebra.

Notice, too, that in the equations $x + x = x$ and $x^2 = x$ the only coefficient to occur is 1. Furthermore if we search for roots of the equation $x^2 - x = 0$, we find that the only roots are the numbers 0 and 1. In other words, these equations enable us to restrict our attention to just two numbers.

The latter half of the 19th century saw several further extensions and refinements of Boole's algebra, but Boole's central concepts remained valid. Of special interest was the work of the German mathematician Ernst Schröder (1841–1902), who developed a complete set of axioms for Boolean algebra. (Boole's axioms, listed previously, were incomplete.)

OR	T	F
T	T	T
F	T	F

Truth table for the modern version of the operation OR

Axioms are a "bare bones" description of a mathematical system in the sense that everything that can be learned about a mathematical system, whether that system is, for example, Boolean algebra or Euclidean geometry, is a logical consequence of the axioms. In this sense mathematicians concern themselves with revealing facts that are, logically speaking, right before their eyes. The axioms always contain all of the information that one can learn about a system. The problem is that the information is not displayed in an obvious way. Any nonobvious statement that can be deduced from a set of axioms is called a theorem. Most mathematicians occupy themselves with deducing new theorems from theorems that have already been proved; this is the art of mathematical discovery. Unfortunately knowing that statement B is, for example, a logical consequence of statement A gives no insight into whether or not statement A is true.

Fortunately there is a final reason why any theorem is true. The ultimate reason that each theorem in a mathematical system is true is that it can be deduced as a logical consequence of the axioms that define the system. The axioms *are* the subject. It is no exaggeration that, mathematically speaking, one can never be completely sure of anything until a set of axioms that define the subject has been stated. Finally, a logically consistent set of axioms for any branch of mathematics is important because it ensures that it is possible to develop the mathematics in such a way that no statement can be proved both true and false. Placing Boole's algebra on a firmer logical foundation was Schröder's contribution.

In the sense that he axiomatized Boolean algebra, Schröder completed the subject: He put Boolean algebra into the logical form that we know today. Of course, many new theorems have been proved in the intervening century or so since Schröder's death, but the theorems were proved in the context of Schröder's axioms. Logicians, philosophers, and a few mathematicians were quick to recognize the value of Boole's insights. His ideas provided a conceptual foundation that enabled the user to examine more closely the relationships that exist between logic and mathematics. Many mathematicians, inspired by Boole's work, went on to do just that. This is one implication of Boole's discoveries, but Boolean algebra has had a more immediate impact on our lives through its use in the design of computer circuitry.

Boolean Algebra and Computers

Boole knew nothing about computers, of course. He died 15 years before the invention of the lightbulb, and the first electronic digital computer began operation in 1946—more than 80 years after his death. Nevertheless the design of computer chips is one of the most important applications of Boolean algebra. Boolean expressions produce one of the values, which we can call true and false or zero and one. Two digits are all that are necessary to express ideas in binary code. (*Binary code* is a way of coding information that depends on precisely two symbols, which, for convenience, are often represented by the digits 0 and 1.)

To appreciate how this works, we can imagine a computer that performs three functions. First, the computer reads an input file consisting of a string of binary digits. (The input file is the information that the computer has been programmed to process.) Second, the computer processes, or alters, the input file in accordance with some preprogrammed set of instructions. Third, the

Computer chips represent Boolean algebra in silicon. (Georgia Tech)

computer displays the results of these manipulations. This is the output file, which we can imagine as consisting of binary code as well. (Of course, the binary output is usually rewritten in a more user-friendly format, but the details of this reformatting process do not concern us here.) The output file is the reason we buy the computer. It represents the answer, the work performed by the machine on the input file.

The middle step, the processing part of this sequence, is the step in which we are interested. The processing takes place via a set of electronic circuits. By an *electronic circuit* we mean any structure through which electricity can flow. Circuits are manufactured in a variety of sizes and can be made of a variety of materials. What is important is that each circuit is capable of modifying or regulating the flow of electrical current in certain very specific ways. The actual control function of a circuit is affected through a set of switches or *gates*. The gates themselves are easily described in terms of Boolean operators.

There are several types of gates. They either correspond to or can be described in terms of the three common Boolean operators, the AND, OR, and NOT operators. (The NOT operator reverses the value of the variable. The value of NOT-true is false, for example, and the value of NOTx is $(1-x)$ as discussed in the section *Boolean Algebra.*) The names of the gates are even derived from Boolean algebra: There are AND-gates, OR-gates, and NOT-gates. By combining Boolean operators we also obtain two other common types of gates: NAND-gates and NOR-gates. Each type of gate regulates the flow of electric current subject to certain conditions.

The idea is that there is a very low-level current flowing through each circuit. This current is constant and has no effect on whether the gate is "open" or "closed." When, however, the voltage of the input current rises above a certain prespecified level, the gate is activated. The level of voltage required to activate the gate is called the threshold voltage.

To see how Boolean algebra comes into play, we describe the AND-gate and the OR-gate. An AND-gate has two inputs, just as the Boolean operator AND has two arguments or independent variables. In the case of the AND-gate, we can let x represent the voltage at one input and *y* represent the voltage at the other input. When the voltage *x* *and* the voltage *y* *simultaneously* exceed the threshold voltage, the AND-gate allows current to pass from one side to the other. If, however, the voltage in either or both of the inputs falls below the threshold voltage, current does not pass to the other side of the gate. It is in this sense that the AND-gate is a physical representation of Boole's own AND operator. Instead of classes of objects, or binary digits, however, the AND-gate operates on electric current.

Similarly the OR-gate is designed to be inclusive, just as the more modern version of the Boolean operator OR is defined inclusively: If either *x* or *y* is true, then *x*OR*y* is also true. The OR-gate operates on two inputs. We can represent the voltage in one input with the letter *x* and the voltage in the second input with the letter *y*. If either *x* or *y* is at or above the threshold voltage then the OR-gate allows the current to pass. Otherwise, the current does not pass.

The five gates—the AND-, OR-, NOT-, NOR-, and NAND-gates—modify input in the form of electric currents to produce a new pattern of electrical currents as output. One can interpret the changes in current pattern—or what is the same thing, the state of the gates—as information, but this is an additional interpretation that is placed on the configuration of circuits. There could be no better physical representation of Boolean algebra than the logic circuits of a computer.

George Boole's exposition of Boolean algebra is contained in a pamphlet, "Mathematical Analysis of Logic" (1847), and a book, *An Investigation into the Laws of Thought on Which Are Founded the Mathematical Theories of Logic and Probabilities* (1854). In these works we find not just a new branch of mathematics, but also a new way of thinking about mathematics. Boole's approach was deliberately more abstract than that of his predecessors. This highly abstract approach, far from making his algebra useless, made his algebra one of the most useful of all mathematical innovations. The most important practical applications of Boole's philosophical and mathematical investigations would not be apparent, however, until about a century after "Mathematical Analysis of Logic" was published.

8

THE THEORY OF MATRICES AND DETERMINANTS

Many new types of algebraic structures have been defined and studied since the time of Galois. Today, in addition to groups and fields, mathematicians study algebraic structures called rings, semigroups, and algebras to name a few. (Here *algebra* refers to a particular type of mathematical object.) Each structure is composed of one or more sets of objects on which one or more operations are defined. The *operations* are rules for combining objects in the sets. The sets and operations together form a *structure*, and it is the goal of the mathematician to discover as much as possible about the logical relationships that exist among different parts of the structure.

In this modern approach to algebra the nature of the objects in the set is usually not specified. The objects are represented by letters. The letters may represent numbers, polynomials, or something else entirely, but usually no interpretation is placed on the letters at all. It is only the relationships that exist *among* objects and sets of objects—not the objects and sets themselves—that are of interest to the mathematician.

One of the first and most important of these "new" mathematical structures to receive the attention of mathematicians was the algebra of matrices. Matrices are tables of numbers or symbols. They combine according to some of the same rules that numbers obey, but some of the relationships that exist between matrices are different from the analogous relationships between numbers.

An important part of the theory of matrices is the theory of determinants. Today a determinant is often described as a function of a matrix. For example, if the elements in the matrix are numbers, then—provided the matrix has as many rows as columns—we can, in theory, use those elements to compute a number called the determinant. The determinant reveals a great deal of useful information about the (square) matrix. If a matrix represents a system of equations, for example, then the determinant can tell us whether or not there exists a single solution to the system. In theory we can even use determinants to compute solutions to systems of equations (although, as we will soon see, the work involved in doing so is usually enormous—too much work to make it a practical approach to problem solving).

The theory of matrices and determinants has proved to be one of the most useful of all branches of mathematics. Not only is the theory an important tool in the solution of many problems within the field of mathematics, it is also one of the most useful in science and engineering. One reason is that this is the type of mathematics that one must know in order to solve systems of linear equations. (A *linear equation* is an equation in which every term is either the product of a number and a variable of the first power or simply a number. For example, $x + y = 1$ is a linear equation, but $x^2 + y = 1$ is not because the x term is raised to the second power.)

Most of us are introduced to systems of linear equations while we are still in junior or senior high school. These are "small" systems, usually involving two or three independent variables. We begin with small systems because the amount of work involved in solving systems of linear equations increases rapidly as the number of variables increases. Unfortunately these small systems fail to convey the tremendous scope of the subject. Today many mathematicians, scientists, and engineers are engaged in solving systems of equations involving many thousands of independent variables. The rush to develop computer algorithms that quickly and accurately solve ever-larger systems of equations has attracted the attention of many mathematicians around the world. The history of matrices, determinants, and related parts of mathematics, however, begins long before the advent of the computer.

Early Ideas

Today when determinants and matrices are taught, matrices are introduced first, and determinants are described as functions of matrices. But historically determinants were discovered almost 200 years before mathematicians began to study matrices.

The Japanese mathematician Seki Kōwa, also known as Takakazu (1642–1708), was the first person to discover the idea of a determinant and investigate some of the mathematics associated with this concept. Seki was born into a samurai warrior family, but at an early age he was adopted by a family of the ruling class. When Seki was age nine a family servant who knew mathematics introduced him to the subject. He demonstrated mathematical talent almost immediately, and later in life Seki became known as the Arithmetical Sage. Today he is often described as the founder of Japanese mathematics. This is something of an exaggeration. There was mathematics in Japan before Seki. Nevertheless he was certainly an important person in the history of Japanese mathematics.

The Arithmetical Sage published very little work during his life. In fact, as was the custom in Japan at the time, he disclosed much of his work to only a select few. As a consequence much of what we know about his discoveries is secondhand or third-hand. Some scholars attribute a great many accomplishments to him: an (unproved) version of the fundamental theorem of algebra, discoveries in the field of calculus, complex algorithms for discovering solutions to algebraic equations, and more. Other scholars attribute quite a bit less. It is certain, however, that Seki discovered determinants, because his writings on this subject are well known.

Seki's ideas on determinants are fairly complex, and he used them in ways that would be difficult to describe here. A simpler approach to determinants was discovered independently in Europe about 10 years after Seki made his initial discovery. The German mathematician and philosopher Gottfried Leibniz (1646–1716) was the second person to discover what we now call determinants.

Leibniz was one of the more important figures in the history of mathematics. He is one of two creators of the calculus, and he made a number of important discoveries in other areas of mathematics. He discovered, for example, the base 2 number system.

With respect to determinants, Leibniz indicated that he was sometimes required to solve a set of three linear equations involving two variables. Such a system may or may not have a solution. Leibniz discovered that the determinant could be used to establish a criterion for the existence or nonexistence of a solution.

In modern notation we might represent a system of three equations in two unknowns as follows:

$$a_{11} + a_{12}x + a_{13}y = 0$$
$$a_{21} + a_{22}x + a_{23}y = 0$$
$$a_{31} + a_{32}x + a_{33}y = 0$$

The letters a_{ij} represent numbers called coefficients. All the coefficients on the left side of the column of "equals" signs can be viewed as part of a table. In that case the first index—the i-index—indicates the row in which the number appears, and the second index—the j-index—indicates the column. (The coefficient a_{12}, for example, belongs to the first row and the second column.) Notice that the first column contains no variables.

Leibniz's system of equations contains more equations—there are three of them—than there are variables; there are only two variables. When the number of equations exceeds the number of variables, the possibility exists that there are simply too many constraints on the variables and that no values for x and y can simultaneously satisfy all the equations. Mathematicians today call such a system—a system for which there are no solutions—overdetermined. But even when there are more equations than there are variables, it is still possible that solutions exist. What Leibniz discovered is a criterion for determining whether such a system of equations is overdetermined. His criterion is very general, and it does not involve computing the solutions to the equations themselves. Instead it places a constraint on the numbers in the table of coefficients. The key is using the a_{ij} to compute a number that we now call the determinant. Leibniz wrote that when the determinant of this type of system is 0, a solution exists. When the determinant is not 0, there are no values of x and y that can simultaneously satisfy all three equations. In addition Leibniz understood how to use determinants to calculate the values of the variables that would satisfy the system of equations.

Leibniz had made an important discovery: He had found a way to investigate the existence of solutions for an entire class of problems. He did this with a new type of function, the determinant, that depends only on the coefficients appearing in the equations themselves. He described his discoveries in letters to a colleague, but for whatever reason he did not publish these results for a wider audience. In fact Leibniz's ideas were not published for more than 150 years after his death. As a consequence his ideas on determinants were not widely known among the mathematicians of his time and had little impact on the development of the subject.

Mathematicians again began to look at determinants as a tool in understanding systems of equations about 50 years after Leibniz first described his discoveries. Initially these ideas were stated and proved only for small systems of variables that in modern notation might be written like this:

$$a_{11}x + a_{12}y + a_{13}z = b_1$$
$$a_{21}x + a_{22}y + a_{23}z = b_2$$
$$a_{31}x + a_{32}y + a_{33}z = b_3$$

The notation is similar to what Leibniz used. The differences are that (1) here there are three equations in three variables and (2) the b_i represent any numbers.

Part of the difficulty that these mathematicians had in applying their insights about determinants to larger systems of equations is that their algebraic notation was not good enough. Determinants can be difficult to describe without very good notation. The calculation of determinants—even for small systems—involves quite a bit of arithmetic, and the algebraic notation needed to describe the procedure can be very complicated as well. For example, the determinant of the system in the previous paragraph is

$$a_{11}a_{22}a_{33} + a_{12}a_{23}a_{31} + a_{13}a_{21}a_{32}$$
$$- a_{31}a_{22}a_{13} - a_{32}a_{23}a_{11} - a_{33}a_{21}a_{12}$$

(The general formula for computing determinants of square matrices of any size is too complicated to describe here. It can, however, be found in any textbook on linear algebra.)

Notice that for a general system of three equations in three unknowns the formula for the 3 × 3 matrix given in the preceding paragraph involves 17 arithmetic operations, that is, 17 additions, subtractions, and multiplications. Computing the determinant of a general system of four equations and four unknowns involves several times as much work when measured by the number of arithmetic operations involved.

In 1750 the Swiss mathematician Gabriel Cramer (1704–50) published the method now known as Cramer's rule, a method for using determinants to solve any system of n linear equations in n unknowns, where n represents any positive integer greater than 1. Essentially Cramer's rule involves computing multiple determinants. Theoretically it is a very important insight into the relationships between determinants and systems of linear equations. Practically speaking Cramer's rule is of little use, because it requires far too many computations. The idea is simple enough, however. For example, in the system of equations given three paragraphs previous, the solution for each variable can be expressed as a fraction in which the numerator and the denominator are both determinants. The denominator of the fraction is the determinant of the system of equations. The numerator of the fraction is the determinant obtained from the original system of equations by replacing the column of coefficients associated with the variable of interest with the column consisting of (b_1, b_2, b_3). In modern notation we might write the value of x as

$$x = \frac{\begin{vmatrix} b_1 & a_{12} & a_{13} \\ b_2 & a_{22} & a_{23} \\ b_3 & a_{32} & a_{33} \end{vmatrix}}{\begin{vmatrix} a_{11} & a_{12} & a_{13} \\ a_{21} & a_{22} & a_{23} \\ a_{31} & a_{32} & a_{33} \end{vmatrix}}$$

where the vertical lines indicate the determinant of the table of numbers inside. (One consequence of this formula is that it fails when the denominator is zero. But the denominator is the determinant of the original system of equations. It can be shown that a unique solution for the system exists if and only if the denominator is not zero.) A more computational approach to expressing the value of x looks like this:

$$x = \frac{b_1a_{22}a_{33} + a_{12}a_{23}b_3 + a_{13}b_2a_{32} - b_3a_{22}a_{13} - a_{32}a_{23}b_1 - a_{33}b_2a_{12}}{a_{11}a_{22}a_{33} + a_{12}a_{23}a_{31} + a_{13}a_{21}a_{32} - a_{31}a_{22}a_{13} - a_{32}a_{23}a_{13} - a_{33}a_{21}a_{12}}$$

Writing out the solution in this way for a system of five equations with five unknowns would take up much of this page.

Putting the computation aside, it can be helpful to think of Cramer's rule as a function of coefficients. The values of the function are the solutions to the equations. In concept, Cramer's rule is similar to the quadratic formula, which enables the user to compute the solutions to equations of the form $ax^2 + bx + c = 0$ using only the numbers a, b, and c, the coefficients that appear in the equation. In other words, the quadratic formula allows one to identify x in terms of a, b, and c. Large systems of linear equations—and here we confine our attention to n linear equations in n unknowns—have many more coefficients than a quadratic equation and so the function is more complicated, but Cramer's rule enables the user to identify $x_1, x_2, \ldots, x_n$ in terms of $a_{11}, a_{12}, a_{13}, \ldots, a_{21}, a_{22}, a_{23}, \ldots, a_{nn}, b_1, b_2, b_3, \ldots, b_n$.

In the years immediately following the publication of this method of solution, mathematicians expended a great deal of effort seeking easier ways to compute determinants for certain special cases as well as applications for these ideas. These ideas became increasingly important as they found their way into physics.

Spectral Theory

New insights into the mathematics of systems of linear equations arose as mathematicians sought to apply analysis, that branch of mathematics that arose out of the discovery of calculus, to the study of problems in physics. The three mathematicians who pointed the way to these new discoveries were all French: Jean le Rond d'Alembert (1717–83), Joseph-Louis Lagrange (1736–1813), and Pierre-Simon Laplace (1749–1827). All three mathematicians contributed to the ideas about to be described; of the three d'Alembert was the first.

Jean le Rond d'Alembert was the child of an artillery officer and a marquise. His birth mother abandoned him on the steps of the Parisian church of Saint Jean le Rond (where he got his name). He was adopted by the wife of an artisan who raised him as her own. He lived with her until her death—48 years later. When he later achieved prominence as a scientist, he spurned his biological mother's attempts to make contact with him.

D'Alembert's biological father never acknowledged his paternity, but he made sure that his son had sufficient money for a first-rate education. In college d'Alembert studied theology, medicine, and law, but he eventually settled on mathematics. Surprisingly, d'Alembert taught himself mathematics while pursuing his other studies, and except for a few private lessons he was entirely self-taught.

Soon after beginning his mathematical studies, d'Alembert distinguished himself as a mathematician, a physicist, and a personality. Never hesitant to criticize the work of others, d'Alembert lived a life marked by almost continuous controversy as well as mathematical and philosophical accomplishments. In his own day, d'Alembert was probably best known as the coeditor of the *Encyclopédie* with Denis Diderot. This was one of the great works of the Enlightenment.

With respect to systems of linear equations, d'Alembert was interested in developing and solving a set of equations that would represent the motion of a very thin, very light string along which several weights were arrayed. With one end tied to a support, the weighted string is allowed to swing back and forth. Under these conditions the motion of the string is quite irregular. Some mathematicians of the time believed that the motion was too complicated to predict. D'Alembert, however, solved the problem for small motions of the string.

As mathematicians uncovered connections between the eigenvalues of matrices and problems in astronomy, progress was accelerated in both algebra and astronomy. (NASA-JPL)

The analytical details of d'Alembert's solution are too complicated to describe here, but the algebra is not. Broadly speaking d'Alembert reduced his problem to what is now known as an eigenvalue problem. A general eigenvalue problem looks like this:

$$a_{11}x + a_{12}y + a_{13}z = \lambda x$$
$$a_{21}x + a_{22}y + a_{23}z = \lambda y$$
$$a_{31}x + a_{32}y + a_{33}z = \lambda z$$

In d'Alembert's problem the unknowns, here represented by the letters x, y, and z, represented functions rather than numbers, but this distinction has no bearing on the algebra that we are interested in discussing.

This system of linear equations is different from the others we have considered in three important ways:

1. The unknowns, x, y, and z, appear on both sides of each equation.
2. The number represented by the Greek letter λ, or lambda, is also an unknown. It is called an eigenvalue of the system of equations.
3. The goal of the mathematician is to find all eigenvalues as well as the solutions for x, y, and z that are associated with each eigenvalue. Each eigenvalue determines a different set of values for x, y, and z.

The problem of determining the eigenvalues associated with each system of equations is important because eigenvalues often have important physical interpretations.

D'Alembert discovered that the only reasonable solutions to his equations were associated with negative eigenvalues. Solutions associated with positive eigenvalues—that is, solutions associated with values of λ greater than 0—were not physically realistic. A solution associated with a positive eigenvalue predicted that once the string was set in motion, the arc along which it swung would become larger and larger instead of slowly "dying down" as actually occurs. The observation that eigenvalues had interesting physical interpretations was also made by other scientists at about the same time.

Pierre-Simon Laplace and Joseph-Louis Lagrange reached similar sorts of conclusions in their study of the motion of the planets. Their research also generated systems of linear equations, where each unknown represented a function (as opposed to a number). They studied systems of six equations in six unknowns because at the time there were only six known planets. Laplace and Lagrange discovered that solutions associated with positive eigenvalues predicted that small perturbations in planetary motion would become ever larger over time. One consequence of Laplace's and Lagrange's observation is that any solution associated with a positive eigenvalue predicted that over time the solar system would eventually fly apart. Lagrange rejected positive eigenvalues on the basis of physical reasoning: The solar system had not already flown apart. Laplace ruled out the existence of positive eigenvalues associated with his system of equations on mathematical grounds. He proved that in a system in which all the planets moved in the same direction, the eigenvalues must all be negative. He concluded that the solar system is stable—that is, that it would not fly apart over time.

Notice the similarities between the model of the solar system and the weighted string problem of d'Alembert. In each case solutions associated with positive eigenvalues were shown to be "nonphysical" in the sense that they did not occur in nature. The connection between algebraic ideas (eigenvalues) and physical ones (weighted strings and planetary orbits) spurred further research into both.

To convey the flavor of the type of algebraic insights that Lagrange and Laplace were pursuing, consider the following eigenvalue problem taken from one of the preceding paragraphs. It is reproduced here for ease of reference.

$$a_{11}x + a_{12}y + a_{13}z = \lambda x$$
$$a_{21}x + a_{22}y + a_{23}z = \lambda y$$
$$a_{31}x + a_{32}y + a_{33}z = \lambda z$$

Notice that we can subtract away each term on the right from both sides of each equation. The result is

$$(a_{11} - \lambda)x + a_{12}y + a_{13}z = 0$$
$$a_{21}x + (a_{22} - \lambda)y + a_{23}z = 0$$
$$a_{31}x + a_{32}y + (a_{33} - \lambda)z = 0$$

This is a new set of coefficients. The coefficient in the upper left corner is now $a_{11} - \lambda$ instead of simply a_{11}. The middle coefficient is now $a_{22} - \lambda$ instead of a_{22}, and similarly the coefficient in the lower right corner is now $a_{33} - \lambda$ instead of a_{33}. The other coefficients are unchanged. From this new set of coefficients a new determinant can be computed. (We can, in fact, use the formula already given for 3 × 3-matrices on page 146: Simply substitute $a_{11} - \lambda$, $a_{22} - \lambda$, $a_{33} - \lambda$ for a_{11}, a_{22}, and a_{33} in the formula.) The result is a third-degree polynomial in the variable λ. This polynomial is called the characteristic polynomial, and its roots are exactly the eigenvalues of the original system of linear equations.

The discovery of the characteristic polynomial established an important connection between two very important branches of algebra: the theory of determinants and the theory of algebraic equations. Laplace and Legendre had discovered the results for particular systems of equations, but there was as yet no general theory of either determinants or eigenvalues. Their work, however, pointed to complex and interesting connections among the theory of determinants, the theory of algebraic equations, and physics. This very rich interplay of different areas of mathematics and science is such a frequent feature of discovery in both fields that today we sometimes take it for granted. At the time of Laplace and Legendre, however, the existence of these interconnections was a discovery in itself.

The work of d'Alembert, Laplace, and Legendre gave a great impetus to the study of eigenvalue problems. Investigators wanted to understand the relationships that existed between the coefficients in the equations and the eigenvalues. The result of these inquiries was the beginning of a branch of mathematics called spectral theory—the eigenvalues are sometimes called the spectral values of the system—and the pioneer in the theoretical study of these types of questions was the French mathematician Augustin-Louis Cauchy (1789–1857).

Cauchy was born at a time when France was politically unstable. This instability profoundly affected both his personal and his professional life. First the French Revolution of 1789 occurred. While Cauchy was still a boy, the revolution was supplanted by a period called the Reign of Terror (1793–94), a period when approximately 17,000 French citizens were executed and many more were imprisoned. In search of safety, Cauchy's family fled Paris, the city of his birth, to a village called Arcueil. It was in Arcueil that Cauchy first met Laplace. Lagrange and Laplace were friends of Cauchy's father, and Lagrange advised the elder Cauchy that his son could best prepare himself for mathematics by studying languages. Dutifully Cauchy studied languages for two years before beginning his study of mathematics.

By the age of 21, Cauchy was working as a military engineer—at this time Napoléon was leading wars against his European neighbors—and pursuing research in mathematics in his spare time. Cauchy wanted to work in an academic environment, but this goal proved difficult for him. He was passed over for appointments by several colleges and worked briefly at others. He eventually secured a position at the Académie des Sciences in 1816. There he replaced the distinguished professor of geometry Gaspard Monge, who lost the position for political reasons.

In July of 1830 there was another revolution in France. This time King Charles X was replaced by Louis-Philippe. As a condition of employment Cauchy was required to swear an oath of allegiance to the new king, but this he refused to do. The result was that he lost the academic post that had meant so much to him. Cauchy found a position in Turin, Italy, and later in Prague in what is now the Czech Republic. By 1838 he was able to return to his old position in Paris as a researcher but not as a teacher: The requirement of the oath was still in effect, and Cauchy still refused to swear his allegiance. It was not until 1848, when Louis-Philippe was overthrown, that Cauchy, who never did swear allegiance, was able to teach again.

All biographies of Cauchy indicate that he was a difficult man, brusque and preachy. This is another reason that he often did not obtain academic appointments that he very much desired. This pattern proved a source of frustration throughout his life. Even toward the end of his career, after producing one of the largest and most creative bodies of work in the history of mathematics, he still failed to gain an appointment that he sought at the Collège de France.

Today there are theorems and problems in many branches of mathematics that bear Cauchy's name. His ideas are fundamental to the fields of analysis, group theory, and geometry as well as spectral theory. After his death his papers were collected and published. They fill 27 volumes.

One of Cauchy's earliest papers was about the theory of determinants. He revisited the problems associated with determinants and eigenvalues several times during his life, each time adding something different. We concentrate on two of his contributions, which we sometimes express in a more modern and convenient notation than Cauchy used.

Cauchy wrote determinants as tables of numbers. For example, he would write the determinant of the system of linear equations

$$a_{11}x + a_{12}y + a_{13}z = b_1$$
$$a_{21}x + a_{22}y + a_{23}z = b_2$$
$$a_{31}x + a_{32}y + a_{33}z = b_3$$

as the table of numbers

$$\begin{matrix} a_{11} & a_{12} & a_{13} \\ a_{21} & a_{22} & a_{23} \\ a_{31} & a_{32} & a_{33} \end{matrix}$$

The coefficients a_{11}, a_{22}, a_{33} lie along what is called the main diagonal. Cauchy proved that when all the numbers in the table are real and when the table itself is symmetric with respect to the main diagonal—so that, for example, $a_{12} = a_{21}$—the eigenvalues, or roots of the characteristic equation, are real numbers. This was the first such observation relating the eigenvalues of the equations to the structure of the table of numbers from which the determinant is calculated. Part of the value of this observation is that it enables the user to describe various properties of the eigenvalues without actually computing them. (As a practical matter, computing eigenvalues for large systems of equations involves a great deal of work.)

Cauchy also discovered a kind of "determinant arithmetic." In modern language he discovered that when two matrices are multiplied together in a certain way—see the sidebar *Matrix Multiplication*—the determinant of the product matrix is the product of the determinants of the two matrices. In other words, if A and B are two square arrays of numbers, then $\det(A \times B) = \det(A) \times \det(B)$, where $\det(A)$ is shorthand for "the determinant of A." These theorems are important because they hint at the existence of a deeper logical structure.

To better appreciate what "deeper logical structure" means, it helps to think about Euclidean geometry and what it means to study Euclidean geometry. What do students study when they study Euclidean geometry? A part of Euclidean geometry is devoted, for example, to the study of triangles, but not every property of a triangle is geometric. Euclidean geometry is not concerned, for example, with how far from the edge of a blackboard or computer screen the triangle appears; nor does it matter how the triangle is tilted. These properties are not geometric. Instead, Euclidean geometry is concerned solely with those properties of a triangle that do not change when, for example, the triangle is rotated or translated. (A translation means that the triangle is moved so that its sides remain parallel to their original orientation.) These motions are called Euclidean motions (or Euclidean transformations), and so one can characterize Euclidean geometry as the study of those properties of figures that are invariant under the set of all Euclidean motions. Lengths are preserved under translations and rotations, for example, and so are the measures of angles. These properties of the triangle are, therefore, geometric in Euclidean geometry. In non-Euclidean geometries these properties may not be geometric.

In a similar sort of way, Cauchy had begun to identify properties of matrices and determinants that are invariant when the matrices undergo certain kinds of transformations. This idea is important because it is often the case that two algebraic descriptions of the same object look different but are in some sense the same. An equation, for example, can be imagined as a kind of description. It describes a set of numbers—the set of numbers that satisfy the equation. The equation is not invariant under multiplication because multiplication of each term by any number (different from zero or one) will change every term in the equation. The solutions, however, are invariant under multiplication, because multiplying both sides of the equation by any nonzero number will leave the solution set unchanged. Invariant properties are fundamental properties, and Cauchy had begun the process of enumerating those properties of matrices, their determinants, and their associated characteristic polynomials that are invariant when the coefficients of the matrices are transformed in certain ways. To describe the specifics of these transformations would take us too far afield, but identifying the transformations and the properties that they leave unchanged is always a major step forward in any program of algebraic research—then as now. The mathematician studies exactly those properties of a system that are invariant under a specified class of transformations. Cauchy's insights into these matters form an important part of the foundations of spectral theory.

The Theory of Matrices

Credit for founding the theory of matrices is often given to the English mathematician Arthur Cayley (1821–95) and his close friend the English mathematician James Joseph Sylvester (1814–97), but others had essentially the same ideas at roughly the same time. The German mathematicians Ferdinand Georg Frobenius (1849–1917) and Ferdinand Gotthold Max Eisenstein (1823–52) and the French mathematician Charles Hermite (1822–1901) are three mathematicians who also made discoveries similar to those of Cayley and Sylvester. Eisenstein, in fact, seems to have been the first to think of developing an algebra of matrices. He had been studying systems of linear equations of the form

$$a_{11}x + a_{12}y + a_{13}z = b_1$$
$$a_{21}x + a_{22}y + a_{23}z = b_2$$
$$a_{31}x + a_{32}y + a_{33}z = b_3$$

and began to consider the possibility of analyzing the mathematical properties of what is essentially the "skeleton" of the equation, the table of coefficients that today we would write as

$$\begin{bmatrix} a_{11} & a_{12} & a_{13} \\ a_{21} & a_{22} & a_{23} \\ a_{31} & a_{32} & a_{33} \end{bmatrix}$$

Although this idea may seem similar to that of Cauchy's tables of coefficients, it is not. It is true that Cauchy used tables of numbers, but he used them as an alternate way of representing the determinant function. Eisenstein contemplated the possibility of developing an algebra in which the objects of interest were not numbers, or the determinant function, or even polynomials, but rather matrices. Unfortunately he died before he could follow up on these ideas. In this discussion we follow the usual practice of emphasizing Cayley's and Sylvester's contributions, but it would, for example, be possible to describe the history of matrix algebra from the point of view of Frobenius, Eisenstein, and Hermite as well.

A great deal has been written about Cayley and Sylvester as researchers and as friends. Each distinguished himself in mathematics at university, Cayley at Trinity College, Cambridge, and Sylvester at Saint John's College, Cambridge. Cayley's early academic successes led to increased opportunities at college as well as a stipend. Sylvester's early successes proved to be a source of frustration. He was barred from a number of opportunities because of discrimination—he was Jewish—and he left Saint John's without graduating. He would eventually receive his degrees in 1841 from Trinity College, Dublin.

When it came time to find employment as mathematicians, neither Cayley nor Sylvester found much in the way of work. Cayley solved the problem by becoming a lawyer. Sylvester, too, became a lawyer, but his route to the legal profession was more circuitous. In 1841 he left Great Britain and worked briefly on the faculty at the University of Virginia. He left the university several months later after an altercation with a student. Unable to find another position in the United States, he returned to London in 1843 to work as an actuary. While working as an actuary, Sylvester tutored private pupils in mathematics, and it was during this time that he tutored the medical pioneer Florence Nightingale in mathematics. (Nightingale was a firm and early believer in the use of statistics to evaluate medical protocols.) Finally, in 1850 Sylvester turned to the legal profession to earn a living. That same year while they were both working as lawyers Cayley and Sylvester met and formed a lifelong friendship.

Cayley worked as a lawyer for 14 years before he joined the faculty at Cambridge in 1863. Sylvester worked as a lawyer for five years until he found a position at the Royal Military Academy, Woolwich. Cayley, a contemplative man, remained at Cambridge for most of the rest of his working life. The exception occurred when he spent a year at Johns Hopkins University in Baltimore, Maryland, at Sylvester's invitation. By contrast Sylvester remained at Woolwich for 15 years and then, in 1876, moved back to the United States to work at Johns Hopkins University. (Sylvester played an important role in establishing advanced mathematical research in the United States.) In 1883 Sylvester returned to the United Kingdom to work at Oxford University.

Although they were both creative mathematicians their approaches to mathematics were quite different. Cayley spoke carefully and produced mathematical papers that were well reasoned and rigorous. By contrast, Sylvester was excitable and talkative and did not hesitate to substitute his intuition for a rigorous proof. He sometimes produced mathematical papers that contained a great deal of elegant and poetic description but were decidedly short on mathematical rigor. Nevertheless his intuition could usually be shown to be correct.

The theory of determinants, spectral theory, and the theory of linear equations had already revealed many of the basic properties of matrices before anyone conceived of the idea of a matrix. Arthur Cayley remarked that logically the theory of matrices precedes the theory of determinants, but historically these theories were developed in just the opposite order. It was Cayley, author of "A Memoir on the Theory of Matrices," published in 1858, who first described the properties of matrices as mathematical objects. He had been studying systems of equations of the form

$$a_{11}x + a_{12}y = u$$
$$a_{21}x + a_{22}y = v$$

(Notice that in this set of equations the variables x and y are the independent variables and the variables u and v are the dependent variables.) Apparently in an effort to streamline his notation he simply wrote $\begin{bmatrix} a_{11} & a_{12} \\ a_{21} & a_{22} \end{bmatrix}$, a shorthand form of the same equation that preserves all of the information.

Having defined a matrix he began to study the set of all such matrices as a mathematical system. The most useful and richest part of the theory concerns the mathematical properties of square matrices of a fixed size—they are called the set of all $n \times n$ matrices, where n represents a fixed natural number greater than 1. In what follows we restrict our attention to 2×2 matrices for simplicity, but similar definitions and results exist for square matrices of any size.

Matrix addition is defined elementwise. Given a pair of 2×2 matrices, which we can represent with the symbols $\begin{bmatrix} a_{11} & a_{12} \\ a_{21} & a_{22} \end{bmatrix}$ and $\begin{bmatrix} b_{11} & b_{12} \\ b_{21} & b_{22} \end{bmatrix}$, the sum of these two matrices is $\begin{bmatrix} a_{11}+b_{11} & a_{12}+b_{12} \\ a_{21}+b_{21} & a_{22}+b_{22} \end{bmatrix}$. (The difference of the two matrices is simply obtained by writing a subtraction sign in place of the addition sign.) With this definition of matrix addition the matrix $\begin{bmatrix} 0 & 0 \\ 0 & 0 \end{bmatrix}$ plays the same role as the number 0 in the real number system. (Notice that the set of matrices of the same size—the set of 2×2 matrices, for example—form a group under the operation of addition.)

Cayley also defined matrix multiplication. The definition of matrix multiplication is not especially obvious to most of us, but to Cayley it was a simple matter because he and others had used this definition in the study of other mathematics problems, even before he had begun the study of matrices (see the sidebar Matrix Multiplication).

There are differences between matrix arithmetic and the arithmetic of numbers that we learn in grade school. One significant difference is that multiplication is not commutative: That is, the order in which we multiply two matrices makes a difference. Given two matrices, which we represent with the letters A and B, it is generally false that AB and BA are equal. Ordinary multiplication of numbers, by contrast, is commutative: 3×4 and 4×3, for example, represent the same number.

Another significant difference between matrix multiplication and the multiplication of numbers is that every *number* (except zero) has a multiplicative inverse. To appreciate what this means, recall that if the letter x represents a number other than zero, there is always another number, which we can write as x^{-1}, such that $x \times x^{-1} = 1$. To be sure, the set of all $n \times n$ matrices has a matrix that corresponds to the number 1. This matrix, which is called the identity matrix and is usually represented by the capital letter I, has 1s along its main diagonal and zeros elsewhere. (Recall the main diagonal consists of those entries of the form $a_{1,1}, a_{2,2}, \ldots, a_{n,n}$.) When the identity is multiplied by any matrix A, the result is A—in other words, $IA = A$ and $AI = A$—and this, of course, is exactly what happens when a number is multiplied by the number 1. But there are *some* (nonzero) $n \times n$ matrices with no multiplicative inverse, or to put it another way, for some $n \times n$ matrices A, there is no matrix A^{-1} such that $A \times A^{-1} = I$. (Notice that because not every $n \times n$ matrix has an inverse, the set of all nonzero $n \times n$ matrices cannot form a group under multiplication because one of the group axioms is that every element in the set must have an inverse.)

There is an additional operation that one can perform in matrix arithmetic that connects the theory of matrices with ordinary numbers: Not only can one compute the product of two $n \times n$ matrices; it is also possible to multiply any matrix by a number. For example, if the letter c represents any number then the product of c and the matrix $\begin{bmatrix} a_{11} & a_{12} \\ a_{21} & a_{22} \end{bmatrix}$ is $\begin{bmatrix} ca_{11} & ca_{12} \\ ca_{21} & ca_{22} \end{bmatrix}$.

Cayley also investigated polynomials in which the variables that appear in the polynomial represent matrices instead of numbers. His most famous result is the relationship between a square matrix and its characteristic polynomial. On page 153, we described the characteristic polynomial of a matrix. Cayley showed that if the matrix is written in place of the variable in the characteristic polynomial and the indicated operations are performed, the result is always the 0 matrix. In other words, each matrix satisfies the equation obtained by setting its characteristic polynomial to 0, so it is sometimes said that every matrix is a root of its own characteristic polynomial. This is called the Hamilton-Cayley theorem after Cayley and the Irish mathematician and astronomer Sir William Henry Rowan Hamilton (1805–65), who discovered the same theorem but from a different point of view.

In symbols, the Hamilton-Cayley theorem looks like this: Let A be a square $n \times n$ matrix. Let $a_n\lambda^n + a_{n-1}\lambda^{n-1} + \ldots + a_1\lambda + a_0$ be its characteristic polynomial, then $a_nA^n + a_{n-1}A^{n-1} + \ldots a_1A + a_0I = 0$, where the symbol zero represents the $n \times n$ matrix that has zeros for all its entries.

Cayley was a prominent and prolific mathematician, but his work on matrices did not attract much attention inside Great Britain. Outside Great Britain it was unknown. Consequently many of his ideas were later rediscovered elsewhere. In the 1880s, James Joseph Sylvester, who in the intervening years had become one of the most prominent mathematicians of his time, turned his attention to the same questions that Cayley had addressed about three decades earlier. Whether Sylvester had read Cayley's old monograph or rediscovered these ideas independently is not clear. In any case Sylvester's work had the effect of drawing attention to Cayley's earlier discoveries—a fact that seemed to please Sylvester. He always spoke highly of his friend—he once described Cayley's memoir on matrices as "the foundation stone" of the subject—but in this case Sylvester's prominence and his emphasis on the contributions of Cayley had the effect of obscuring the work of Frobenius, Eisenstein, Hermite, and others.

Sylvester did more than rediscover Cayley's work, however. Sylvester had made important contributions to the theory of

MATRIX MULTIPLICATION

Matrix multiplication is defined for square matrices of a fixed but arbitrary size in such a manner that many of the laws that govern the arithmetic of ordinary numbers carry over to the matrix case. In what follows we restrict our attention to 2 × 2 matrices, but similar definitions apply to any $n \times n$ matrix.

Let the matrix $\begin{bmatrix} c_{11} & c_{12} \\ c_{21} & c_{22} \end{bmatrix}$ represent the product of the matrices $\begin{bmatrix} a_{11} & a_{12} \\ a_{21} & a_{22} \end{bmatrix}$ and $\begin{bmatrix} b_{11} & b_{12} \\ b_{21} & b_{22} \end{bmatrix}$. Each number c_{ij} is obtained by combining the ith row of the "a-matrix" with the jth column of the "b-matrix" in the following way:

$$\begin{bmatrix} a_{11} & a_{12} \\ a_{21} & a_{22} \end{bmatrix} \times \begin{bmatrix} b_{11} & b_{12} \\ b_{21} & b_{22} \end{bmatrix} = \begin{bmatrix} a_{11}b_{11} + a_{12}b_{21} & a_{11}b_{12} + a_{12}b_{22} \\ a_{21}b_{11} + a_{22}b_{21} & a_{21}b_{21} + a_{22}b_{22} \end{bmatrix}$$

For example, to compute c_{12}, which is equal to $a_{11}b_{12} + a_{12}b_{22}$, multiply the first entry of the first row of the a-matrix by the first entry of the second column of the b-matrix, then add this to the product of the second entry of the first row of the a-matrix multiplied by the second entry of the second column of the b-matrix. Here are some consequences of this definition of multiplication:

1. The matrix $\begin{bmatrix} 1 & 0 \\ 0 & 1 \end{bmatrix}$, usually denoted as I, plays the same role as the number 1 in the real number system in the sense that $AI = IA = A$ for all 2 × 2 matrices.
2. If we let A, B, and C represent any three $n \times n$ matrices, then multiplication "distributes" over addition, just as in ordinary arithmetic: $A(B + C) = AB + AC$.
3. If we let A, B, and C represent any three $n \times n$ matrices, the associative property applies: $A(BC) = (AB)C$.
4. Matrix multiplication is not usually commutative: $AB \neq BA$.

How did this definition of multiplication arise? Mathematicians, Cayley among them, had already studied functions of the form $z = \dfrac{a_{11}y + a_{12}}{a_{21}y + a_{22}}$.

(continues)

MATRIX MULTIPLICATION
(continued)

If we take a second function of the same form, say, $y = \dfrac{b_{11}x + b_{12}}{b_{21}x + b_{22}}$, and we write $\dfrac{b_{11}y + b_{12}}{b_{21}x + b_{22}}$ in place of y in the expression for z, and finally perform all of the arithmetic, we obtain the following expression: $z = \dfrac{(a_{11}b_{11} + a_{12}b_{21})x + (a_{11}b_{12} + a_{12}b_{22})}{(a_{21}b_{11} + a_{22}b_{21})x + (a_{21}b_{21} + a_{22}b_{22})}$. Compare this with the entries in the product matrix already given. The corresponding entries are identical. It is in this sense that matrix multiplication was discovered before matrices were discovered!

determinants for decades, and he had learned how to use determinants to investigate a number of problems. In a sense he had become familiar with many of the problems that are important to the theory of matrices before discovering matrices themselves. (Sylvester himself coined the term *matrix.*)

Sylvester was interested in the relationships that exist between a matrix and its eigenvalues. He discovered, for example, that if A represents an $n \times n$ matrix and λ is an eigenvalue of A, then λ^j is an eigenvalue of the matrix A^j, where A^j represents the matrix product of A multiplied by itself j times. He produced other results in a similar vein. For example, suppose the matrix A has an inverse. Let A^{-1}, represent the inverse of the matrix A so that $A^{-1} \times A$ equals the identity matrix, that is, the matrix with 1s down the main diagonal and 0s elsewhere. Let λ represent an eigenvalue of A; then λ^{-1}—also written as $1/\lambda$—is an eigenvalue of A^{-1}.

The work of Cayley, Sylvester, and others led to the development of a branch of mathematics that proved to be very useful in ways that they could not possibly have predicted. For example, in the early years of the 20th century, physicists were searching for a way of mathematically expressing new ideas about the inner workings of the atom. These were the early years of that branch of physics called quantum mechanics. It turned out that the theory of matrices—developed by Cayley, Sylvester, Frobenius, Hermite, and Eisenstein in the preceding century—was exactly the right language for expressing the ideas of quantum mechanics. All the physicists needed to do was use the mathematics that had been previously developed. The theory of matrices proved useful in other ways as well.

Emmy Noether and the Theory of Rings

Mathematicians' understanding of algebra as a mathematical discipline changed radically in the 20th century. For millennia equations were the subject matter of algebra. Mathematicians used algebra to state equations, and they used algebra to develop algorithms for solving equations. During the 19th century, they developed concepts and techniques that revealed when certain algorithms did not exist. (See the accounts of the work of Abel and Galois in chapter 6.) As the 19th century drew to a close, mathematicians became increasingly preoccupied with using algebra to make broad statements about the properties of systems of polynomials, systems of matrices, and the properties of specific number systems. Today, however, a different conception of what it means to do algebraic research prevails. One of the most influential pioneers in the new algebra was the German mathematician Emmy Noether (1882–1935).

Emmy Noether demonstrated a talent for languages and mathematics. Her original intention was to teach foreign languages at a secondary school, and she received certifications in French and English. Eventually, however, she turned her attention to mathematics. (Her father, Max Noether, was a prominent mathematician who taught at Erlangen University.) Emmy received a Ph.D. from Erlangen University, but at that time women were barred from becoming university faculty members. Occasionally, she taught at Erlangen in place of her father, but she did so without pay. Her mathematical talents had, however, been recognized by David Hilbert and Felix Klein, two of the most prominent mathematicians in the

Emmy Noether, whose abstract approach to the study of algebra revolutionized how mathematicians everywhere understood the subject (Special Collections Department, Bryn Mawr College)

world at the time. Both were on the faculty at Göttingen University, which for generations had been home to some of the world's most successful mathematicians. Although Hilbert and Klein petitioned the university to waive its rules and hire Noether as a faculty member, they were initially unsuccessful. Instead, she taught classes under Hilbert's name. Her stature as a mathematician continued to grow, however, and soon mathematicians from around the world went to Göttingen to hear her lectures. Finally, the university relented, and Noether became a faculty member at Göttingen. She remained at Göttingen until the National Socialist German Workers' Party (Nazi Party) came to power. Noether, who was Jewish, left Germany for the United States. She found a position at Bryn Mawr College in late 1933, and in 1935 she died from complications following surgery.

Early in her career, Noether's best-known contribution was in the area of mathematical physics. (At Hilbert's request she had investigated a question relating to the theory of relativity.) But her first interest was algebra. To appreciate part of her contribution in this area, it helps to know that some of her earliest research involved the decomposition of polynomials—that is, she studied the problem of representing higher degree polynomials as products of lower order polynomials. (The polynomial $x^3 - 1$ can, for example, be factored using only real numbers as $(x - 1)(x^2 + x + 1)$. If complex numbers are allowed then it can be factored as $(x - 1)$ $(x + [1 + 3i]/2)(x + [1 - 3i/2]$.) The decompositions that are possible depend upon the numbers that one is willing to accept as roots and the numbers that appear as coefficients. She was also interested in the factorization of integers. (Every integer can be written [factored] as a product of prime numbers. The number 24, for example, can be factored as $2^3 \times 3$, and the number 33 can be factored as 3×11. Prime numbers are their own factors.) Factoring integers and factoring polynomials have a lot in common. One of Noether's great insights was to identify what these two different-looking systems (and others!) have in common and to study an abstract model of all such systems. The name for this abstract model is a ring. Noether did not "discover" rings. Other mathematicians before her had used the concept to study integers and polynomials. What Noether did was to introduce the study of abstract rings, which were generalized models of classes of specific mathematical systems.

To understand Noether's insight, consider, for example, the set of polynomials with rational numbers as coefficients, and consider the set of integers. We will represent the set of all such polynomials with the letter *P*, and we will represent the set of all integers with the letter *Z*. Both sets have a lot in common. The following is a list of some of their commonalities:

1. The set *P* forms a group under the operation of addition and so does the set *Z*. (The group axioms as they apply to the symmetry group of a square are listed on pages 116–117, or see the glossary.)

2. For both *P* and *Z*, addition is commutative, which is another way of saying that it is always true that $a + b = b + a$ whether a and b represent integers or polynomials.

3. If two elements of *P* are multiplied together, the result is another element of *P*, and if two elements of *Z* are multiplied together, the result is another element in *Z*.

4. Multiplication of elements in both sets is associative, which is another way of saying that it is always true that $a \times (b \times c) = (a \times b) \times c$, whether a, b, and c represent integers or polynomials.

5. Multiplication distributes over addition, which is another way of saying that it is always true that $a \times (b + c) = a \times b + a \times c$ whether a, b, and c represent integers or polynomials.

Noether's idea was to consider an arbitrary set, which we can represent as $\{a, b, c, \ldots\}$, and two operations defined on this set. Let the "+" sign denote one of the operations and the "×" sign denote the other operation. We can say that these signs represent addition and multiplication, although this is just a matter of convenience since we do not know what, if anything, the letters in the set represent. We suppose that under these operations, conditions 1–5 are satisfied. The resulting mathematical "structure" is called a ring, and conditions 1–5 are the axioms that define a particular type of ring called an "associative ring." (We can, if we choose, change the axioms somewhat to obtain different types of rings—other axioms can be added depending on the needs of the researcher—but as a matter of definition conditions 1 through 5 must be satisfied in order for the object to be called a ring.)

Noether discovered that rings have a rich mathematical structure in the sense that she could prove many theorems about any set that satisfies the ring axioms. At first glance, this might not seem to be significant since she was proving theorems about relationships among sets of letters. This activity requires a certain mental agility, but it might seem to be no more significant than solving a crossword puzzle. What makes Noether's approach important is that what is true of an abstract ring—for example, an abstract associative ring—is true for every mathematical system that conforms to the axioms that define an abstract associative ring. The set P of all polynomials with rational coefficients, the set Z of integers, the set of all $n \times n$ matrices with the operations defined earlier (the n in "$n \times n$ matrices" is fixed but arbitrary so, for example, the list includes the set of all 2×2 matrices and the set of all $1{,}000 \times 1{,}000$ matrices), and the set of all even integers under the usual definitions of addition and multiplication are examples of associative rings. Many other examples exist. By studying an abstract associative ring, Noether was able to make important statements about the logical structure of all associative rings.

Early in the 20th century, mathematicians were faced with an incredible array of different-looking systems. A discovery about the property of one system might also apply to other similar systems, but this was not always clear because there was no rigorous definition of what "similar" meant. What Emmy Noether did—at least in algebra, at least for rings—was to introduce a different way of looking at mathematical systems. She demonstrated that many concrete systems of numbers, polynomials, and matrices, for example, had enough in common that they could be adequately represented by a single abstract model. Theorems that were proved within the context of the model also applied to all of the systems that the model represented. She did not just generate new knowledge about algebra, she introduced a new conception of algebra. Algebra became concerned with the study of abstract "structures," models that represented classes of concrete objects.

Nicolas Bourbaki and Mathematical Structure

Modern mathematics has been heavily influenced by Nicolas Bourbaki. Paul Halmos, an important mathematician in his own right, called Bourbaki ". . . one of the most influential mathematicians of the 20th century." But Halmos's tribute was only partly serious. Although Bourbaki produced a number of important and influential books, including a landmark text on algebra, Bourbaki did not exist. Nicolas Bourbaki is a pseudonym for a group of French mathematicians. Despite—or perhaps because of—changes in membership, the group continued to publish for decades. (Members were supposed to quit at age 50.)

The Nicolas Bourbaki group began to meet in the mid-1930s with the goal of producing a text on calculus that would ". . . define for 25 years the syllabus for the certificate in differential and integral calculus . . ." Driven by the perception that France had fallen behind Germany in mathematical research, the original Bourbaki project included some of the best mathematicians of the era, including Henri Cartan (1904–2008), Jean Dieudonné

Faced with an increasingly diverse collection of mathematical objects and logical structures, members of the Nicolas Bourbaki group sought to identify unifying concepts. (Stewart Dickson)

(1906–92), and Andre Weil (1906–98) among others. Nicolas Bourbaki was supposed to be a collaborative effort at producing mathematical works that emphasized modern concepts. Members would occasionally meet face-to-face, and they regularly published an in-house newsletter that enabled the participants to exchange ideas and debate about the best way to proceed on each project.

The group produced articles and books. Among the books was a 10-volume set called *Eléments de mathématique*, which included the following titles: *Theory of Sets*, *Algebra*, *General Topology*, *Functions of a Real Variable*, *Topological Vector Spaces*, *Integration*, *Lie Groups and Lie Algebras*, *Commutative Algebra*, *Spectral Theories*, and *Differential and Analytic Manifolds*. Among the articles was an oft-quoted work entitled, "The Architecture of Mathematics." This article raised a question that is still debated today: Is mathematics a single branch of knowledge or has it become a "tower of Babel" comprised of numerous disciplines isolated from each other by disparate methods, goals, and vocabulary? Most mathematicians would agree that it is better if experts in different branches of mathematics can communicate across their areas of expertise. In this way, they can exchange ideas, recognize common problems, and share common solutions, but as mathematics has become increasingly specialized this has become harder to accomplish. The unity of the field could be maintained, according to Bourbaki, by emphasizing the axiomatic method in the discovery of mathematical truth and by placing a greater emphasis on the role of mathematical structures. The existence of such structures enables mathematicians to simplify.

One sometimes hears criticism of the abstract nature of mathematics—that it is unnecessarily complicated. Mathematicians' emphasis on symbolic language as opposed to plain prose, for example, is sometimes characterized as confusing. Why use a symbol when a word will do? But this criticism stems from a misunderstanding about the nature of mathematics. Mathematicians must be precise when they say something if they are to say anything meaningful, and they have learned through experience that an abstract language *is* the simplest way to present mathematical ideas precisely. One page of algebraic symbols may be worth 100 pages of plain prose—assuming that the ideas can be expressed in plain prose at all.

An abstract approach to mathematics is not just simpler; it is also conceptually more powerful. Theorems proved about one abstract mathematical structure apply to all instances of that structure. What Noether had done with rings, Bourbaki attempted to impose across all of mathematics. By attempting to identify (or create) a mathematical architecture that would reveal how different branches of mathematics are related to each other, they hoped to demonstrate that mathematics is a conceptually unified whole, a universe of the mind, and not just a jumbled collection of isolated results.

Not only was it necessary, therefore, to create and exploit models of classes of systems—groups, rings, fields, and so on—it was also necessary to specify how these conceptual models were related to each other. Some mathematical groups, for example, are more general than others. Recall that in the discussion of the symmetry transformations of the square (see chapter 6) there was a subgroup, the group of rotations, that was "nested" within the larger group. One could say, therefore, that the group of symmetry transformations of the square has a richer structure than the group of rotations of the square. Many other examples of this nesting phenomenon exist within the class of groups, within the class of rings, and within the class of fields. But every ring contains a group and every field contains a ring. Consequently, there are important relations between the class of groups and the class of rings, between the class of groups and the class of fields, and between the class of rings and the class of fields. And there are other algebraic structures in addition to the ones mentioned here. (There is not enough space in this book to discuss them all.) Explicitly identifying how these structures and classes of structures are related to each other is an enormous task, but if it can be accomplished, it will be easier to understand how ideas in one branch of mathematics apply to other branches of mathematics. It will be easier to see mathematics as a unified whole.

Although Bourbaki emphasized the role of structure and even developed criteria for delineating relationships between structures, their program was not especially successful in developing an overarching theory of mathematical structure. Another theory, called category theory, has superseded Bourbaki's ideas about structure. (A description of category theory requires too much mathematical background to be given here.) It is too soon to say, however, whether or not category theory will reveal a unifying structure or set of structures that apply across all of mathematics; whether or not it will prove to be an all-embracing theory in which each discipline has its own place and its own set of relationships with other disciplines.

The Problem Solvers

Because a great deal of academic algebraic research is concerned with the study of abstract groups, rings, fields, and other algebraic objects, it is easy to lose sight of the fact that many mathematicians continue to work full-time developing better algorithms. They are the problem solvers, the intellectual descendants of Brahmagupta, al-Khwārizmī, and Tartaglia. The difference is that the equations (and the algorithms) have changed.

Computation has always been an important part of mathematics. It remains important today, but now most computations are performed by computers. Consequently, the search for better algorithms usually means the search for better computer algorithms. Computers process and store information in ways that are quite different from the ways that people perform these same tasks, and the best algorithms make essential use of these differences. In addition, the types of problems that computers solve are different from the types of problems that people solve.

Systems of linear equations are of special interest to researchers who concentrate on so-called numerical methods. In concept, the linear systems studied by researchers have the same properties as the linear systems that students learn to solve in junior and senior high school. A model for a three-variable system can be written like this:

$$\begin{aligned} a_{11}x + a_{12}y + a_{13}z &= b_1 \\ a_{21}x + a_{22}y + a_{23}z &= b_2 \\ a_{31}x + a_{32}y + a_{33}z &= b_3 \end{aligned}$$

or it can be written in matrix form like this:

$$\begin{bmatrix} a_{11} & a_{12} & a_{13} \\ a_{21} & a_{22} & a_{23} \\ a_{31} & a_{32} & a_{33} \end{bmatrix} \begin{bmatrix} x \\ y \\ z \end{bmatrix} = \begin{bmatrix} b_1 \\ b_2 \\ b_3 \end{bmatrix} \tag{8.1}$$

The a_{11}, a_{12}, . . . appearing in the 3×3 matrix are called the coefficients of the matrix. They are numbers. The b_1, b_2, and b_3 also represent numbers. The x, y, and z are the "unknowns." What is new about the systems of linear equations used in modern scientific computations is their size. The systems studied by contemporary researchers are so large that no person could ever write one down—never mind solve it. Such large systems require their own specialized algorithms. Not that the techniques used to solve small systems of equations—the types we learn in school—are "wrong." In theory, the same techniques that are used to solve small systems can also be used to solve large ones, but while the techniques used to solve small systems are conceptually correct, they are seldom useful in solving larger systems. The techniques that are used to solve systems of equations by hand are just too slow and too labor intensive—even for a computer—when they are applied to very large systems of equations.

Any engineer, scientist, or mathematician who requires a state-of-the-art computer to solve a problem is usually allocated a fixed amount of time to obtain a solution. The reason is that other researchers are waiting to use the same computer. The supply of powerful computers is limited, and demand for access to these machines is strong. If an algorithm is inefficient, then a researcher's problem may not be solved in the allocated time—in which case it might not be solved at all—or one problem may be solved when two could have been solved. Solving problems as efficiently as possible can mean the difference between success and failure for a program of research.

One important class of algorithms, for example, is the class of *sparse matrix techniques*. Many scientific and engineering problems produce very large systems of linear equations in which most of the coefficients are zero. Sparse matrix techniques have no meaning for the systems of linear equations that one encounters in high school because those systems have so few coefficients that it hardly matters whether some of the coefficients are zero or not. But when there are many millions of coefficients, sparse matrix techniques can mean the difference between solving the problem and not solving it.

Sparse matrix techniques are used both to store and to manipulate matrices. The storage techniques are called compression techniques. A simple example of a matrix compression technique would be to store only the numbers a_{11}, a_{22}, and a_{33} instead of the entire sparse matrix

$$\begin{bmatrix} a_{11} & 0 & 0 \\ 0 & a_{22} & 0 \\ 0 & 0 & a_{33} \end{bmatrix}.$$

The techniques become more sophisticated as the sparse matrices become less ordered.

There are also a variety of methods for solving problems involving sparse matrices. The goal of these algorithms is to exploit the sparse structure of the matrix. To better appreciate why specialized algorithms are necessary, consider the problem of solving a system of linear equations in which the matrix has n columns and n rows. The number of entries in such a matrix is n^2. If we count the number of simple arithmetic operations—that is, the number of additions, subtractions, multiplications, and divisions—needed to compute the inverse of an $n \times n$ matrix, the number of operations is roughly proportional to n^3. When n equals 3, as it does in equation 8.1, we might compute the inverse to the matrix, and then multiply both sides of the equation by the inverse. The result would be an equation with the variables isolated on one side and the answers on the other. By hand, this is a lot of work, but students all around the world do this everyday in linear algebra classes. (To be clear about the procedure just described, in one dimension it looks like this: The equation corresponding to equation 8.1 would be $ax = b$, and the solution, which is obtained by multiplying both sides of the equation by the "1×1 matrix" a^{-1}, is $x = b / a$. A similar sort of procedure is possible when a represents an $n \times n$ matrix with a nonzero determinant and b and x are $n \times 1$ matrices. We omit the details of how this is done.)

Now consider a problem that involves a matrix with 1,000 rows and 1,000 columns. If this problem were written out, it would look like equation 8.1 but much bigger. It would, for example, have 1,000 variables instead of three. (A problem involving 1,000 variables is a small problem by contemporary standards.) The number of entries in a 1,000 × 1,000 matrix is 1 million, and the number of operations needed to compute the inverse of such a matrix is roughly proportional to 1,000 × 1,000 × 1,000, or 1 billion. If the number of variables again increased by a factor of 10, the number of operations needed to compute the inverse would increase by a factor of about 1,000. It would, therefore, be proportional to 1 trillion. The larger the problem, the more critical it is, therefore, to use the best possible algorithms. The work increases rapidly as the number of variables increases, and if the matrix is sparse, the use of sparse matrix techniques becomes critical to obtaining a solution in the allotted time.

Another aspect of contemporary computational mathematics that extends the ideas described in this book concerns the development of new types of coordinate systems. To see why new coordinate systems are necessary, recall that a great deal of classical mathematics is concerned with the visualization of functions defined over sets containing infinitely many points. Part of the reason for mathematicians' preference for infinite sets over finite ones is that infinite point sets often simplify the study of functions. But computers do not make use of the infinite. They only require that there are "enough" points to complete the computation, and a finite number of points is always sufficient for computation. Often the only points that the computers use are "grid points," those points where coordinate lines cross. (Imagine a plane with a Cartesian coordinate system, then the grid points could be exactly those points with integer coordinates.)

If the grid points with integer coordinates are too far apart for a particular application, they can be supplemented by drawing additional lines parallel to the y-axis and passing through the points $(m/n, 0)$ and with lines parallel to the x-axis and passing through the points $(0, m/n)$, where n is a fixed natural number greater than 1 and m is any integer. At the points where these lines intersect each other, additional grid points are created. In theory, this method of introducing more grid points always works provided n is large enough. One can always create enough grid points to enable the computer to produce an accurate picture of the function in question. In practice, however, the method fails because it is terribly inefficient. As more points are introduced, the amount of computation required to evaluate the function at all of the grid points rises rapidly—too rapidly for the method to be practical.

The key to creating an efficient coordinate system is to introduce additional grid points only where they are needed. In a region where a function changes very slowly, a few grid points are all that is required to create an accurate visualization of the graph of the function. By contrast, in a region where a function changes rapidly—its graph may contain, for example, many oscillations in a small region or a single sharp isolated cusp—the computed graph will fail to reveal the presence of these features if the grid points are spaced too far apart. The computation will fail for the same reason one cannot convey the tempo of a piece of music with a single snap of the fingers—there is not enough information to draw firm conclusions. Consequently, the coordinate system must be adapted to the specific function under consideration. Grid points must be placed far apart in regions where the function changes slowly, and they must be densely packed in regions where the function changes rapidly. Coordinate systems with this property are called adaptive, and the techniques used to create the adaptive systems are called numerical grid generation techniques.

Today, there is a branch of mathematics specifically concerned with the creation of adaptive coordinate systems. Individuals concerned with establishing the necessary algorithms (and proving that the algorithms have the necessary properties) are the intellectual successors to Descartes and Fermat, the first mathematicians to recognize the utility of coordinate systems in the study of functions. (See, for example, the interview with Dr. Bonita Saunders in the Afterword.)

Numerical algorithms for computers are highly specialized. The language, the techniques, and the concepts that researchers use in the development of these algorithms are understood by a relatively small number of individuals, but most of us are somewhat familiar with the numerical simulations that result when these techniques are used in conjunction with powerful computers. Sometimes we see the simulations on television or on the Web or we see photographs in books. Combustion engineers, meteorologists, biochemists, aerospace engineers, and many others use these simulations in two essential ways. First, the simulations are used to gain insight into the phenomena that each researcher seeks to

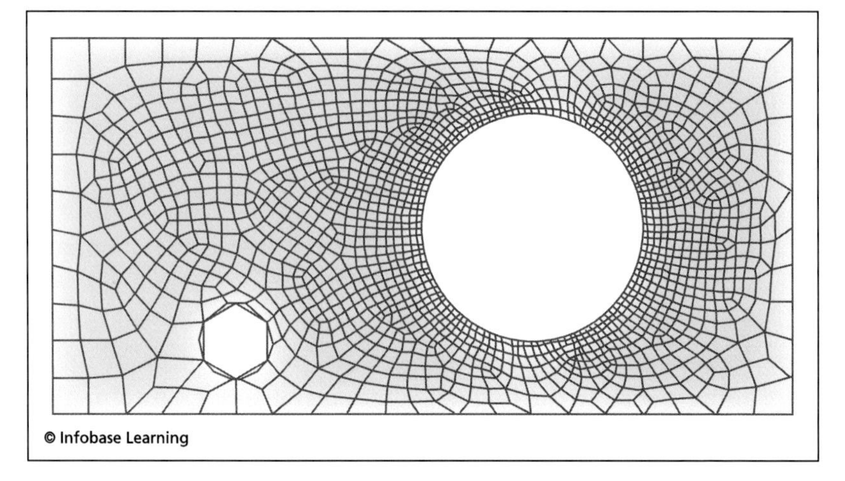

An adaptive coordinate system on a rectangular region punctured by a large and a small hole. It is used to calculate the effects of heating around the edge of the larger hole. Notice the coordinate system is highly irregular. In particular, the coordinate lines are dense in areas of rapid change and less dense elsewhere. This coordinate system enables the investigator to efficiently use the power of the computer. The creation of such coordinate systems is an important discipline within applied mathematics. (www.sandia.gov)

model. Second, the simulations also enable engineers and scientists to better understand the limitations of their models by comparing the outcomes of their simulations to experimental results. It is no exaggeration to say that contemporary scientific progress would be impossible without these highly specialized algorithms.

CONCLUSION

Mathematics is unique in the way that it progresses. To see the difference, compare mathematics with science. Contemporary astronomers do not much concern themselves with what ancient Greek astronomers thought, because they know that most of what ancient Greek astronomers thought was true is false. Modern astronomy has replaced ancient astronomy. By contrast, what Greek mathematicians proved true 2 millennia ago is still true today. It has not been replaced; nor will it be replaced. The reason is that in mathematics logic is the only criteria by which truth is judged. Mathematically speaking, if something is logical it is correct. A future logical deduction cannot prove a previous logical deduction was incorrect. The situation is different for scientists. They may produce logical results, but their discoveries must also agree with the world as it is revealed by experiment. Mathematics is under no such restriction. Science grows in a (more or less) linear way. New results replace old ones. Mathematics grows like a quilt. One patch is added to another—no replacement, just accumulation.

Algebra is as old as civilization. Every civilization that has left written records has left records of algebraic calculations, some more sophisticated than others, and for most of history algebra was concerned with equations. Mathematicians sought better ways to state equations and to solve them. This aspect of algebra is still important. Many mathematicians are developing software that efficiently solves huge or complicated systems of equations. New breakthroughs in hardware require innovations in software and vice versa. The mathematics is complex and the results are astonishing. The scale of these calculations would have been inconceivable to mathematicians even a few generations ago, and much of modern life depends on the continued success of these mathematicians. Algebra as a collection of algorithms is a mathematical tradition that dates back to the mathematicians of Mesopotamia who carefully pressed the results of their computations into slabs of wet clay and the mathematicians of ancient China who recorded their results in the form of elegantly drawn characters.

In the 20th century, many mathematicians turned their attention to the development and study of generalized models of mathematical systems. They sought to identify what was essential in a class of systems and to discard the rest. The essential parts were stated in the form of axioms, and once the axioms were determined, the act of mathematical discovery was "reduced" to drawing logical conclusions from the axioms. Examples of these axiomatic systems include abstract groups, rings, and fields. So successful was this approach that it has become the major preoccupation of academic mathematicians. The creation and study of abstract models have led to a simplification of mathematics. Many formerly disparate branches of mathematics are now perceived as applications of certain basic principles applied to specific systems. Without the simplification made possible by this higher level of abstraction, contemporary mathematics would just be a jumble of disconnected results.

Today some have turned their attention to the development of abstract models that would be powerful enough to reveal the structure of mathematics as a whole. The search for these larger structures has, so far, only been partly successful, but the fact that mathematicians can conceive of the question and achieve some success in formulating an answer shows just how far algebra has progressed. Progress in algebra is, as it has always been, an essential component of progress in science, engineering, and mathematics.

AFTERWORD

AN INTERVIEW WITH DR. BONITA SAUNDERS ON THE DIGITAL LIBRARY OF MATHEMATICAL FUNCTIONS

Dr. Bonita Saunders (Dr. Bonita Saunders)

Dr. Bonita Saunders is a research mathematician at the National Institute of Standards and Technology (NIST), one of the primary federal centers for research in the physical sciences. She is a graduate of the College of William and Mary, and she received her Ph.D. from Old Dominion University. While studying for her Ph.D., she performed research at NASA's Langley Research Center on the application of numerical grid generation to problems in computational fluid dynamics. She has published extensively on the subjects of numerical grid generation and on techniques for the visualization of functions. This interview took place on February 11, 2010.

Tabak: Could you tell me a little bit about NIST and the work that you do there?

Saunders: NIST is a governmental agency that falls under the Department of Commerce. Its mission is to develop measurement standards and technology that help U.S. industries compete—whether it's within the country or globally. I'm a research mathematician in the Mathematical and Computational Sciences Division, which comes under the Information Technology Laboratory. My duties are basically to develop software and design techniques that facilitate the solving of problems in the mathematical and physical sciences and to consult with the other laboratories at NIST.

Tabak: What caught my eye about the work that you do at NIST was the research you have done on visualizing functions—in particular, the Digital Library of Mathematical Functions (DLMF). What was the motivation for compiling this "dictionary" of functions?

Saunders: The DLMF is actually a handbook containing the formulas, properties, and graphs of mathematical functions important in the physical sciences. There are two parts to it. There's the digital library part of it, which will be freely available on the web, and the book version, which will be called the *NIST Handbook of Mathematical Functions*. It is designed to replace another handbook of mathematical functions that's still widely used—the *NBS Handbook of Mathematical Functions*, edited by Abramowitz and Stegun and published in 1964 by the National Bureau of Standards (NBS), the former name of NIST.

Tabak: I'm familiar with it.

Saunders: And even though the book goes all the way back to 1964 and really hasn't been changed much over the years, it's still one of the most cited publications to come out of NIST or NBS. It's still used by the physical sciences community. There was a push within the community—especially within the physics community—for an update of the book. It was decided that we would not only update it, but completely rewrite it, expand it, and also put it on the Web as a digital library.

The authors of the chapters—just as in the original handbook—come from many different places. They are mathematicians, physicists, and engineers from all over the United States and abroad who are key experts on the definitions and properties of the functions. The project is a pretty big endeavor. The editing, organization, and initial validation of the chapters—and the design and construction of the digital library, which includes a mathematical search engine and interactive visualizations—were done at NIST. An international team of experts is responsible for the final validation of the chapters.

Tabak: Could you talk a little bit about the applications for these functions and how you chose the functions that are included?

Saunders: It's kind of interesting to look at the original handbook. You have these high-level mathematical functions that arise as solutions to physical problems—

Tabak: First, could you explain what you mean by a "high-level" mathematical function?

Saunders: High-level mathematical functions are also called special functions. Some examples would be Bessel functions, Airy functions, hypergeometric functions, or orthogonal polynomials. When trying to solve partial differential equations that model some kind of physical process, a researcher might discover that the equations reduce to simpler equations. The form or structure of the new equations may suggest that a particular mathematical function is the solution.

By looking at the definition and properties of the function in the *Handbook* the researcher can determine whether that is the case. If correct, the scientist may not only have an exact solution to his problem, but also gain qualitative information about how the solution behaves—where its value is zero, if and where it blows up, where its maximum and minimum values are. This may give the scientist more information than simply solving the initial equations on a computer. Even if the new equations don't exactly match those of the special function, this initial analysis may still give the scientist clues about what to look for when trying to solve the original equations numerically, that is, on a computer.

Tabak: So the functions are used in meteorology? Fluid dynamics? Aeronautics?

Saunders: Practically any area where physical phenomena are modeled by equations. For instance, the Airy functions come up in the field of optics—solving problems in that area. Bessel functions can be found in applications involving heat conduction, hydrodynamics, and computational fluid dynamics.

Tabak: Have there been many changes in the list of functions between the 1964 edition of the *Handbook* and today's?

Saunders: Most of the functions that are in the '64 edition are in the current edition, but some new functions have been added. In the original edition there were a lot of tables of values. Now that you can use many computer packages to get as many values of a function as you want, most of the tables have been taken out. While the original edition was about 1,000 pages, the new edition was expected to be much smaller, but there was so much additional information that the new edition is also close to 1,000 pages.

Tabak: When they computed values for the functions in 1964, they computed values at regularly spaced grid points. I gather that is not what you are doing now. Could you talk a little about numerical grid generation—what that involves and why you do it?

Saunders: Probably most computations would still be done over a regularly spaced grid, but I used numerical grid generation to create accurate graphs of the functions. While the original handbook contains very few graphs, the new handbook will have more than 400 color graphs and the digital version, close to 500. The digital library will also have more than 200 interactive 3-D visualizations of function surfaces.

So the grid generation comes up in trying to make the graphs and visualizations as accurate as possible. You want the data to be accurate, but you also want the graph or the plot to be accurate. To ensure data accuracy, we compute the function several different ways using whatever software or available information that we have—commercial codes, FORTRAN code, or even the chapter author's personal code. We then compare the results and try to determine the cause of any discrepancies.

Once we know the data are accurate, we also want the plot to be realistic.

Tabak: Yes—

Saunders: Because you can have accurate data and still not have a graph that really shows the key features of that function—

Tabak: And how would that happen? How could you have accurate data and yet have a graph that was inaccurate?

Saunders: Suppose you try to plot a function over variables x and y with a standard commercial package. If you select the function and simply ask for a plot over a given x-range and y-range, you may get a surface plot that misses key areas. For example, if the function being plotted represents the absolute value of another function it might have a sharp cusp where the value is zero. If there are not enough data points near the cusp, you may not see it. If there is a point where the function has a pole—that is, the function goes off to infinity—how does the package show that? It can't show infinity. It has to stop somewhere. The area around a pole usually looks similar to a stove pipe that goes on forever. Some packages automatically choose a height, cut off the stove pipe, and cover it. This makes it look like the covered area is part of the function. Other packages just connect whatever points are in the vicinity of the pole. When there are several poles this can create a jagged effect that looks like a mountain range.

We wanted to make sure we brought out the key features of each function and made the graphs look as accurate as possible—showing cusps, zeros, poles, branch cuts (where the function splits into multiple values). So basically, that's where the grid generation comes in. I design a grid that brings out those key features, and I compute the value of the function at the grid points to obtain the graph. And so how do I know where the key features are?

Tabak: Yes, I was going to ask—

Saunders: A variety of ways. I may do an initial computation of the function using a package. If I see an area where it looks like there might be something, then I investigate that area more carefully. I may also talk to the chapter author or read the author's draft to get information about poles, branch cuts or whatever. It's an exploration. You're trying to get as much information as you possibly can.

Once I have that information I have an idea for how the grid should look. If I am going to capture a cusp, I have to find the location of the zeros—where the function is equal to zero. I make sure those points are on the grid. If there's a pole, or if the values of the function become very large (or small) on the region over which I want to compute the function, I try to decide at what height I want to stop the function. Do I want it to go up to 5, or 20, or 30? Then I compute the contour of the function for that height. For example, if I choose a maximum height of, let's say, 10 then I compute the locations where that function will have a value of 10, and I use that contour to create a boundary for the grid that I will use to compute the function over.

Tabak: So the grid lies on the plane like curves on a topographical map?

Saunders: That's exactly what you have. For example, to bring out the features of the Airy function I chose a very short height—

about 3. So I compute the contour curves for a height of 3, and then I connect those curves so that I have a closed boundary. So instead of computing the function over a rectangular boundary, I compute it over a boundary that has the contour shapes in it. It gives me a nice clipping of the function, a nice smooth cut, and I don't have to worry about a ragged edge at the top. The contour for a pole will usually look like a small circle and probably be in the interior of the domain. That means my grid will have a hole in it. When I compute the function over the grid, the function surface will be clipped so that I get a nice smooth cut at the height I selected. This produces a graph that is more accurate, less confusing, and aesthetically more pleasing.

Tabak: So just to be sure that I understand: The coordinate system that you use for a function will look like a topographical map. Some lines run around the sides of features—like lines running around the sides of mountains on a map—and these are lines of constant height. And some lines run down the sides of the features like lines running down the sides of the mountains. Then you compute the values of the function at the points where the lines intersect?

Saunders: That's the general idea. It's a little more complex than that. What you are saying would probably take more time. What I'm doing is something called numerical grid generation. Basically, that's the development of curvilinear coordinate systems. Do you know about polar coordinate systems?

Tabak: Sure.

Saunders: A polar coordinate system is basically a curvilinear coordinate system, where you have a mapping from a rectangle to a circle. The rectangle has coordinates r and θ (theta)—that is, $x = r \cos \theta$ and $y = r \sin \theta$—so on the rectangle you have the axis of the r and θ, and on the other side, on the physical side, you have x and y. You can think of this rectangle as being mapped to the circle. Your lines of constant r map to concentric circles on the other side, and your lines of constant θ are mapped to rays emanating from the center of the circle. Numerical grid generation is a generalization of that idea. You have a mapping, that is, a function, where you have a rectangle or square, mapped to something that has an arbitrary shape. So I set up my mapping so that I'm mapping my rectangle to whatever shape I want—the domain with the contour boundary, for instance.

Now as I set up this mapping the question is, "How do I fill in the coordinate lines on the contour-shaped domain?" Well, my mapping is made of splines, which are functions that are piecewise polynomials. It maps a square to the contour-shaped domain over which I want to compute the special function. Let's say I call the coordinates for the square domain, *c* and *d*, and the coordinates for the contour-shaped domain, *x* and *y*. In a similar fashion as the polar coordinate system, lines of constant *c* or *d* values on the square are mapped to lines (or they may look like curves) on the contour-shaped domain. The lines will cross each other to produce what looks like a net, or grid. The points of intersection are called the grid points. Each *(c,d)* ordered pair on the square is mapped to an *(x,y)* ordered pair on the contour-shaped domain. The grid generation mapping must be carefully defined so that the grid lines do not intersect adjacent lines or overlap the domain boundary.

Once I have the grid on the contour-shaped domain, I compute the special function at the grid points to obtain the plot data for the graph. So I will have a list of *x*, *y*, and *z* values where *z* is the value of the special function at the ordered pair *(x,y)*.

Tabak: If this were the 1960s, you would stop here and publish a table of numbers, right? This is your table. But this is an adaptive table—adapted to the individual function. So in an area where the function behaves in a complicated way the table of values would cluster in that area, and then the data would spread out over areas where the function varies less?

Saunders: It's somewhat adaptive now, and I'm trying to make it more so. That is basically the idea. Where I have a cusp—a place where the surface comes to a sharp point at a zero—I will have more points there so you can really see that and try to make sure that a grid point actually hits the location of the zero. Near a pole the surface may rise or decrease very quickly, so you probably need more points around there. Yes, that's the idea.

Tabak: OK.

Saunders: This concept of numerical grid generation actually comes from computational fluid dynamics. Numerical grid generation was my area of research for my Ph.D. dissertation. I studied this as a graduate researcher at NASA Langley Research Center. This technique is actually used to develop boundary fitted grids to solve partial differential equations for aerodynamic problems. For example, numerical grid generation has commonly been used to study the design of airfoils—you know, a cross section of a wing?

Tabak: Right.

Saunders: For the two-dimensional study of flow past a wing they designed a grid with an airfoil shape in the middle. The grid points are concentrated near the boundary of the airfoil. They use the grid to solve partial differential equations that model the motion of the air as the wing moves through it.

When we started looking at the problem of trying to develop accurate grids for graphing, I realized that it was actually just a numerical grid generation problem—that I could use the same techniques that I used for problems in computational fluid dynamics in this problem of graphing. This was a win-win situation for me. I was able to do more research in grid generation and also do work that helped NIST at the same time.

Tabak: Now you have the grid. The next step is to turn the values on the grid into a surface?

Saunders: I used the grid to produce a table of ordered triples (x,y,z). So I now have a table of values that can be plotted to produce a surface.

Tabak: At this point in the process, we can imagine it as a three-dimensional space with a finite number of points in it, right? The points on the surface?

Saunders: Right. Our next step was to figure out a way to put the data into a file format that could be viewed on the Web. After doing some research and talking to people at NIST who knew something about developing Web-based graphics, we decided to use VRML (Virtual Reality Modeling Language) and later X3D (Extensible 3D). These formats are specially made for displaying and manipulating 3-D objects on the Web. We collaborated with NIST computer scientists who were able to develop routines to translate our data into the VRML and X3D file formats. These formats are recognized by special browsers and plug-ins that users can download for free. With the VRML/X3D browser the user can not only display the surface, but also perform several interactive maneuvers—rotate the surface, zoom in and out, or use any of the custom features we have included. For example, the user can change the scale of the surface in the x-, y-, or vertical direction. Scaling down close to 0 in the vertical direction yields a special type of contour plot called a density plot. And our cutting plane feature allows the user to see the intersection curves that develop when a plane moves through the surface.

Tabak: I had read about that. It reminds me of the way that the ancient Greeks generated their conic sections. They imagined the cone and then they would cut it with a plane to get their curve, right?

Saunders: Right.

Tabak: These software packages allow you, essentially, to connect the points and produce a surface. Now you have a surface. But these are transcendental functions so the values are approximate and the surface is a continuum, but you've calculated the surface from a finite collection of points. That's a lot of approximations. So after you've done all of this work—and I know that you have produced many of these visualizations—how do you know that the surfaces are correct? How do you check your work?

Saunders: The same way that I figure out what I'm looking for. I look at all the information that I have about the function. Even though high-level math functions can be very complicated, they tend to be fairly smooth in most regions. Anomalies tend to stand out. Also, you always have an approximation; you just want to make sure that the accuracy of the approximation is acceptable. You actually have a little more freedom when it comes to a graph or visualization of a function because the level of detail that your eye can detect is limited. The number of digits needed for a precise graph of a function is usually far fewer than the number of digits of accuracy needed to use the function in calculations to solve a problem. Once you get past a certain point, adding more digits of accuracy to the data points won't give a more accurate picture to the eye. That's why my primary concern is to find out where the key features—poles, zeros, and branch cuts—are and make sure those features are presented clearly.

Tabak: So the finished product will look like a landscape and your software allows you to move over the landscape and look at it from certain points of view or to look at some region more closely or from further away. Is that a fair characterization?

Saunders: Yes, basically. Except in this case it will look like you are moving the surface. We have also put in fixed viewpoints that you can cycle through. And we color the surface two ways. One color map is designed so that the color on the surface changes with the height of the function—

Tabak: I caught that scheme, but I didn't understand the second scheme.

Saunders: The other way we color it is according to the argument of the function. These functions have complex values, and complex numbers can also be written in the form $a + ib = re^{i\theta}$, and

θ is the argument, also called the phase. The color at a particular surface point is then based on the argument of the function at that point. There is also a four-color scheme where the surface color identifies the quadrant in which the phase angle lies.

Tabak: When is this project scheduled to be completed?

Saunders: Very soon.

Tabak: Is that right?

Saunders: The book version is already at the publisher. They're trying to get it ready, and we want the Web site to be ready at the same time—probably within two or three months.

Tabak: Congratulations. It's nice that it will be ready to view when people read this interview. *(laughter)*

Saunders: It is being published by Cambridge University Press.

Tabak: This was very interesting. I especially enjoyed learning how all of these ideas come together to produce a finished product. Thank you very much for your time and your expertise.

Saunders: Thank you.

CHRONOLOGY

ca. 3000 B.C.E.
Hieroglyphic numerals are in use in Egypt.

ca. 2500 B.C.E.
Construction of the Great Pyramid of Khufu takes place.

ca. 2400 B.C.E.
An almost complete system of positional notation is in use in Mesopotamia.

ca. 1650 B.C.E.
The Egyptian scribe Ahmes copies what is now known as the Ahmes (or Rhind) papyrus from an earlier version of the same document.

ca. 585 B.C.E.
Thales of Miletus carries out his research into geometry, marking the beginning of mathematics as a deductive science.

ca. 540 B.C.E.
Pythagoras of Samos establishes the Pythagorean school of philosophy.

ca. 500 B.C.E.
Rod numerals are in use in China.

ca. 420 B.C.E.
Zeno of Elea proposes his philosophical paradoxes.

ca. 399 B.C.E.
Socrates dies.

ca. 360 B.C.E.
Eudoxus, author of the method of exhaustion, carries out his research into mathematics.

ca. 350 B.C.E.

The Greek mathematician Menaechmus writes an important work on conic sections.

ca. 347 B.C.E.

Plato dies.

332 B.C.E.

Alexandria, Egypt, center of Greek mathematics, is established.

ca. 300 B.C.E.

Euclid of Alexandria writes *Elements,* one of the most influential mathematics books of all time.

ca. 260 B.C.E.

Aristarchus of Samos discovers a method for computing the ratio of the Earth-Moon distance to the Earth-Sun distance.

ca. 230 B.C.E.

Eratosthenes of Cyrene computes the circumference of Earth.

Apollonius of Perga writes *Conics.*

Archimedes of Syracuse writes *The Method, On the Equilibrium of Planes,* and other works.

206 B.C.E.

The Han dynasty is established; Chinese mathematics flourishes.

ca. C.E. 150

Ptolemy of Alexandria writes *Almagest,* the most influential astronomy text of antiquity.

ca. C.E. 250

Diophantus of Alexandria writes *Arithmetica,* an important step forward for algebra.

ca. 320

Pappus of Alexandria writes his *Collection,* one of the last influential Greek mathematical treatises.

415

The death of the Alexandrian philosopher and mathematician Hypatia marks the end of the Greek mathematical tradition.

ca. 476

The astronomer and mathematician Aryabhata is born; Indian mathematics flourishes.

ca. 630

The Hindu mathematician and astronomer Brahmagupta writes *Brahma Sphuta Siddhānta*, which contains a description of place-value notation.

ca. 775

Scholars in Baghdad begin to translate Hindu and Greek works into Arabic.

ca. 830

Mohammed ibn-Mūsā al-Khwārizmī writes *Hisāb al-jabr wa'l muqābala*, a new approach to algebra.

833

Al-Ma'mūn, founder of the House of Wisdom in Baghdad, Iraq, dies.

ca. 840

The Jainist mathematician Mahavira writes *Ganita Sara Samgraha*, an important mathematical textbook.

1086

An intensive survey of the wealth of England is carried out and summarized in the tables and lists of the *Domesday Book.*

1123

Omar Khayyám, the author of *Al-jabr w'al muqābala* and the *Rubáiyát*, the last great classical Islamic mathematician, dies.

ca. 1144

Bhaskara II writes the *Lilavati* and the *Vija-Ganita*, two of the last great works in the classical Indian mathematical tradition.

ca. 1202

Leonardo of Pisa (Fibonacci), author of *Liber abaci*, arrives in Europe.

1360

Nicholas Oresme, a French mathematician and Roman Catholic bishop, represents distance as the area beneath a velocity line.

1471

The German artist Albrecht Dürer is born.

1482

Leonardo da Vinci begins to keep his diaries.

ca. 1541

Niccolò Fontana, an Italian mathematician, also known as Tartaglia, discovers a general method for factoring third-degree algebraic equations.

1543

Copernicus publishes *De revolutionibus*, marking the start of the Copernican revolution.

1545

Girolamo Cardano, an Italian mathematician and physician, publishes *Ars magna*, marking the beginning of modern algebra. Later he publishes *Liber de ludo aleae*, the first book on probability.

1579

François Viète, a French mathematician, publishes *Canon mathematicus*, marking the beginning of modern algebraic notation.

1585

The Dutch mathematician and engineer Simon Stevin publishes "La disme."

1609

Johannes Kepler, author of Kepler's laws of planetary motion, publishes *Astronomia nova*.

Galileo Galilei begins his astronomical observations.

1621

The English mathematician and astronomer Thomas Harriot dies. His only work, *Artis analyticae praxis*, is published in 1631.

ca. 1630

The French lawyer and mathematician Pierre de Fermat begins a lifetime of mathematical research. He is the first person to claim to have proved "Fermat's last theorem."

1636

Gérard (or Girard) Desargues, a French mathematician and engineer, publishes *Traité de la section perspective*, which marks the beginning of projective geometry.

1637

René Descartes, a French philosopher and mathematician, publishes *Discours de la méthode*, permanently changing both algebra and geometry.

1638

Galileo Galilei publishes *Dialogues Concerning Two New Sciences* while under arrest.

1640

Blaise Pascal, a French philosopher, scientist, and mathematician, publishes *Essai sur les coniques*, an extension of the work of Desargues.

1642

Blaise Pascal manufactures an early mechanical calculator, the Pascaline.

1654

Pierre de Fermat and Blaise Pascal exchange a series of letters about probability, thereby inspiring many mathematicians to study the subject.

1655

John Wallis, an English mathematician and clergyman, publishes *Arithmetica infinitorum*, an important work that presages calculus.

1657

Christiaan Huygens, a Dutch mathematician, astronomer, and physicist, publishes *De ratiociniis in ludo aleae*, a highly influential text in probability theory.

1662

John Graunt, an English businessman and a pioneer in statistics, publishes his research on the London Bills of Mortality.

1673

Gottfried Leibniz, a German philosopher and mathematician, constructs a mechanical calculator that can perform addition, subtraction, multiplication, division, and extraction of roots.

1683

Seki Kōwa, Japanese mathematician, discovers the theory of determinants.

1684

Gottfried Leibniz publishes the first paper on calculus, *Nova methodus pro maximis et minimis.*

1687

Isaac Newton, a British mathematician and physicist, publishes *Philosophiae naturalis principia mathematica*, beginning a new era in science.

1693

Edmund Halley, a British mathematician and astronomer, undertakes a statistical study of the mortality rate in Breslau, Germany.

1698

Thomas Savery, an English engineer and inventor, patents the first steam engine.

1705

Jacob Bernoulli, a Swiss mathematician, dies. His major work on probability, *Ars conjectandi*, is published in 1713.

1712

The first Newcomen steam engine is installed.

1718

Abraham de Moivre, a French mathematician, publishes *The Doctrine of Chances*, the most advanced text of the time on the theory of probability.

1743

The Anglo-Irish Anglican bishop and philosopher George Berkeley publishes *The Analyst*, an attack on the new mathematics pioneered by Isaac Newton and Gottfried Leibniz.

The French mathematician and philosopher Jean Le Rond d'Alembert begins work on the *Encyclopédie*, one of the great works of the Enlightenment.

1748

Leonhard Euler, a Swiss mathematician, publishes his *Introductio.*

1749

The French mathematician and scientist Georges-Louis Leclerc, comte de Buffon publishes the first volume of *Histoire naturelle.*

1750

Gabriel Cramer, a Swiss mathematician, publishes "Cramer's rule," a procedure for solving systems of linear equations.

1760

Daniel Bernoulli, a Swiss mathematician and scientist, publishes his probabilistic analysis of the risks and benefits of variolation against smallpox.

1761

Thomas Bayes, an English theologian and mathematician, dies. His "Essay Towards Solving a Problem in the Doctrine of Chances" is published two years later.

The English scientist Joseph Black proposes the idea of latent heat.

1769

James Watt obtains his first steam engine patent.

1781

William Herschel, a German-born British musician and astronomer, discovers Uranus.

1789

Unrest in France culminates in the French Revolution.

1793

The Reign of Terror, a period of brutal, state-sanctioned repression, begins in France.

1794

The French mathematician Adrien-Marie Legendre (or Le Gendre) publishes his *Éléments de géométrie*, a text that influences mathematics education for decades.

Antoine-Laurent Lavoisier, a French scientist and discoverer of the law of conservation of mass, is executed by the French government.

1798

Benjamin Thompson (Count Rumford), a British physicist, proposes the equivalence of heat and work.

1799

Napoléon seizes control of the French government.

Caspar Wessel, a Norwegian mathematician and surveyor, publishes the first geometric representation of the complex numbers.

1801

Carl Friedrich Gauss, a German mathematician, publishes *Disquisitiones arithmeticae.*

1805

Adrien-Marie Legendre, a French mathematician, publishes *Nouvelles méthodes pour la détermination des orbites des comètes*, which contains the first description of the method of least squares.

1806

Jean-Robert Argand, a French bookkeeper, accountant, and mathematician, develops the Argand diagram to represent complex numbers.

1812

Pierre-Simon Laplace, a French mathematician, publishes *Théorie analytique des probabilités*, the most influential 19th-century work on the theory of probability.

1815

Napoléon suffers final defeat at the battle of Waterloo.

Jean-Victor Poncelet, a French mathematician and the "father of projective geometry," publishes *Traité des propriétés projectives des figures.*

1824

The French engineer Sadi Carnot publishes *Réflexions sur la puissance motrice du feu*, wherein he describes the Carnot engine.

Niels Henrik Abel, a Norwegian mathematician, publishes his proof of the impossibility of algebraically solving a general fifth-degree equation.

1826

Nikolay Ivanovich Lobachevsky, a Russian mathematician and "the Copernicus of geometry," announces his theory of non-Euclidean geometry.

1828

Robert Brown, a Scottish botanist, publishes the first description of Brownian motion in "A Brief Account of Microscopical Observations."

1830

Charles Babbage, a British mathematician and inventor, begins work on his analytical engine, the first attempt at a modern computer.

1832

János Bolyai, a Hungarian mathematician, publishes *Absolute Science of Space.*

The French mathematician Évariste Galois is killed in a duel.

1843

James Prescott Joule publishes his measurement of the mechanical equivalent of heat.

1846

The planet Neptune is discovered by the French mathematician Urbain-Jean-Joseph Le Verrier from a mathematical analysis of the orbit of Uranus.

1847

Georg Christian von Staudt publishes *Geometrie der Lage*, which shows that projective geometry can be expressed without any concept of length.

1848

Bernhard Bolzano, a Czech mathematician and theologian, dies. His study of infinite sets, *Paradoxien des Unendlichen*, is first published in 1851.

1850

Rudolph Clausius, a German mathematician and physicist, publishes his first paper on the theory of heat.

1851

William Thomson (Lord Kelvin), a British scientist, publishes "On the Dynamical Theory of Heat."

1854

George Boole, a British mathematician, publishes *Laws of Thought*. The mathematics contained therein makes possible the later design of computer logic circuits.

The German mathematician Bernhard Riemann gives the historic lecture "On the Hypotheses That Form the Foundations of Geometry." The ideas therein play an integral part in the theory of relativity.

1855

John Snow, a British physician, publishes "On the Mode of Communication of Cholera," the first successful epidemiological study of a disease.

1859

James Clerk Maxwell, a British physicist, proposes a probabilistic model for the distribution of molecular velocities in a gas.

Charles Darwin, a British biologist, publishes *On the Origin of Species by Means of Natural Selection*.

1861

Karl Weierstrass creates a continuous nowhere differentiable function.

1866

The Austrian biologist and monk Gregor Mendel publishes his ideas on the theory of heredity in "Versuche über Pflanzenhybriden."

1872

The German mathematician Felix Klein announces his Erlanger Programm, an attempt to categorize all geometries with the use of group theory.

Lord Kelvin (William Thomson) develops an early analog computer to predict tides.

Richard Dedekind, a German mathematician, rigorously establishes the connection between real numbers and the real number line.

1874

Georg Cantor, a German mathematician, publishes "Über eine Eigenschaft des Inbegriffes aller reelen algebraischen Zahlen," a pioneering paper that shows that all infinite sets are not the same size.

1890

The Hollerith tabulator, an important innovation in calculating machines, is installed at the United States Census for use in the 1890 census.

Giuseppe Peano publishes his example of a space-filling curve.

1894

Oliver Heaviside describes his operational calculus in his text *Electromagnetic Theory.*

1895

Henri Poincaré publishes *Analysis situs*, a landmark paper in the history of topology, in which he introduces a number of ideas that would occupy the attention of mathematicians for generations.

1898

Émile Borel begins to develop a theory of measure of abstract sets that takes into account the topology of the sets on which the measure is defined.

1899

The German mathematician David Hilbert publishes the definitive axiomatic treatment of Euclidean geometry.

1900

David Hilbert announces his list of mathematics problems for the 20th century.

The Russian mathematician Andrey Andreyevich Markov begins his research into the theory of probability.

1901

Henri-Léon Lebesgue, a French mathematician, develops his theory of integration.

1905

Ernst Zermelo, a German mathematician, undertakes the task of axiomatizing set theory.

Albert Einstein, a German-born American physicist, begins to publish his discoveries in physics.

1906

Marian Smoluchowski, a Polish scientist, publishes his insights into Brownian motion.

1908

The Hardy-Weinberg law, containing ideas fundamental to population genetics, is published.

1910

Bertrand Russell, a British logician and philosopher, and Alfred North Whitehead, a British mathematician and philosopher, publish *Principia mathematica*, an important work on the foundations of mathematics.

1913

Luitzen E. J. Brouwer publishes his recursive definition of the concept of dimension.

1914

Felix Hausdorff publishes *Grundzüge der Mengenlehre.*

1915

Wacław Sierpiński publishes his description of the now-famous curve called the Sierpiński gasket.

1917

Vladimir Ilyich Lenin leads a revolution that results in the founding of the Union of Soviet Socialist Republics.

1918

World War I ends.

The German mathematician Emmy Noether presents her ideas on the roles of symmetries in physics.

1920

Zygmunt Janiszewski, founder of the Polish school of topology, dies.

1923

Stefan Banach begins to develop the theory of Banach spaces.

Karl Menger publishes his first paper on dimension theory.

1924

Pavel Samuilovich Urysohn dies in a swimming accident at the age of 25 after making several important contributions to topology.

1928

Maurice Frechet publishes his *Les espaces abstraits et leur théorie considérée comme introduction à l'analyse générale*, which places topological concepts at the foundation of the field of analysis.

1929

Andrey Nikolayevich Kolmogorov, a Russian mathematician, publishes *General Theory of Measure and Probability Theory*, establishing the theory of probability on a firm axiomatic basis for the first time.

1930

Ronald Aylmer Fisher, a British geneticist and statistician, publishes *Genetical Theory of Natural Selection*, an important early attempt to express the theory of natural selection in mathematical language.

1931

Kurt Gödel, an Austrian-born American mathematician, publishes his incompleteness proof.

The Differential Analyzer, an important development in analog computers, is developed at Massachusetts Institute of Technology.

1933

Karl Pearson, a British innovator in statistics, retires from University College, London.

Kazimierz Kuratowski publishes the first volume of *Topologie*, which extends the boundaries of set theoretic topology (still an important text).

1935

George Horace Gallup, a U.S. statistician, founds the American Institute of Public Opinion.

1937

The British mathematician Alan Turing publishes his insights on the limits of computability.

Topologist and teacher Robert Lee Moore begins serving as president of the American Mathematical Society.

1939

World War II begins.

William Edwards Deming joins the United States Census Bureau.

The Nicolas Bourbaki group publishes the first volume of its *Éléments de mathématique.*

Sergey Sobolev elected to the USSR Academy of Sciences after publishing a long series of papers describing a generalization of the concept of function and a generalization of the concept of derivative. His work forms the foundation for a new branch of analysis.

1941

Witold Hurewicz and Henry Wallman publish their classic text *Dimension Theory.*

1945

Samuel Eilenberg and Saunders Mac Lane found the discipline of category theory.

1946

The Electronic Numerical Integrator and Calculator (ENIAC) computer begins operation at the University of Pennsylvania.

1948

While working at Bell Telephone Labs in the United States, Claude Shannon publishes "A Mathematical Theory of Communication," marking the beginning of the Information Age.

1951

The Universal Automatic Computer (UNIVAC I) is installed at U.S. Bureau of the Census.

1954

FORmula TRANslator (FORTRAN), one of the first high-level computer languages, is introduced.

1956

The American Walter Shewhart, an innovator in the field of quality control, retires from Bell Telephone Laboratories.

1957

Olga Oleinik publishes "Discontinuous Solutions to Nonlinear Differential Equations," a milestone in mathematical physics.

1965

Andrey Nikolayevich Kolmogorov establishes the branch of mathematics now known as Kolmogorov complexity.

1972

Amid much fanfare, the French mathematician and philosopher René Thom establishes a new field of mathematics called catastrophe theory.

1973

The C computer language, developed at Bell Laboratories, is essentially completed.

1975

The French geophysicist Jean Morlet helps develop a new kind of analysis based on what he calls "wavelets."

1980

Kiiti Morita, the founder of the Japanese school of topology, publishes a paper that further extends the concept of dimension to general topological spaces.

1982

Benoît Mandelbrot publishes his highly influential *The Fractal Geometry of Nature.*

1989

The Belgian mathematician Ingrid Daubechies develops what has become the mathematical foundation for today's wavelet research.

1995

The British mathematician Andrew Wiles publishes the first proof of Fermat's last theorem.

JAVA computer language is introduced commercially by Sun Microsystems.

1997

René Thom declares the mathematical field of catastrophe theory "dead."

2002

Experimental Mathematics celebrates its 10th anniversary. It is a refereed journal dedicated to the experimental aspects of mathematical research.

Manindra Agrawal, Neeraj Kayal, and Nitin Saxena create a brief, elegant algorithm to test whether a number is prime, thereby solving an important centuries-old problem.

2003

Grigory Perelman produces the first complete proof of the Poincaré conjecture, a statement about some of the most fundamental properties of three-dimensional shapes.

2007

The international financial system, heavily dependent on so-called sophisticated mathematical models, finds itself on the edge of collapse, calling into question the value of the mathematical models.

2008

Henri Cartan, one of the founding members of the Nicolas Bourbaki group, dies at the age of 104.

GLOSSARY

algebra (1) a mathematical system that is a generalization of arithmetic, in which letters or other symbols are used to represent numbers; (2) the study of the formal relations between symbols belonging to sets on which one or more operations has been defined

algebraic equation an equation of the form $a_n x^n + a_{n-1} x^{n-1} + \ldots + a_1 x + a_0 = 0$ where n can represent any natural number, x represents the variable raised to the power indicated, and a_j, which (in this book) always represents a rational number, is the coefficient by which x^j is multiplied

algorithm a formula or procedure used to solve a mathematical problem

analytic geometry the branch of mathematics that studies geometry via algebraic methods and coordinate systems

axiom a statement accepted as true that serves as a basis for deductive reasoning

characteristic equation an algebraic equation associated with the determinant of a given matrix with the additional property that the matrix acts as a root of the equation

coefficient a number or symbol representing a number used to multiply a variable

combinatorics the branch of mathematics concerned with the selection of elements from finite sets and the operations that are performed with those sets

commensurable evenly divisible by a common measure. Two lengths (or numbers representing those lengths) are commensurable when they are evenly divisible by a common unit

complex number any number of the form $a + bi$ where a and b are real numbers and i has the property that $i^2 = -1$

composite number a whole number greater than 1 that is not prime

conic section any member of the family of curves obtained from the intersection of a double cone and a plane

coordinate system a method of establishing a one-to-one correspondence between points in space and sets of numbers

deduction a conclusion obtained by logically reasoning from general principles to particular statements

degree of an equation for an algebraic equation of one variable, the largest exponent appearing in the equation

determinant a particular function defined on the set of square matrices. The value of the determinant is a real or complex number

determinant equation an equation or system of equations for which there exists a unique solution

eigenvalue the root of a characteristic equation

ellipse a closed curve formed by the intersection of a right circular cone and a plane

field a set of numbers with the property that however two numbers are combined via the operations of addition, subtraction, multiplication, and division (except by 0), the result is another number in the set

fifth-degree equation an algebraic equation in which the highest exponent appearing in the equation is 5

fourth-degree equation an algebraic equation in which the highest exponent appearing in the equation is 4

fundamental principle of analytic geometry the observation that under fairly general conditions one equation in two variables defines a curve

fundamental principle of solid analytic geometry the observation that under fairly general conditions one equation in three variables defines a surface

fundamental theorem of algebra the statement that any polynomial of degree n has n roots

geometric algebra a method of expressing ideas usually associated with algebra via the concepts and techniques of Euclidean geometry

group a set of objects together with an operation analogous to multiplication such that (1) the "product" of any two elements in the set is an element in the set; (2) the operation is associative, that is, for any three elements, a, b, and c in the group $(ab)c = a(bc)$; (3) there is an element in the set, usually denoted with the letter e, such that $ea = ae = a$ where a is any element in the set; and (4) every element in the set has an inverse, so that if a is an element in the set, there is an element called a^{-1} such that $aa^{-1} = e$

group theory the branch of mathematics devoted to the study of groups

hyperbola a curve composed of the intersection of a plane and both parts of a double, right circular cone

identity the element, usually denoted with the letter e, in a group with the property that if g is any element in the group, then $eg = ge = g$

indeterminate equation an equation or set of equations for which there exist infinitely many solutions

integer any whole number

invariant a property of a mathematical system that remains unchanged when the elements that comprise the system are transformed according to some rule. Whether a property is invariant or not depends on the transformation rule. The discovery and exploitation of invariants is an important part of mathematical research

irrational number any real number that cannot be expressed as a/b, where a and b are integers and b is not 0

linear equation an algebraic equation in which every term is a number or a variable of degree 1 multiplied by a number

matrix a rectangular array or table of numbers or other quantities

natural number the number 1, or any number obtained by adding 1 to itself sufficiently many times

one-to-one correspondence the pairing of elements between two sets, *A* and *B*, such that each element of *A* is paired with a unique element of *B* and to each element of *B* is paired a unique element of *A*

parabola the curve formed by the intersection of a right circular cone and a plane that is parallel to a line that generates the cone

polynomial a mathematical expression consisting of the sum of terms of the form ax^n, where a represents a number, x represents a variable, and n represents a nonnegative integer

prime number a natural number greater than 1 that is—among the set of all natural numbers—evenly divisible only by itself and 1

Pythagorean theorem the statement that for a right triangle the square of the length of the hypotenuse equals the sum of the squares of the lengths of the remaining sides

Pythagorean triple three numbers each of which is a natural number such that the sum of the squares of the two smaller numbers equals the square of the largest number.

quadratic equation See SECOND-DEGREE EQUATION.

quadratic formula a mathematical formula for computing the roots of any second-degree algebraic equation by using the coefficients that appear in the equation

rational number any number of the form a/b, where a and b are integers and b is not 0

real number any rational number or any number that can be approximated to an arbitrarily high degree of accuracy by a rational number

rhetorical algebra algebra that is expressed in words only, without specialized algebraic symbols

ring a set of objects on which two operations, often called addition and multiplication, are defined. Under addition, the set forms a commutative GROUP. Under multiplication, the elements in the set satisfy the following two properties: (1) the product of any two elements in the set is another element in the set, and (2) multiplication is associative. Addition and multiplication are related through the distributive law: $a \times (b + c) = a \times b + a \times c$. Sometimes other conditions are imposed in addition to these.

root for any algebraic equation any number that satisfies the equation

second-degree equation an algebraic equation in which the highest exponent appearing in the equation is 2

sparse matrix a matrix in which most of the coefficients appearing in the matrix are zero

spectral theory the study that seeks to relate the properties of a matrix to the properties of its eigenvalues

syllogism a type of formal logical argument described in detail by Aristotle in the collection of his writings known as *The Organon*

symmetry transformation a change, such as a rotation or reflection, of a geometric or physical object with the property that the spatial configuration of the object is the same before and after the transformation

syncopated algebra a method of expressing algebra that uses some abbreviations but does not employ a fully symbolic system of algebraic notation

third-degree equation an algebraic equation in which the highest exponent appearing in the equation is 3

unit fraction a fraction of the form $1/a$, where a is any integer except zero

FURTHER RESOURCES

MODERN WORKS

Adler, Irving. *Thinking Machines, a Layman's Introduction to Logic, Boolean Algebra, and Computers.* New York: John Day, 1961. This old book is still the best nontechnical introduction to computer arithmetic. It begins with fingers and ends with Boolean logic circuits.

Bashmakova, Izabella G. *The Beginnings and Evolution of Algebra.* Washington, D.C.: Mathematical Association of America, 2000. Aimed at older high school students, this is a detailed look at algebra from the age of Mesopotamia to the end of the 19th century. It presupposes a strong background in high school-level algebra.

Boyer, Carl B., and Uta C. Merzbach. *A History of Mathematics.* New York: John Wiley & Sons, 1991. Boyer was one of the preeminent mathematics historians of the 20th century. This work contains much interesting biographical information. The mathematical information assumes a fairly strong background.

Bunt, Lucas Nicolaas Hendrik, Phillip S. Jones, Jack D. Bedient. *The Historical Roots of Elementary Mathematics.* Englewood Cliffs, N.J.: Prentice-Hall, 1976. A highly detailed examination—complete with numerous exercises—of how ancient cultures added, subtracted, divided, multiplied, and reasoned.

Carroll, L. *Symbolic Logic and The Game of Logic.* New York: Dover Publications, 1958. Better known as the author of Alice in Wonderland, Lewis Carroll was also an accomplished mathematician. The language in these two books (bound as one) is old-fashioned but very accessible.

Courant, Richard, and Herbert Robbins. *What Is Mathematics? An Elementary Approach to Ideas and Mathematics.* New York: Oxford University Press, 1941. A classic and exhaustive answer to the question posed in the title. Courant was an influential 20th-century mathematician.

Danzig, Tobias. *Number, the Language of Science.* New York: Macmillan, 1954. First published in 1930, this book is painfully elitist; the author's many prejudices are on display in every chapter. Yet it is one of the best nontechnical histories of the concept of number ever written. Apparently it was also Albert Einstein's favorite book on the history of mathematics.

Dewdney, Alexander K. *200% of Nothing: An Eye-Opening Tour through the Twists and Turns of Math Abuse and Innumeracy.* New York: John Wiley & Sons, 1993. A critical look at how mathematical reasoning has been abused to distort truth.

Eastaway, Robert, and Jeremy Wyndham. *Why Do Buses Come in Threes? The Hidden Mathematics of Everyday Life.* New York: John Wiley & Sons, 1998. Nineteen lighthearted essays on the mathematics underlying everything from luck to scheduling problems.

Eves, Howard. *An Introduction to the History of Mathematics.* New York: Holt, Rinehart & Winston, 1953. This well-written history of mathematics places special emphasis on early mathematics. It is unusual because the history is accompanied by numerous mathematical problems. (The solutions are in the back of the book.)

Freudenthal, Hans. *Mathematics Observed.* New York: McGraw-Hill, 1967. A collection of seven survey articles about math topics from computability to geometry to physics (some more technical than others).

Gardner, Martin. *The Colossal Book of Mathematics.* New York: Norton, 2001. Martin Gardner had a gift for seeing things mathematically. This "colossal" book contains sections on geometry, algebra, probability, logic, and more.

———. *Logic Machines and Diagrams.* Chicago: University of Chicago Press, 1982. An excellent book on logic and its uses in computers.

Guillen, Michael. *Bridges to Infinity: The Human Side of Mathematics.* Los Angeles: Jeremy P. Tarcher, 1983. This book consists of an engaging nontechnical set of essays on mathematical topics, including non-Euclidean geometry, transfinite numbers, and catastrophe theory.

Heath, Thomas L. *A History of Greek Mathematics.* New York: Dover Publications, 1981. First published early in the 20th century and reprinted numerous times, this book is still one of the main references on the subject.

Hoffman, Paul. *Archimedes' Revenge: The Joys and Perils of Mathematics.* New York: Ballantine, 1989. A relaxed, sometimes silly look at an interesting and diverse set of math problems ranging from prime numbers and cryptography to Turing machines and the mathematics of democratic processes.

Hogben, L. *Mathematics for the Million.* New York: W. W. Norton, 1968. This is a classic text that has been in print for many decades. Written by a creative scientist, it reveals a view of mathematics, its history, and its applications that is both challenging and entertaining. Highly recommended.

Jacquette, D. *On Boole.* Belmont, Calif.: Wadsworth/Thompson Learning, 2002. This book gives a good overview of Aristotelian syllogisms, Boolean algebra, and the uses of Boolean algebra in the design of computer logic circuits.

Joseph, George G. *The Crest of the Peacock: The Non-European Roots of Mathematics.* Princeton, N.J.: Princeton University Press, 1991. One of the best of a new crop of books devoted to this important topic.

Kline, Morris. *Mathematics and the Physical World.* New York: Thomas Y. Crowell, 1959. The history of mathematics as it relates to the history of science, and vice versa.

———. *Mathematics for the Nonmathematician.* New York: Dover Publications, 1985. An articulate, not very technical overview of many important mathematical ideas.

———. *Mathematics in Western Culture.* New York: Oxford University Press, 1953. An excellent overview of the development of

Western mathematics in its cultural context, this book is aimed at an audience with a firm grasp of high school-level mathematics.

McLeish, John. *Number.* New York: Fawcett Columbine, 1992. A history of the concept of number from Mesopotamia to modern times.

Pappas, Theoni. *The Joy of Mathematics.* San Carlos, Calif.: World Wide/Tetra, 1986. Aimed at a younger audience, this work searches for interesting applications of mathematics in the world around us.

Pierce, John R. *An Introduction to Information Theory: Symbols, Signals and Noise.* New York: Dover Publications, 1961. Despite the sound of the title, this is not a textbook. Among other topics, Pierce, formerly of Bell Laboratories, describes some of the mathematics involved in encoding numbers and text for digital transmission or storage—a lucid introduction to the topics of information and algebraic coding theory.

Reid, Constance. *From Zero to Infinity: What Makes Numbers Interesting.* New York: Thomas Y. Crowell, 1960. A well-written overview of numbers and the algebra that stimulated their development.

Schiffer, M. and Leon Bowden. *The Role of Mathematics in Science.* Washington, D.C.: Mathematical Association of America, 1984. The first few chapters of this book, ostensibly written for high school students, will be accessible to many students; the last few chapters will find a much narrower audience.

Smith, David E., and Yoshio Mikami. *A History of Japanese Mathematics.* Chicago: Open Court, 1914. Copies of this book are still around, and it is frequently quoted. The first half is an informative nontechnical survey. The second half is written more for the expert.

Stewart, Ian. *From Here to Infinity.* New York: Oxford University Press, 1996. A well-written, very readable overview of several important contemporary ideas in geometry, algebra, computability, chaos, and mathematics in nature.

Swetz, Frank J., editor. *From Five Fingers to Infinity: A Journey through the History of Mathematics.* Chicago: Open Court, 1994.

This is a fascinating, though not especially focused, look at the history of mathematics. Highly recommended.

Swetz, Frank. *Sea Island Mathematical Manual: Surveying and Mathematics in Ancient China.* University Park: Pennsylvania State University Press, 1992. The book contains many ancient problems in mathematics and measurement and illustrates how problems in measurement often inspired the development of geometric ideas and techniques.

Tabak, John. *Numbers: Computers, Philosophers, and the Search for Meaning.* New York: Facts On File, 2004. More information about how the concept of number and ideas about the nature of algebra evolved together.

Thomas, David A. *Math Projects for Young Scientists.* New York: Franklin Watts, 1988. This project-oriented text is an introduction to several historically important geometry problems.

Yaglom, Isaac M. *Geometric Transformations*, translated by Allen Shields. New York: Random House, 1962. Aimed at high school students, this is a very sophisticated treatment of "simple" geometry and an excellent introduction to higher mathematics. It is also an excellent introduction to the concept of invariance.

Zippin, Leo. *The Uses of Infinity.* New York: Random House, 1962. Contains lots of equations—perhaps too many for the uninitiated—but none of the equations is very difficult. The book is worth the effort required to read it.

ORIGINAL SOURCES

It can sometimes deepen our appreciation of an important mathematical discovery to read the discoverer's own description. Often this is not possible, because the description is too technical. Fortunately there are exceptions. Sometimes the discovery is accessible because the idea does not require a lot of technical background to appreciate it. Sometimes the discoverer writes a nontechnical account of the technical idea that she or he has discovered. Here are some classic papers:

Ahmes. *The Rhind Mathematical Papyrus: Free Translation, Commentary, and Selected Photographs, Transcription, Literal Translations,* translated by Arnold B. Chace. Reston, Va.: National Council of Teachers of Mathematics, 1979. This is a translation of the biggest and best of extant Egyptian mathematical texts, the Rhind papyrus (also known as the Ahmes papyrus). It provides insight into the types of problems and methods of solution known to one of humanity's oldest cultures.

Al-Khwārizmī, Mohammed ibn-Mūsā. *Robert of Chester's Latin translation of the Algebra of al-Khwārizmī, with an introduction, critical notes, and an English version of Louis Charles Karpinski.* Norwood, Mass.: Norwood Press, 1915. Various versions of this important original work are still around. It can be found in academic libraries and on the Internet. The algebra is simple, but it is still hard to read because of the absence of algebraic symbols. Al-Khwārizmī's book changed history.

Boole, George. *Mathematical Analysis of Logic.* In *The World of Mathematics*, vol. 3, edited by James R. Newman. New York: Dover Publications, 1956. This is a nontechnical excerpt from one of Boole's most famous works. Although there is no "Boolean algebra" in this article, it contains Boole's own explanation for what he hoped to gain from studying the laws of thought.

Brahmagupta and Bhascara. *Algebra, with Arithmetic and Mensuration.* Translated by Henry Colebrook. London: John Murray, 1819. Although it was first published almost 200 years ago, copies of this book can be found in most academic libraries, and it is also available on the Internet. It is interesting to see these great minds struggle to create a new way of thinking.

Descartes, René. *The Geometry.* In *The World of Mathematics,* vol. 1, edited by James Newman. New York: Dover Publications, 1956. This is a readable translation of an excerpt from Descartes's own revolutionary work *La Géométrie.*

Euclid. *Elements.* Translated by Sir Thomas L. Heath. *Great Books of the Western World.* Vol. 11. Chicago: Encyclopaedia Britannica, 1952. See especially book I for Euclid's own exposition of the axiomatic method, and read some of the early propositions in this volume to see how the Greeks investigated mathematics without equations.

Galilei, Galileo. *Dialogues Concerning Two New Sciences.* Translated by Henry Crew and Alfonso de Salvio. New York: Dover Publications, 1954. An interesting literary work as well as a pioneering physics text. Many regard the publication of this text as the beginning of the modern scientific tradition. And in it one can find Galileo seeking to use algebraic language.

Hardy, Godfrey H. *A Mathematician's Apology.* Cambridge, England: Cambridge University Press, 1940. Hardy was an excellent mathematician and a good writer. In this oft-quoted and very brief book Hardy seeks to explain and sometimes justify his life as a mathematician.

Keyser, Cassius J. *The Group Concept.* In *The World of Mathematics,* vol. 4, edited by James R. Newman. New York: Dover Publications, 1956. Keyser was a successful 20th-century American mathematician. The language in this article is stilted and old-fashioned, but look past matters of style to read his beautiful description of geometric "shape" as that property that is invariant under a group of similarity transformations.

Mathematics: Its Content, Methods, and Meaning. Vols. 1 and 3, edited by A. D. Aleksandrov, A. N. Kolmogorov, and M. A. Lavrent'ev, and translated by S. H. Gould and T. Bartha. Mineola, N.Y.: Dover Publications, 1999. This three-volume set, now available as a single enormous volume, consists of a collection of survey articles written for the well-informed layperson by some of the great mathematicians of the 20th century. Chapters 3 and 4 in volume 1 are written by the Russian mathematician B. N. Delone and discuss analytic geometry and the theory of algebraic equations. Chapter 16, which is written by the Russian mathematician D. K. Faddeev discusses linear algebra. Highly recommended.

Mikami, Yoshio. *The Development of Mathematics in China and Japan.* New York: Chelsea Publishing, 1913. This book can still be found in most academic libraries and because it is now in the public domain it is freely available on the Web. It is a well-written book about an important topic.

Poincaré, Henri. *Mathematical Creation.* In *The World of Mathematics,* vol. 4, edited by James R. Newman. New York: Dover Publications, 1956. Poincaré was one of history's most successful mathematicians. In particular, he made a number of important contributions to group theory, mathematical physics, topology, and philosophy. He also liked to write popular articles about mathematics. This is one of them.

Russell, Bertrand. *Mathematics and the Metaphysicians.* In *The World of Mathematics.* Vol. 3, edited by James Newman. New York: Dover Publications, 1956. An introduction to the philosophical ideas upon which mathematics is founded written by a major contributor to this field.

INTERNET RESOURCES

Mathematical ideas are often subtle and expressed in an unfamiliar vocabulary. Without long periods of quiet reflection, mathematical concepts are sometimes difficult to appreciate. This is exactly the type of environment one does not usually find on the Web. To develop a real appreciation for mathematical thought, books are better. That said, the following Web sites are some good resources.

Eric Weisstein's World of Mathematics. Available online. URL: http://mathworld.wolfram.com/. Accessed January 15, 2010. This site has brief overviews of a great many topics in mathematics. The level of presentation varies substantially from topic to topic.

Fife, Earl, and Larry Husch. Math Archives. "History of Mathematics." Available online. URL: http://archives.math.utk.edu/topics/history.html. Accessed January 15, 2010. Information on mathematics, mathematicians, and mathematical organizations.

Gangolli, Ramesh. *Asian Contributions to Mathematics.* Available online. URL: http://www.pps.k12.or.us/depts-c/mc-me/be-as-ma.pdf. Accessed January 15, 2010. As its name implies, this well-written online book focuses on the history of mathematics in Asia and its effect on the world history of mathematics. It also includes information on the work of Asian Americans, a welcome contribution to the field.

Heinlow, Lance, and Karen Pagel. "Math History." Online Resource. Available online. URL: http://or.amatyc.org. Accessed January 15, 2010. Created under the auspices of the American Mathematical Association of Two-Year Colleges, this site is a very extensive collection of links to mathematical and math-related topics.

Howard, Mike. *Introduction to Crystallography and Mineral Crystal Systems*. Available online. URL: http://www.rockhounds.com/rockshop/xtal/. Accessed January 15, 2010. The author has designed a nice introduction to the use of group theory in the study of crystals through an interesting mix of geometry, algebra, and mineralogy.

The Math Forum @ Drexel. The Math Forum Student Center. Available online. URL: http://mathforum.org/students/. Accessed January 15, 2010. Probably the best Web site for information about the kinds of mathematics that students encounter in their school-related studies. You will find interesting and challenging problems and solutions for students in grades K-12 as well as a fair amount of college-level information.

O'Connor, John L., and Edmund F. Robertson. The MacTutor History of Mathematics Archive. Available online. URL: http://www.gap.dcs.st-and.ac.uk/~history/index.html. Accessed January 15, 2010. This is a valuable resource for anyone interested in learning more about the history of mathematics. It contains an extraordinary collection of biographies of mathematicians in different cultures and times. In addition it provides information about the historical development of certain key mathematical ideas.

PERIODICALS

+Plus

URL: http://pass.maths.org.uk

A site with numerous interesting articles about all aspects of high school math. They send an e-mail every few weeks to their subscribers to keep them informed about new articles at the site.

Parabola: A Mathematics Magazine for Secondary Students

Australian Mathematics Trust
University of Canberra
ACT 2601
Australia

Published twice a year by the Australian Mathematics Trust in association with the University of New South Wales, *Parabola* is a source of short high-quality articles on many aspects of mathematics. Some back issues are also available free online. See URL: http://www.maths.unsw.edu.au/Parabola/index.html.

Scientific American

415 Madison Avenue
New York, NY 10017

A serious and widely read monthly magazine, *Scientific American* regularly carries high-quality articles on mathematics and mathematically intensive branches of science. This is the one "popular" source of high-quality mathematical information that you will find at a newsstand.

INDEX

Page numbers in *italic* indicate illustrations;
page numbers followed by *c* indicate chronology entries.

T

THE FUTURE OF GENETICS

THE FUTURE OF GENETICS

Beyond the Human Genome Project

RUSS HODGE

FOREWORD BY NADIA ROSENTHAL, PH.D.

This book is dedicated to the memory of my grandparents, E. J. and Mabel Evens and Irene Hodge; to my parents, Ed and Jo Hodge; and especially to my wife, Gabi, and my children, Jesper, Sharon, and Lisa, with love.

THE FUTURE OF GENETICS: Beyond the Human Genome Project

For information contact:

Facts On File, Inc.
An imprint of Infobase Publishing
132 West 31st Street
New York NY 10001

Library of Congress Cataloging-in-Publication Data

Hodge, Russ.
The future of genetics : beyond the human genome project / Russ Hodge ; foreword by Nadia Rosenthal.
p. ; cm. — (Genetics and evolution)
Includes bibliographical references and index.
ISBN 978-0-8160-6684-1 (alk. paper)
1. Genetics—Popular works. 2. Genetics—Forecasting. I. Title. II. Series: Genetics and evolution.
[DNLM: 1. Genetics, Medical. 2. Genome, Human. 3. Molecular Biology—methods. QZ 50 H688f 2010]

QH437.H63 2010
303.48'3—dc22 2009018297

Facts On File books are available at special discounts when purchased in bulk quantities for businesses, associations, institutions, or sales promotions. Please call our Special Sales Department in New York at (212) 967-8800 or (800) 322-8755.

You can find Facts On File on the World Wide Web at http://www.factsonfile.com

Text design by Kerry Casey
Illustrations by Sholto Ainslie
Photo research by Elizabeth H. Oakes
Composition by Hermitage Publishing Services
Cover printed by Bang Printing, Brainerd, MN
Book printed and bound by Bang Printing, Brainerd, MN
Date printed: March 2010
Printed in the United States of America

This book is printed on acid-free paper.

I say that it touches a man that his blood is sea water and his tears are salt, that the seed of his loins is scarcely different from the same cells in a seaweed, and that of stuff like his bones coral is made. I say that the physical and biologic law lies down with him, and wakes when a child stirs in the womb, and that the sap in a tree, uprushing in the spring, and the smell of the loam, where the bacteria bestir themselves in darkness, and the path of the sun in the heaven, these are facts of first importance to his mental conclusions, and that a man who goes in no consciousness of them is a drifter and a dreamer, without a home or any contact with reality.

—*An Almanac for Moderns: A Daybook of Nature,*
Donald Culross Peattie, copyright ©1935
(renewed 1963) by Donald Culross Peattie

Contents

Foreword

Science has played an increasingly central role in human affairs over the last two centuries, affecting every aspect of our daily lives. From the time of the Industrial Revolution, we have somewhat taken for granted our increasing dependence on the products of mechanical ingenuity that drive agriculture, manufacturing, mining, transportation, and communications. At the beginning of the 19th century, our great-grandparents could not easily have imagined our modern, engine-powered world. When we try to conceive of the advances that our great-grandchildren will witness, it is likely to be the fruits of genetic research that will change lives most dramatically.

The Future of Genetics offers an exciting, sometimes startling survey of the mechanics of today's genetic research at the dawn of a new century and provides a glimpse into its future. In chapter 1, Russ Hodge shows how the last century of genetic research has paved the way for the rapid progress we are witnessing in the field. Increasingly sophisticated tools of genetic analysis and manipulation are currently being developed in laboratories across the world. Drawing from other research streams, such as advances in molecular and cell biology, imaging, and informatics, today's genetics is offering a view of nature that is richer and more varied than we could have imagined previously.

The new tools of genetics also suggest prospects for changing the course of natural selection, sometimes in ways that are every bit as bizarre as science fiction. In chapter 2, Hodge explores how current advances in genetics have infiltrated and influenced our societal views and values. In a world where the communication of information is not only instantaneous but can also be manipulated to suit the needs of politicians, journalists, and the marketplace, a little information can be a dangerous thing.

In these accounts, geneticist are often morphed into the monsters they are supposedly creating, and the relevance of their extraordinary discoveries is sometimes lost on a public craving sensationalism.

Despite the view that putting the power of genetics in the hands of amoral inventors is hazardous, human curiosity is insatiable, and the outcome of genetic research is already beginning to make a major difference in our lives. Whether in regard to the early detection and prevention of congenital disease, the promise of a longer and healthier life span through a better understanding of the role of genes in aging, or the creation of new drugs based on a person's unique genetic makeup, the impact of genetics on medicine will continue to grow. Chapter 3 reviews the progress that has already been made in harnessing new knowledge gleaned from genetics to control our personal destinies and looks forward to the ways we may handle these changes as a society.

Other areas of human society, as explored in chapter 4, are just as affected by the new genetics as they were by the mechanics of the Industrial Revolution. Experimentation with genetically modified crops has not been a popular aspect of this progress, but in times of food shortages and global climate change, genetically modified crops are already affecting agriculture in our lifetime. Other directions that genetic research may take may be just as inevitable, but their cost might be too great: The creation of new designer organisms or the resuscitation of extinct species, for example, might be exciting grist for science fiction writers but could have disastrous consequences similar to the ill-advised introduction of cane toads into Australia in the 1930s for agricultural pest control that has resulted in a plague. The more genetics teaches about biodiversity, the more we appreciate how our own population explosion and the environmental devastation we create in the name of sustaining our way of life threatens Earth's ecosystem, upon which we vitally depend. Just as we have evolved from the days of coal-driven engines by developing new, quieter, and less polluting versions, the mechanics of tomorrow will need to harness the power of genetics for the benefit of society and the protection of the environment. This is a responsibility we cannot afford to ignore.

—Nadia Rosenthal, Ph.D.
Head of Outstation
European Molecular Biology Laboratory
Rome, Italy

Preface

In laboratories, clinics, and companies around the world, an amazing revolution is taking place in our understanding of life. It will dramatically change the way medicine is practiced and have other effects on nearly everyone alive today. This revolution makes the news nearly every day, but the headlines often seem mysterious and scary. Discoveries are being made at such a dizzying pace that even scientists, let alone the public, can barely keep up.

The six-volume Genetics and Evolution set aims to explain what is happening in biological research and put things into perspective for high-school students and the general public. The themes are the main fields of current research devoted to four volumes: *Evolution, The Molecules of Life, Genetic Engineering,* and *Developmental Biology.* A fifth volume is devoted to *Human Genetics,* and the sixth, *The Future of Genetics,* takes a look at how these sciences are likely to shape science and society in the future. The books aim to fill an important need by connecting the history of scientific ideas and methods to their impact on today's research. *Evolution,* for example, begins by explaining why a new theory of life was necessary in the 19th century. It goes on to show how the theory is helping create new animal models of human diseases and is shedding light on the genomes of humans, other animals, and plants.

Most of what is happening in the life sciences today can be traced back to a series of discoveries made in the mid-19th century. Evolution, cell biology, heredity, chemistry, embryology, and modern medicine were born during that era. At first these fields approached life from different points of view, using different methods. But they have steadily grown closer, and today they are all coming together in a view of life that stretches from single molecules to whole organisms, complex interactions between species, and the environment.

The meeting point of these traditions is the cell. Over the last 50 years biochemists have learned how DNA, RNA, and proteins carry out a complex dialogue with the environment to manage the cell's daily business and to build complex organisms. Medicine is also focusing on cells: Bacteria and viruses cause damage by invading cells and disrupting what is going on inside. Other diseases—such as cancer or Alzheimer's disease—arise from inherent defects in cells that we may soon learn to repair.

This is a change in orientation. Modern medicine arose when scientists learned to fight some of the worst infectious diseases with vaccines and drugs. This strategy has not worked with AIDS, malaria, and a range of other diseases because of their complexity and the way they infiltrate processes in cells. Curing such infectious diseases, cancer, and the health problems that arise from defective genes will require a new type of medicine based on a thorough understanding of how cells work and the development of new methods to manipulate what happens inside them.

Today's research is painting a picture of life that is much richer and more complex than anyone imagined just a few decades ago. Modern science has given us new insights into human nature that bring along a great many questions and many new responsibilities. Discoveries are being made at an amazing pace, but they usually concern tiny details of biochemistry or the functions of networks of molecules within cells that are hard to explain in headlines or short newspaper articles. So the communication gap between the worlds of research, schools, and the public is widening at the worst possible time. In the near future young people will be called on to make decisions—large political ones and very personal ones—about how science is practiced and how its findings are applied. Should there be limits on research into stem cells or other types of human cells? What kinds of diagnostic tests should be performed on embryos or children? How should information about a person's genes be used? How can privacy be protected in an age when everyone carries a readout of his or her personal genome on a memory card? These questions will be difficult to answer, and decisions should not be made without a good understanding of the issues.

I was largely unaware of this amazing scientific revolution until 12 years ago, when I was hired to create a public information office at one of the world's most renowned research laboratories. Since that time I have had the great privilege of working alongside some of today's greatest researchers, talking to them daily, writing about their work, and picking their brains about the world that today's science is creating. These books aim to share those experiences with the young people who will shape tomorrow's science and live in the world that it makes possible.

Acknowledgments

This book would not have been possible without the help of many people. First I want to thank the dozens of scientists with whom I have worked over the past 12 years, who have spent a great amount of time introducing me to the world of molecular biology. In particular I thank Volker Wiersdorff, Patricia Kahn, Eric Karsenti, Thomas Graf, Nadia Rosenthal, Walter Rosenthal, and Walter Birchmeier. My agent, Jodie Rhodes, was instrumental in planning and launching the project. Frank K. Darmstadt, executive editor, kept things on track and made great contributions to the quality of the text. Sincere thanks go as well to the production and art departments for their invaluable contributions. I am very grateful to Beth Oakes for locating the photographs for the entire set. Finally, I thank my family for all their support. That begins with my parents, Ed and Jo Hodge, who somehow figured out how to raise a young writer, and extends to my wife and children, Gabi, Jesper, Sharon, and Lisa, who are still learning how to live with one.

Introduction

The Future of Genetics was inspired by a few facts and experiences that most people have shared in one way or another. My grandfather was born in the year 1900 and died toward the end of the 20th century, having witnessed such amazing transformations that the world must have seemed almost alien. For example, discoveries had led to airplanes, televisions, computers, and the Internet, to name only a few. Many of the changes were comforting to a middle-class family in the developed world: electricity everywhere, all kinds of luxuries to make daily life easier, vast amounts of information and entertainment transmitted directly into the home, and great advances in medicine, including the insulin that extended his life.

However, these positive changes came at a pace that was difficult to adjust to, with some worrisome side effects. The hunger for energy led to widespread strip-mining of coal in his home state, and burning the coal caused smog and global warming. Modern physics had produced weapons capable of wiping out human civilization and nuclear power plants that dotted the countryside—were they really safe? Industrial chemicals and other types of pollution ate at the ozone layer and contaminated the soil. The technology that permitted the creation of cell phones and the Internet could be used to eavesdrop on people. And the same methods that produced insulin were being used to insert new genes into crops grown by his neighbors, with consequences that might bring risks; it was hard to be sure.

Many people seem to feel that scientific discoveries and progress are inevitably accompanied by unforeseen dangers. This can be witnessed anywhere. Start a conversation about genetic engineering at any café, classroom, or dinner party. Phrases such as "scientists should not meddle with nature," or "the Earth will take revenge on humans," or "as doctors get smarter, nature will invent smarter diseases" are bound to be heard.

Where do these beliefs come from? Are they a valid appreciation of real risks, an impression given by highly exaggerated scenarios of science fiction, have people's feelings about the biological sciences been unfairly contaminated by experiences with other types of research and technology? Is cloning really anything to be scared of or have people simply seen too many bad science fiction films? Is biological research itself dangerous or do problems arise from the pressure to turn science into products that can be catapulted onto global markets? Are people afraid of science or are they afraid of change in general? This book aims to give readers the facts they need to find sensible answers to these questions and to distinguish serious issues from hype and unrealistic fears. That is difficult at a time when the line between information, entertainment, and ideologies has become blurred. Optimistic press releases are released by companies trying to find investors; science fiction films sensationalize, exaggerate, and dramatize rather than explain, and scientific data is being misused by people with political or religious agendas.

It is important to try to form a realistic picture of how science might change the future, even though speculations about the distant future of technology often turn out to be wrong—where are the starships or time machines imagined just a few decades ago? On the other hand, some visions do come true. Today's biology is having an enormous impact on medicine, particularly in diagnosing disease and understanding how drugs work, and researchers strongly believe that it will soon change the way doctors cope with complex illnesses like cancer. It may even provide treatments for genetic diseases. Many of the possibilities for therapies or technologies foreseen by scientists already exist in some form. The question is whether they can be extended to humans safely and in a controlled way. Change may come more rapidly than anyone anticipates, and there may be intense pressure to sell technologies before they have been adequately tested. This can already be seen in the case of some drugs. Some medications for attention-deficit hyperactivity disorder (ADHD) were prescribed to children before their long-term effects on children's development were fully understood.

The ethical issues need to be confronted now, in preparation for a future in which biotechnology is more powerful. Yet making intelligent decisions will require a realistic understanding of how science works. Research has become thoroughly global, which means that the only way to regulate certain practices may be for the global research community as a whole to agree that some things should not be done.

Currently, there are areas of widespread agreement. Few scientists see the sense in cloning a person—which means taking material from a person's cells and producing an identical twin at a later date. On the other hand, it would be useful to know how to stimulate embryonic stem cells into rebuilding damaged nerves, muscles, or heart tissue—and accomplishing one of these things might require learning to do the other.

Most scientists also agree that a person's genes should not be altered in a way that can be passed along to his or her children. But just a century ago, many scientists—and others—felt differently. They promoted misguided efforts to improve humanity in the first half of the 20th century in programs that forcibly sterilized thousands of mental patients and other "undesirables" throughout the United States. This thinking culminated in the Holocaust in Nazi Germany, whose perpetrators also claimed to be trying to improve mankind. Ever since, society has rejected the idea of tampering with the human genome. But people might find it hard to turn their backs on the possibility of ridding their own families of a mutation that causes the symptoms and suffering of Huntington's or Alzheimer's, cancer, or some other devastating disease.

The Future of Genetics considers where research in genetics, molecular biology, and medicine is headed while trying to cleanly separate facts from fiction and ideologies. The first chapter sets the stage by showing how over the last 150 years different strands of biological research have become interwoven to create a new kind of interdisciplinary science. These trends have been accompanied by works of fiction—from *Frankenstein* to *Brave New World* to *Jurassic Park*—in which authors have explored the social impacts and ethical implications of discoveries. Chapter 2 explores how some of these famous works—and other science-related events—have shaped people's perceptions of science. Chapter 3 presents a range of very new technologies that are giving scientists a broader view of life and providing new ways to manipulate organisms and the environment. The final chapter focuses on some of the most fascinating questions that scientists are posing about the future: the causes of aging and death, the nature of the brain and mind, and the future of life on Earth. Genetics is playing a key role in research into all of these areas; it may also be the gateway to improving people's lives and ensuring that the Earth remains a hospitable place to live.

1

The Origins of Twenty-first-Century Biology

The 21st century opened with the completion of a working draft of the human *genome sequence.* After more than 15 years of intense effort that involved thousands of people working day and night around the globe, researchers finally finished cracking the entire human genetic code. The text consists of only four letters—the chemical subunits called *nucleotides,* or bases, that make up *deoxyribose nucleic acid,* or *DNA.* When 3 billion of such subunits are strung together, they contain enough information to build a person. Just 50 years after the discovery that genes were made of DNA, the code was deciphered. At the moment, the state of that information is more like the contents of a kitchen's shelves than a recipe book. The current challenge is to learn how cells use the ingredients to create human beings and every other form of life on Earth.

The human genome encodes a vast amount of information about human evolution and the way a fertilized egg cell develops into a body. It also holds clues to the origins of diseases such as cancer and the mechanisms behind processes like aging. To make use of this information, scientists will have to learn how the genome produces other types of molecules—*RNAs* and *proteins*—and discover how they control the structure and behavior of cells.

Today's biology is a mix of methods and concepts from fields such as chemistry, physics, genetics, embryology, evolution, mathematics, and medicine. Interestingly, nearly all of these modern fields can be traced back to a series of breakthroughs that occurred within a few decades of the mid-19th century. The best way to understand today's science and its implications for the future is to take a brief look backward and see how these disciplines have become interwoven in the intervening years.

PLANTS, ANIMALS, AND CELLS

In 1833, Professor Johannes Müller (1801–58) moved from the city of Bonn, Germany, to take up a new position at the Humboldt University of Berlin. Until his death 25 years later, he carried out research into human senses and the nervous system while training an entire generation of young scientists. In the early 19th century, Germany, England, and a few other European countries were the world's hotspots of scientific discovery. Several of Müller's students went on to revolutionize—and in some cases create—the modern fields of cell biology, embryology, and medicine.

When Müller moved to Berlin, he brought along a talented student named Theodor Schwann (1810–82). One day at the Berlin train station, Schwann struck up a conversation with Mathias Schleiden (1804–81), a fellow student. Schleiden had studied law in Heidelberg in southern Germany, then moved north to Hamburg to open a law practice, but there he suffered bouts of depression that led to an unsuccessful attempt at suicide. As a result, he carried a bullet in his brain for the rest of his life. Now he had moved to Berlin to start over, this time as a botanist. He too was taking classes with Müller.

Their professor had become an enthusiastic user of a new type of microscope that had just been invented by the Englishman Joseph Jackson Lister. Up to that time microscopes had been limited by problems of blurring and distortion. Lister's innovation was to mount two lenses at fixed distances from each other in a tube, a construction that dramatically increased

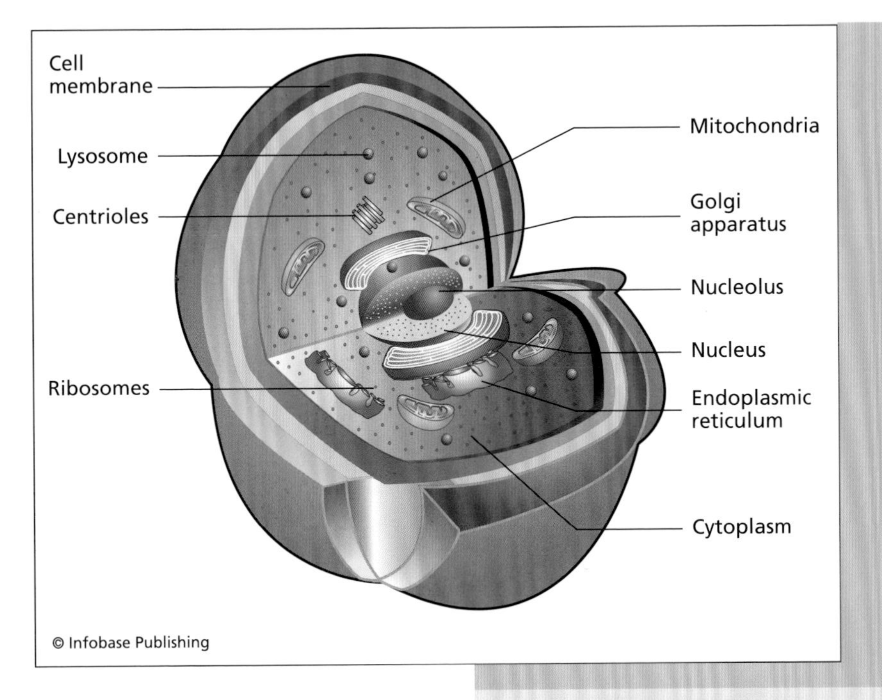

Theodor Schwann and Mathias Schleiden played a key role in the birth of modern biology with their discovery that both plants and animals were made up of cells. New staining techniques soon began to reveal "organelles" and other substructures within the cells.

resolution and gave scientists their sharpest view ever of the microscopic world. Over the course of hundreds of hours studying plant specimens on carefully prepared slides, Schleiden discovered that plants were built entirely of fundamental units—single cells—which somehow formed from the nuclei of other cells.

One night at dinner with Schwann, Schleiden mentioned what he had found. Schwann was less interested in plants, but Schleiden's idea might explain what he had been seeing in animal tissue. The two men left their meal half-eaten and rushed over to Schwann's laboratory. Previously, anatomists had believed that animal tissues were made of fibers, grains, tubes, and other objects. They had been looking at cells, Schwann realized, without knowing it.

Discovering that bodies were made of more fundamental units was a huge leap for science. It gave researchers a new way to look at the formation of embryos, for example—as a collection of cells that developed in different ways to form various types of tissues. But neither man followed the idea to its logical conclusion. That would be the accomplishment of another of Müller's students, Rudolf Virchow (1821–1902), one of the greatest physicians of the 19th century.

Upon graduation, Virchow held a double professorship at the University of Berlin and the Charité hospital. There, he treated patients and carried out research related to fundamental questions about cells. In 1858, he took Schleiden and Schwann's observations a step further with his statement of the doctrine *Omnis cellula e cellula,* "Every cell originates from a similar, previously existing cell." Today, this is such a basic principle of biology that it seems obvious, but at the time many scientists believed that cells could somehow arise by themselves, in a process called *spontaneous generation,* crystallizing from fluids or more basic substances. Virchow's simple new idea had a huge impact because it changed the way scientists thought about all sorts of questions, from the growth of embryos to the nature of disease.

German physician Rudolf Virchow first proposed the theory that every cell arises from an existing cell, which had a tremendous impact on embryology, the study of cancer and infectious diseases, and the rest of the life sciences.

For example, it gave Virchow a new view of cancer. He realized that tumors arose from small pools of cells that divided too often in the wrong places. Removing the source might stop the spread of the disease. He developed new laboratory methods to diagnose cancer and new surgical procedures to treat it. Some of his ideas

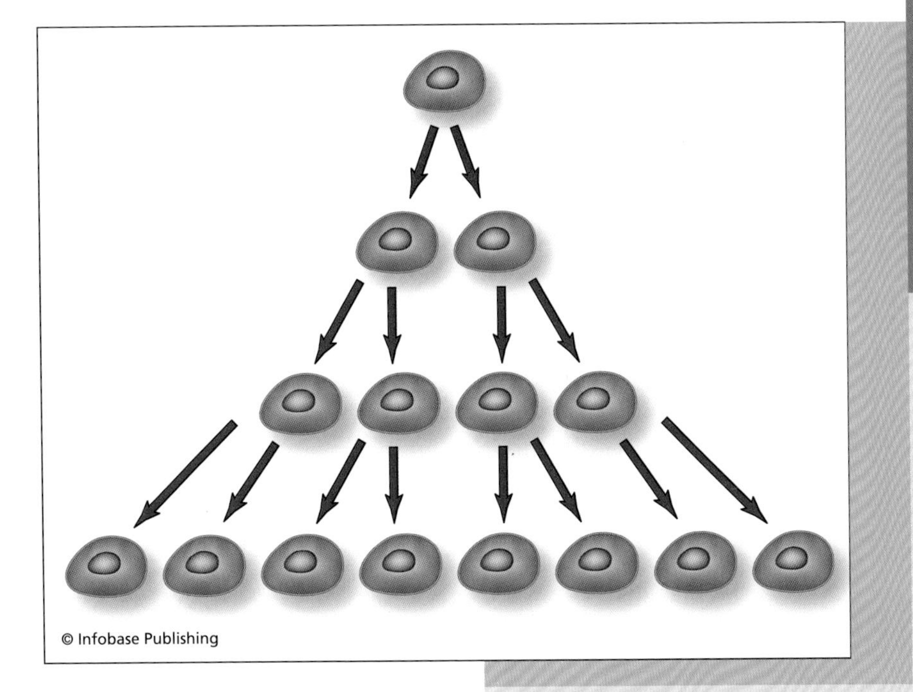

Omnis cellula e cellula—"Every cell originates from a similar, previously existing cell." This applies equally to tissues of the eye, a maggot that suddenly appears in rotting meat, or a tumor cell.

were ahead of their time—his experience with patients, for example, suggested that the disease often arose at sites of injuries or infections. This made him think that tumors might be linked to inflammations—the body's response to injuries. For more than a century, most scientists rejected this idea, but there now is compelling evidence that he was right. Some types of cancer have been linked to infections by viruses, inflammations, and autoimmune diseases in which the body has trouble distinguishing between its own and foreign cells.

Virchow went on to become one of the most famous scientists in the world, a Renaissance man passionately interested in other sciences such as archaeology (he participated in the archaeologist Heinrich Schliemann's first excavations of Troy) and a bold social thinker. Thanks to his efforts, Berlin developed a modern sanitation system that greatly improved the health of the city. He was elected to parliament where he pushed for democratic reforms. This made him such a thorn in the side of Prime Minister Otto von Bismarck that the statesman challenged him to a duel. Virchow's response was to laugh.

CELL BIOLOGY AND MEDICINE

Virchow's discovery of the connection between cancer and cells helped found a new type of medicine. Within a few years, the transformation of medicine into a modern science would be complete with the discovery that bacteria—also cells—were responsible for a wide range of epidemics such as cholera, tuberculosis, and the plague. Two major figures in this revolution were the German physician Robert Koch (1843–1910) and the French scientist Louis Pasteur (1822–95).

The idea that diseases were caused by tiny parasites had a historical precedent. In 36 B.C.E., the Roman scholar Marcus Terrentius Varro had warned people not to build their homes too close to swamps because such areas "breed certain minute creatures which cannot be seen by the eyes, but which float in the air and enter the body through the mouth and nose and cause serious diseases." Arabic physicians of the Middle Ages suggested that microscopic substances were responsible for infectious diseases. But well into the 19th century, most European physicians still held to a *miasma theory of disease,* which suggested that tiny particles of decomposed material floated through the air, accompanied by unpleasant smells, and caused sickness through poisoning. The idea was useful in a way: It encouraged improvements in sanitation that often resulted in better health. Clean air and water usually held far fewer dangerous microorganisms. Attributing disease to bad air, however, was like a sleight of hand in which a magician gets the public to look at the wrong hand.

The widening use of microscopes began to change this situation. In 1840, Friedrich Henle (1809–85), one of Johannes Müller's assistants in Berlin, wrote "On Miasma and Contagia," an essay in which he challenged the prevailing theory. No one had ever proven that miasma existed, he wrote; it was more likely that the air was merely the route taken by tiny living parasites as they moved from one host to another. Disease organisms had not been found, he said, because they looked so similar to the tissues they infected. While Henle had yet to identify such organisms, he was confident that they would be found, and he joined Müller in pushing medical students to spend time at the microscope.

As this new idea of disease was being introduced in Germany, a researcher in northern France was about to perform an important experiment that would help confirm it. Louis Pasteur was a gifted chemist who had become more and more interested in microorganisms. In the early 1850s, he took on the question of *fermentation*—the type of chemical transformation that occurred in the production of wine and other types of alcoholic drinks. People had made use of the phenomenon for thousands of years without understanding why it happened. Pasteur proved that this chemical problem was actually a biological one. Within fermenting liquids he discovered yeast cells. He wrote, "I am of the opinion that alcoholic fermentation never occurs without simultaneous organization, development and multiplication of cells."

Where did the cells come from? Virchow's theory of *Omnis cellula e cellula* was still new. Most researchers were convinced that complex, visible organisms such as maggots came from eggs already found in rotting meat or which had somehow drifted onto it, but the origins of microorganisms were less clear. In 1864, Pasteur published the results of a very careful set of experiments in which he demonstrated that microbes such as bacteria or yeast could not arise in sterile conditions. "Never will the doctrine of spontaneous generation recover from the mortal blow struck by this simple experiment," he wrote.

The next step was to develop a cellular view of disease. Henle had moved to the University of Göttingen in central Germany; one of his students there, Robert Koch, undertook the search for disease organisms in earnest. He began with the bacterium that caused anthrax, a serious disease that affected cows, sheep, and other grazing animals. Humans could catch it by eating meat from infected animals or being exposed to their fur or wool. The bacterium had been discovered in the blood of

This stamp honors the German physician Robert Koch, who was the first to prove a connection between a disease and a specific microbe—the bacterium responsible for anthrax. Koch established criteria that are still fundamental to identifying the causes of infectious diseases. *(Deutsche Reichspost)*

sheep in 1850 by a French researcher named Casimir Davaine, but its connection to the disease was not entirely clear; animals sometimes contracted anthrax without having had contact with other sick animals. Koch's experiments showed that in its normal form the organism could only survive a short time outside of a host, but it was also capable of forming capsulelike spores that could lie dormant on a field for long periods of time. Ingested by grazing animals, the microbes could become active again and trigger the disease. The work established the first definitive link between a bacterium and illness. Thinking that other disease organisms might also survive outside the body, Koch pushed hospitals in Berlin to begin sterilizing their surgical instruments. He went on to improve methods for growing microorganisms in cell cultures and staining them so that they could be seen more easily under the microscope.

Koch's next discovery was *Mycobacterium tuberculosis,* the cause of tuberculosis. Long one of mankind's worst diseases, it was responsible for one of every seven deaths in the mid-19th century. The finding was considered so important that Koch was awarded the 1905 Nobel Prize in physiology or medicine.

He devoted the next several years of his life to trying to develop a sort of vaccine made of extracts from the bacterium. Although it did not work and his career suffered as a result, Koch had a profound impact on the development of modern medicine. He developed a set of conditions, now known as Koch's postulates, which had to be fulfilled to prove that a particular microbe was responsible for a disease. The following principles are still considered fundamental to disease research:

- the microorganism has to be found in every patient or organism suffering from the disease
- researchers must be able to grow it in pure cultures in the laboratory
- even after several generations of growth in the laboratory, it must still be capable of causing the disease
- if it has been artificially introduced in an animal and causes the disease, researchers must be able to extract and culture it again

Armed with Koch's principles and methods, his students went on to find the microbes responsible for typhoid, leprosy, the bubonic plague, and other major diseases.

Louis Pasteur was having much better success in the search for cures. As he worked on a form of cholera that infected chickens, a series of chance events led him to some of the basic principles underlying vaccination. He was growing cholera bacteria in the laboratory and injecting it into the birds. One round of cultures became spoiled. When he tried to use it to infect a new round of chickens, they did not develop the disease. He tried again, using the same birds and a fresh batch of the microbe. They became sick but made a complete recovery. Pasteur reasoned that something about dead or weakened microbes could protect the birds—and possibly people—from future infections.

This phenomenon had been seen before. In the 18th century, physicians had begun inoculating people with bits of tissue taken from the sores of a victim of smallpox, often giving them a milder form of the disease that protected them later. In 1777, concerned that an outbreak of smallpox would threaten the Continental army, George Washington used the method to vaccinate his troops. Historians believe this may have had an important influence on the outcome of the Revolutionary War. Two decades later, Edward Jenner (1749–1823) developed the first true vaccine when he infected people with cowpox. It caused a mild disease but protected them from smallpox, which was much more deadly. The reason that it worked lay in the fact that the viruses that cause the disease are closely related and the immune system does not distinguish between them.

Pasteur began artificially weakening disease organisms in the laboratory for use as vaccines for the prevention of cholera, anthrax, and rabies. In 1885, he made a very risky decision to treat a young boy named Joseph Meister, who had been bitten by a rabid dog. The only treatment available was an experimental vaccine developed by a colleague, and it had only been tested in a few dogs. The boy's cure quickly elevated Pasteur to the status of a national hero. It also convinced the medical world that microorganisms caused illnesses and could sometimes be used to cure them. The germ theory of infectious diseases had come to stay, and it provided doctors with their first effective defense against epidemics that had long haunted mankind.

Still, they are not the solution to every infectious disease, at least not yet. Most modern vaccines are directed against viruses. The development of bacteria-killing antibiotics in the 20th century initially suggested that vaccines against bacteria might not be necessary. This was welcome because attempts to develop vaccines against tuberculosis and many other serious diseases had failed. Researchers have also been unable to develop treatments for HIV and many other viruses because, like some bacteria, they have mechanisms that help them evade the immune system. The AIDS virus slips into the white blood cells that should be fighting it and remains hidden there for a long time; bacteria and other parasites sometimes adopt disguises by changing the molecules on their surfaces. The vaccines needed to fight such clever parasites will have to be more sophisticated.

There is a renewed interest in making bacterial vaccines because many dangerous strains have developed resistance against common antibiotics. Finding cures will require a precise understanding of both the infectious agents and the immune system. The strategies that are being developed to do so may equally be useful in the fight against cancer and some types of genetic diseases. This theme is explored in more detail later in the chapter.

EVOLUTION, GENETICS, AND MATHEMATICS

In 1858, as Virchow was announcing his cell theory in Berlin, an even greater revolution was occurring in Great Britain. Two Englishmen were about to shake the foundations of what nearly everyone thought about the origins of life and its immense diversity around the world. Charles Darwin (1809–82) and Alfred Russel Wallace (1823–1913), living on opposite sides of the globe, had come to nearly identical conclusions. Just as each cell arose from a preexisting cell, each of the Earth's species had evolved from an earlier form of life, in an immensely long, unbroken chain stretching back to the first cell.

Evolution was based on a few logical principles that could be observed nearly everywhere:

- *Variation*—Each member of a species (except for identical twins) has slightly different features
- *Heredity*—Some of these unique characteristics are passed down from plant or animal parents to their offspring.
- *Natural selection*—In a given environment, some features give certain members of a species an advantage at survival and reproduction. If this bias continues over many generations, more and more of the population will be made up of their descendants, until they dominate the entire species.

Different environments—with unique climates, types of food, predators, and other factors—would favor different features. The genius of Darwin and Wallace was to see how, over long periods of time, tiny variations within a species could transform it into new ones.

There was little debate about the single points of the theory; what worried religious leaders and others was that, taken

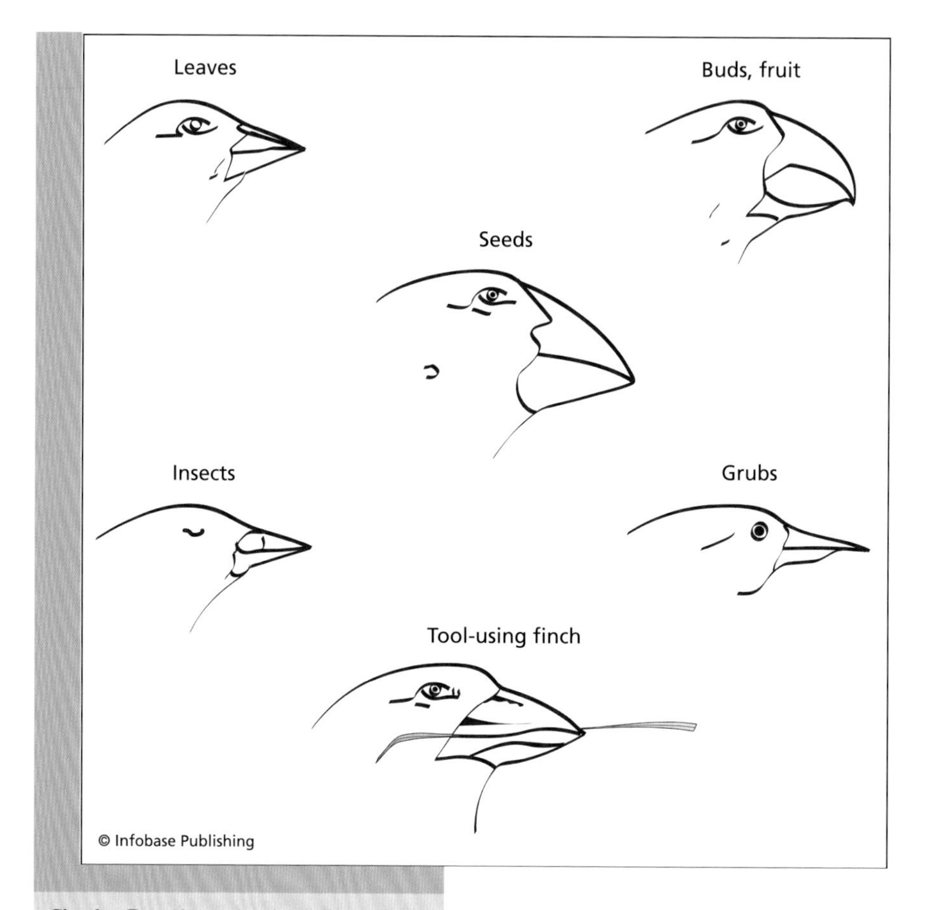

Charles Darwin noticed small differences in several species of finches living on the Galápagos Islands in the Pacific—a perfect example of natural selection. Mutations caused small changes in the shapes of beaks, which allowed the birds to exploit different sources of food. Over time the birds whose beaks were best suited to their lifestyle survived better and had more offspring, and now specific beak-building genes dominate in each population.

together, they threatened the idea of a miraculous creation, in which humans and other species sprang fully formed from the mind of a creator. For the first time, scientists had a coherent theory to explain the diversity of life—and possibly even its origins—that could be tested in various ways.

Heredity was fundamental to the theory, but in the 1850s no one had any idea of how features were transmitted from parents to their offspring. Darwin's own speculations on the subject—that parents' traits were combined in a souplike mixture—were quickly proven wrong by his cousin Francis Galton (1822–1911). It was not a blow to the theory: Any mechanism for heredity would work, as long as it produced children who were neither perfect copies of their parents nor completely unlike them. If those conditions were met, selection would act on a species.

As the powerful framework of evolution began changing the way biologists looked at life, a reclusive monk named Gregor Johann Mendel (1822–84) was quietly going about solving some of the major questions about heredity. He succeeded at a problem that had stumped the greatest scientific minds of history for several reasons. First, he worked with peas and other plants in which reproduction could be carefully controlled. Second, he started from the assumption that various aspects of an organism—such as the shape and color of a seed—were separate features that were inherited independently of each other. Finally, he was talented at statistics, becoming one of the first researchers to apply mathematics to a biological problem. Exact numbers would be necessary to discover the laws that governed inheritance.

Mendel's work revealed that in most species, males and females each contributed one hereditary unit (later called a gene) to their offspring. These units might be the same or different—for example, both parents might pass along a gene that made peas yellow, or one might contribute a gene for yellowness while the other passed along greenness. Different forms of the same gene would be called *alleles.* For example, there was a gene that determined a pea's shape, with wrinkled and round alleles. A single plant might inherit two alleles for one gene. If that happened, Mendel concluded, one of them would be *dominant* and the other *recessive.* In peas, the green allele was dominant over yellow.

Darwin never read Mendel's work. The monk, on the other hand, was aware of evolution, but he did not see its immediate relevance to the questions he was investigating. Given the tools and concepts at hand, he could only look at one type of variation—alleles—and the focus of his work was why plant offspring were so similar to their parents, rather than where new traits might come from. To make a comparison: If the two men had been studying languages, Mendel would have been focused on small differences between the accents of two English speakers, whereas Darwin wanted to understand the relationship between French and Latin, or English and German. Different approaches were necessary, and this gap between genetics and evolution would not be closed for many decades.

Mendel did not live to see it happen. On the advice of another scientist, he tried to reproduce his results in another plant; it turned out to be difficult to handle and the experiments yielded confusing results. He began to doubt his own work just at the time he was appointed head of the abbey where he lived, and he spent the last years of his life immersed in its business.

The importance of his findings remained unappreciated until they were rediscovered at the beginning of the 20th century by three scientists working independently on similar problems in different countries: a Dutchman named Hugo de Vries (1848–1935), a German named Carl Correns (1864–1933), and an Austrian researcher, Erich Tschermak von Seysenegg (1871–1962). William Bateson (1861–1926), a British scientist, also played a vital role in bringing Mendel's ideas to the world and transforming them into a new science, which he called genetics. Bateson first encountered Mendel's name through a paper written by de Vries that he read while on a train to London. Bateson went on to demonstrate that Mendel's principles held true for animals as well as plants. He also believed they might provide a way of linking heredity to evolution.

This turned out to be more difficult than anyone expected, mainly because of the issue of variation. Genetics could explain part of it—for example, how two plants that produced green peas might give rise to a plant with yellow ones. But this happened because the trait for yellowness was already there. The focus of genetics was how existing alleles were shuffled from parents to their offspring. Evolution needed something more: an explanation for the appearance of completely new traits. Human beings were not just a peculiar arrangement of alleles that had once existed in bacteria and were simply being shuffled around in new ways.

De Vries proposed a solution with the concept of *mutations*: mistakes that occurred as genes were copied or during their passage from parent to child. Two decades later, American scientists would discover another type of change that could occur: Genes sometimes were duplicated, and offspring could inherit extra copies. This provided extra genetic material for evolution to work on and could partly explain how very complex organisms might arise from much simpler ones.

This discovery came from the research team of Thomas Hunt Morgan (1866–1945), who had established a laboratory at Columbia University in New York with the intent of catching evolution in the act. He planned to breed a species until mutations occurred, then study how the changes moved through the population. To do so, he needed an organism that reproduced quickly, had lots of offspring, and was easy to take care of. A colleague recommended the simple fruit fly *Drosophila melanogaster,* which could be raised in glass beakers and fed on mashed bananas.

For two years, Morgan and his students raised flies without discovering any mutations. He may have been about to give up when in 1909 they began to appear—subtle differences in the color of the insects' bodies and eyes. The fact that they were transmitted to their offspring along Mendelian patterns proved that genes were responsible. Each new mutation revealed the existence of a new gene—if a fly suddenly developed white eyes, it must have undergone a mutation in a gene that normally made them red. No one knew how many genes an animal had or how they worked, and Morgan's focus quickly shifted away from evolution toward these themes. Over the next three decades, his laboratory found dozens of new genes and identified their positions on chromosomes. It would be a long time before the cause of mutations would be understood—spelling changes in the chemical language of DNA—but the group discovered that genes could be duplicated, inverted, or undergo other types of changes.

Mutations seemed to be abrupt breaks in hereditary information, like computer files that had become corrupted. Bateson believed that a small number of such events could quickly give rise to new species, but initially many of his colleagues disagreed. Darwin had seen evolution as a very slow process that gradually bent, twisted, and stretched the existing features of organisms into new forms, rather than quickly replacing them with something different. For example, in a species of antelopes, some animals would inevitably have slightly longer horns than others. If longer horns made them more attractive to females or otherwise led them to have more offspring, then length would undergo positive selection over many generations. This vision was probably largely due to Darwin's fluid mixing hypothesis of heredity, which offered no explanation for the arrival of completely new features.

The debate arose from a misunderstanding about how genes functioned. Genes could encode quantitative features like an animal's height, weight, or the length of its horns; but a single mutation could also have very dramatic effects—giving a fish two heads or a goat an extra pair of legs. The reasons would not become clear until many years later, with the discovery of the connections between cell chemistry and embryonic development. In the early 20th century, the disagreement had to be resolved by mathematicians. Evolution had attracted their attention because it raised interesting statistical questions—for example, how a mutation that happened in a single organism could spread through a population. Many were skeptical that such single events could spread far enough to become visible to natural selection, especially since many mutations were recessive. This meant that two parents would have to have an allele before it appeared in their offspring. George Udny Yule (1871–1951), a Scottish statistician, predicted that dominant genes would multiply in a population and wipe out recessive traits. But the American geneticist William Castle (1867–1962) calculated that with no natural selection at all, the frequency of alleles in a population would remain stable over many generations.

Darwin had predicted that natural selection would work hardest when the pressure on a population was extreme—when a high number of predators lurked in the neighborhood, when populations exceeded the food supply, or when environmental conditions changed. In animals such as birds, the pressure would be high all the time—he calculated that in just 200 years, eight pairs of swifts could produce 10,000 billion billion billion descendants if nothing kept them under control. But what about animals that were less fertile? Darwin made some calculations based on one of the slowest-breeding creatures on Earth, the elephant. If a pair bred between the ages of 30 and 90 and had only six offspring, they would produce 19 million descendants after only 750 years. That was far beyond the real population, and elephants had been around for much longer. Natural selection was clearly working on them as well. The question was how to detect it.

The mathematicians Godfrey Hardy (1877–1947) and Wilhelm Weinberg (1862–1937) independently came up with the same answer, based on statistics and probability. Their formula to describe the spread of an allele, now called the Hardy-Weinberg rule, consists of the following steps:

1. Determine the frequency of specific alleles among the adults in a species.
2. Find out which types of adults mate with each other.
3. Estimate the frequency of alleles among their offspring using Mendel's ratios.
4. Discover how many of the offspring survive to reproduce.

The formula could be used to test hypotheses about populations and evolution. It verified William Castle's prediction that in a large population with random mating and no natural selection, the genetic makeup would stay the same over many generations. Recessive alleles could survive for a long time; they would not be wiped out by dominant ones. The method could also be used to detect natural selection. If the proportions of alleles in a population changed significantly over many generations, it was a sign that something was favoring some forms of a gene over others.

The Hardy-Weinberg rule still did not address the big question of whether evolution was driven by mutations or by very gradual changes in features. That problem would be tackled by

Ronald Fisher (1890–1962), a talented mathematician who developed an interest in evolution at an early age. As an undergraduate, he began thinking about ways to use mathematics to bring evolution and genetics together. In one project, he showed that evolution could not be primarily driven by new mutations that happened all the time; instead, once a change in a gene occurred, the new allele became part of a species' gene pool and behaved like any other allele. At the beginning it would be rare, but if it offered a reproductive advantage, it might spread quickly.

Calculating rates of species change required numbers, so Fisher invented a value called variance. In his 1930 book, *The Genetical Theory of Natural Selection,* he also introduced a concept of *fitness,* meaning a measurement of how well a species is adapted (or not) to its environment. For most species, which had been shaped by millions of years of natural selection, this value would be high. But it could change because species were molded by the environments of the past, rather than the present. Human fitness, for example, was the product of the hunter-gatherer lifestyle practiced for 99 percent of human history, rather than the circumstances of modern industrial society.

Fisher's contemporary Sewall Wright (1889–1988), an American geneticist, saw selection happening within a fitness landscape, an imaginary place of peaks and valleys. The purpose of this metaphor was to describe how selection could change a species' profile over time. Most individuals in a species would be near the peak—an allele that was best adapted to the environment, with other variants of a gene falling off in a slope, and some vary rare alleles in the valleys. But changes in the environment might suddenly favor different characteristics. This would shift the ideal position of the peak and cause the number of infrequent alleles to rise.

Another important figure in the coalescence of genetics and mathematics was John Burdon Sanderson Haldane (1892–1964), one of the most colorful figures of 20th-century science. During his youth, he carried out experiments with his father, such as studies of the effects of air pressure on the body, and he used himself to experiment on. This damaged some of his vertebrae

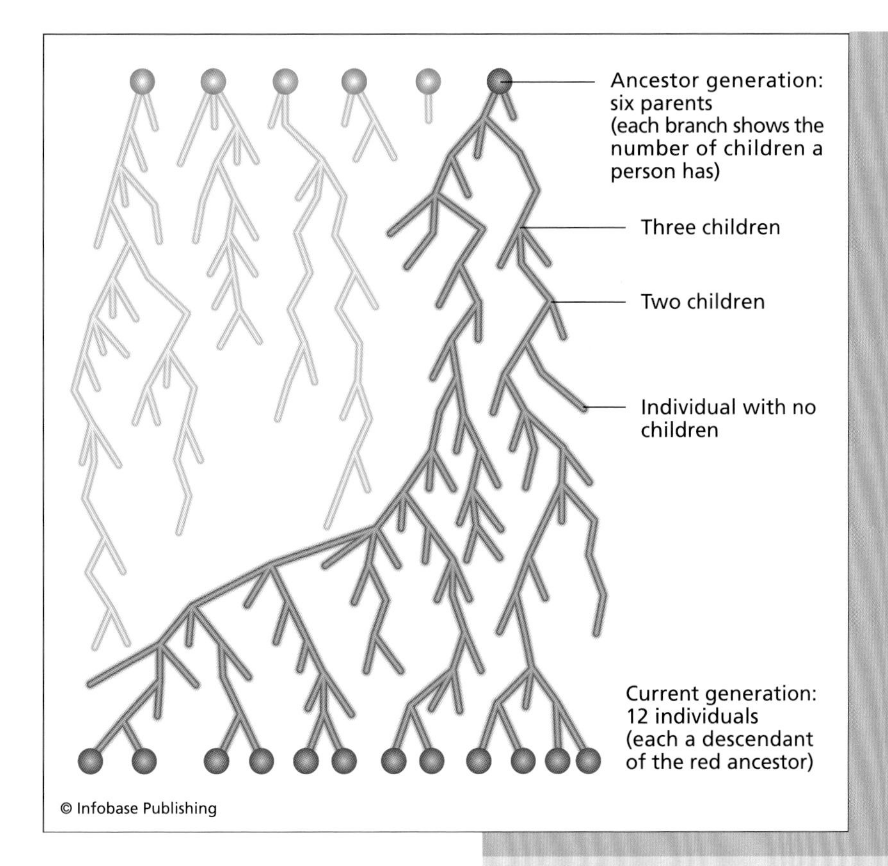

This diagram shows how a small advantage in reproduction of some individuals and their descendants can have a big effect on a species over time. The family line at right (red) consistently has more children than the descendants of the five other ancestors at the top (blue). The differences may not be obvious in a single generation, but over time this family's genes may come to dominate an entire species.

and left him with a perforated eardrum. The latter was not a particular problem, he wrote: "The drum generally heals up; and if a hole remains in it, although one is somewhat deaf, one can blow tobacco smoke out of the ear . . . which is a social accomplishment."

In 1924, Haldane began a series of scientific papers called "A Mathematical Theory of Natural and Artificial Selection." He calculated that even if a trait offered only a very small reproductive advantage to the organism that inherited it, as little as 0.1 of 1 percent, with enough time it could become

The Failed Marriage between Evolution and Sociology

Although 150 years of research into evolution has confirmed evolutionary theory in a lot of ways—including some that Darwin never envisioned—it is still widely misunderstood. The misconceptions are most obvious in the way people have tried to apply Darwin's ideas to human society. Evolution arrived on the scene at a time when industry and technology were rapidly changing the Western world. Many people were obsessed with progress and believed that it would lead to a utopian society. Evolution brought the unwelcome messages that humans were not the ultimate goal of creation and that history was not driving their species toward physical and moral perfection. It did not necessarily produce creatures that were more complex or intelligent than their ancestors, and species could fail—by becoming extinct.

Whereas evolution has been a unifying concept in science, helping to draw biology close to physics, chemistry, mathematics, and all the other disciplines described in this chapter, it has had a much harder time finding common ground with the social sciences. Part of the reason lies with what was happening in the world of research and British society in 1858, the year that *On the Origin of Species* was published.

At that time the word evolution existed, but had a different meaning, referring to the development of embryos into newborns and adults. That process clearly had a destination. A chicken egg produced a chick, and a baby in a human mother's womb gave rise to a human being. Herbert Spencer (1820–1903), a British philosopher and political and sociological theorist who was quickly becoming one of the most important thinkers of the 19th century, felt that human society was undergoing a similar developmental transformation.

Spencer described scientific progress as a method of trying out ideas and discarding those that did not work in favor of better ones. He proposed that the entire universe might behave the same way, moving from a state of simplicity to more complexity, from imperfection to higher order—not in a religious sense, but simply because natural laws worked that way. Since culture was a product of human beings, the laws that governed their biology must also dictate the development of culture. In an 1857 article called "Progress: Its Law and Cause," he wrote, "Whether it be in the development of the Earth, in the development in Life upon its surface, in the development of Society, of Government, of Manufactures, of Commerce, of Language, Literature, Science, Art, this same evolution of the simple into the complex, through a process of continuous differentiation, holds throughout."

When Darwin's book appeared, Spencer quickly became one of its strongest supporters. He believed it provided a firm biological foundation for his own ideas. But like many others, he was unable to give up the idea that humans were the pinnacle of evolution. He coined the phrase *survival of the fittest* to describe natural selection. Darwin was uncomfortable about this because he knew that fitness—the way Spencer meant it—was a loaded word. It assigned human values to the natural world and was a disguised attempt to marry concepts of improvement and progress to natural selection. In society, success meant the acquisition of wealth, power, and prestige. In evolution, success meant something completely different—living long enough to have more fertile offspring than other plants or animals of the same species.

(continues)

(continued)

Many people who read Spencer's books saw the parallels to Darwin's account of biological evolution without understanding the important differences, and this would have an important impact on the relationship between biology and sociology over the next century. Like most other philosophers, Spencer could not bear the idea that humanity's future was a matter of chance. Left on its own, without governments to intervene, human society would progress by favoring stronger and healthier individuals—not necessarily the rich, because he realized that poorer social classes were not really responsible for the conditions in which they lived. On the other hand, some of these people were clearly unfit through idleness or incompetence. Spencer believed that for such people to starve or suffer was a natural process.

It is not hard to see how these ideas could be turned against the poor, the sick, or groups that were considered somehow unfit by those in power. Spencer opposed charities and donations to the poor on the grounds that they ran against the principles of selection and promoted the survival of the unworthy. "The quality of a society is physically lowered by the artificial preservation of its feeblest member," he wrote. Helping defective people survive could harm society and perhaps even the human race. Taken to its extreme, this idea eventually led to *eugenics* programs—initiatives to control human mating. These programs were based on a complete misunderstanding of human genetics, and their disastrous consequences are discussed in the next chapter.

frequent in a population. He plotted what might happen after thousands of generations. A comparison of mathematical predictions to real measurements of allele frequencies would reveal how much of an advantage any particular mutation provided.

Haldane extended his calculations to unusual hypothetical cases, such as the following. Suppose that two recessive genes that affected eyesight were circulating in a population. On its own, each led to poor eyesight, but if someone inherited both genes, there would be an improvement. Haldane could predict whether the trait would survive and how frequent it might become.

By the 1930s, theorists had brought genetics and evolution together in what became known as the modern synthesis. But huge questions remained. No one knew what genes were made of, or why they led organisms to develop red eyes, five fingers, or any of their other features. In the meantime, geneticists and evolutionary researchers had begun to think about the implications of this new type of science on their own species.

CELL BIOLOGY, CHEMISTRY, AND GENES

If one had to pick a single icon to represent modern biology, it would surely be the DNA double helix: the spiral staircase–like ladder of sugars and nucleotides that contains the information needed to make a bacterium, plant, fly, or human being. This model of the molecule, proposed in 1953 by James Watson (1928–) and Francis Crick (1916–2004), revolutionized biology by showing how DNA could be copied and how mutations could arise—essentially proving, overnight, that genes were made of DNA. It also hinted at a way that genes might influence the structure and behavior of cells and, thereby, the formation of animal bodies.

The model could only be built because of a coming-together of chemistry, physics, and biology that had been under way for nearly a century. Until just a few years before Watson and Crick's discovery, many believed that proteins carried species' hereditary information, in spite of evidence in favor of DNA that had been accumulating since the late 19th century.

In 1866, the German biologist Ernst Haeckel (1834–1919) suggested that hereditary material might be found in the cell nucleus, whose function was unknown up to that time. Two years later, while trying to find a way to remove and study the nucleus, a Swiss biologist named Friedrich Miescher (1844–95) achieved the first extraction of DNA from white blood cells. Its chemistry was odd compared to that of other cellular molecules; for example, it contained large amounts of phosphorus, which was virtually unknown in other organic molecules. When Miescher submitted his results to a scientific journal, they were so unusual that the editor insisted that the experiments be repeated before agreeing to publish the discovery.

Microscopists were also gathering evidence that the cell nucleus played a key role in heredity. Oscar Hertwig (1849–1922), professor of zoology at the University of Berlin, looked at the huge, pearly white eggs of sea urchins and discovered that a sperm cell brings a new nucleus into the egg, which then fuses with the egg's own nucleus. The rest of the sperm is discarded. Three years later, his countryman Walther Flemming (1843–1905) stained nuclei with dyes and saw chromosomes for the first time. He watched as cells divided and discovered that the chromosomes were split up among the two new daughters. But Mendel's work was still unknown and without his data—showing that both parents contributed equally to heredity—the importance of these new findings was not immediately clear. Wilhelm Roux (1850–1924) and August Weismann (1834–1914), also German professors, figured out that fertilization is a process of combining chromosomes from each parent. These threads, Roux wrote, must contain the hereditary material, and he proposed that the information they contained was in a linear form, like the words of a text.

(opposite page) An organism's germ cells—sperm or egg—are among the first cells to develop in the embryo. They arise before the development of sex organs and migrate as these organs form. These images show the location of germ cells (yellow spheres) at various stages in the development of a fruit fly larva. Red shows the tissue that will become the fly gut, and green is tissue that will become sex organs. By maturity, the cells have taken up a position next to the future sex organs.

Weismann tried to pull all of these observations into a single theory. He believed that organisms maintained reproductive *germ cells* separate from the rest of their cells (which he called the soma), and this helped explained why organisms did not pass

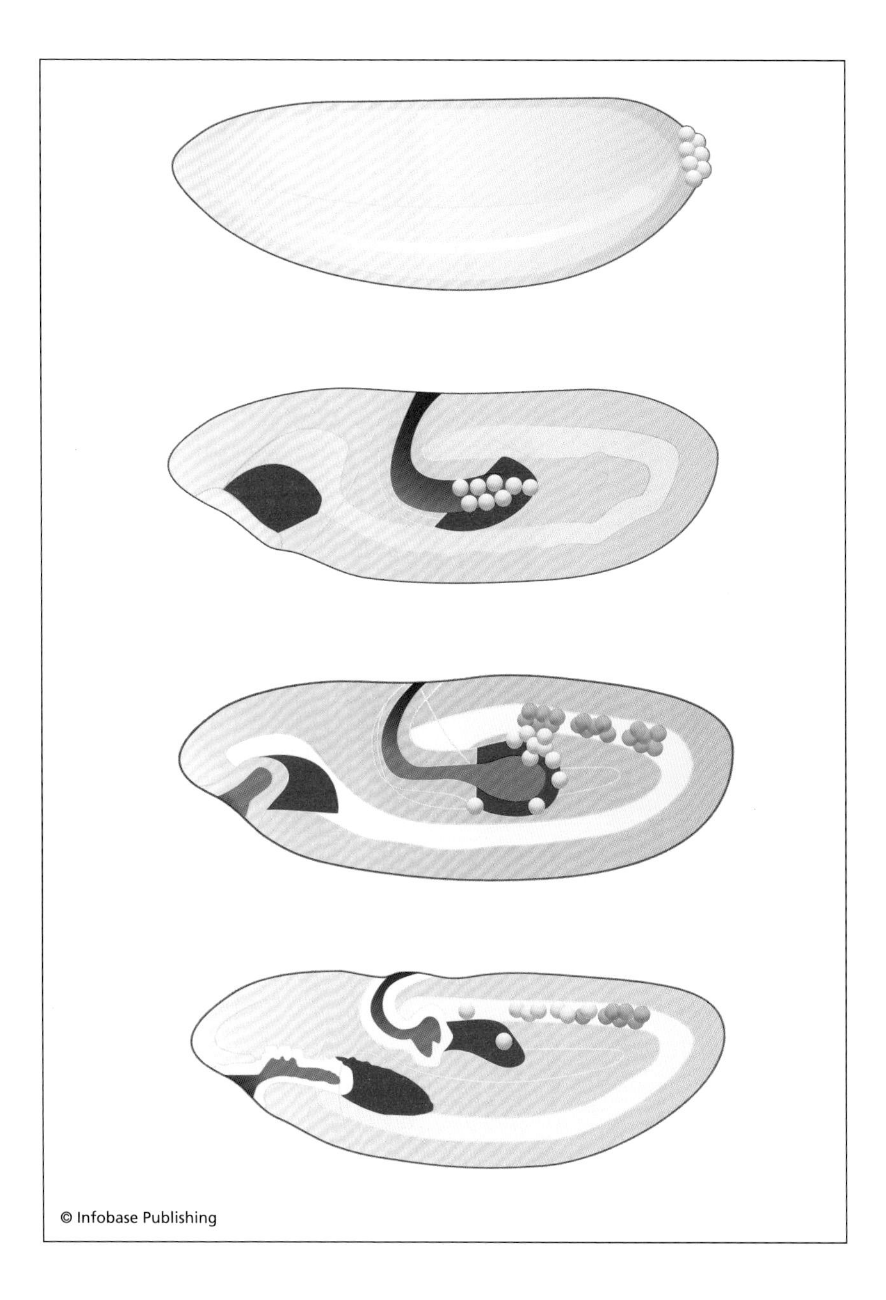

along traits acquired during their lifetimes to their offspring. This idea was central to evolution but was still controversial among scientists, many of whom felt that natural selection was a severe and amoral system. Weismann put it to the test with an experiment in which he cut off the tails of mice for several generations in a row. If Darwin was wrong, he reasoned, the mice would eventually produce offspring with no tails. But this never happened. Neither behavior nor lifetime events affected the protected germ cells.

Weismann believed the material in these cells, which he called the germ plasm, would be the key to understanding heredity. Whatever the substance was, it was passed along intact from generation to generation, separate from the rest of the body. The soma was like a flower which grew and died within a year; the germ plasm was like the body of the plant, which survived season after season. The function of sex was to mix up the germ plasm of separate organisms, ensuring variety within species.

With the rediscovery of Mendel's work, the search for the physical location and the chemical nature of genes was ready to begin in earnest. At the turn of the century, the American geneticist Walter Sutton (1877–1916) proposed that genes were located on chromosomes. Working at the same time in Munich, Theodor Boveri (1862–1915) discovered that if more than one sperm managed to fertilize an egg, the resulting embryo had too many chromosomes, failed to develop, and died at a very early stage. This led him to an important conclusion: An organism needed not only a complete set of chromosomes, but also the right number. Through a series of experiments conducted between 1901 and 1905, he became convinced that each chromosome possessed unique qualities. Each contained a unique subset of the instructions needed to build an organism. Too many chromosomes meant too many instructions, and too few meant that important information was missing. The next task was to try to discover which information was stored where. An important first step came in 1905, when the American biologists Nettie Stevens (1861–1912) and Edmund Wilson (1856–1939) discovered that the X-Y chromosome pair contained the genes that determined an organism's sex.

If chromosomes contained only DNA, scientists would surely have realized much more quickly that genes were made of this molecule. However, the DNA in a cell nucleus is linked to a huge number of proteins in a mixture called chromatin. Many researchers were convinced that the genetic code was contained in the proteins, which were complex molecules built of 20 amino acids. The chemical language of DNA was much simpler, made up of only four nucleotides—too simple, perhaps, to produce complex organisms.

A major step forward came through the work of Frederick Griffith (1879–1941), a medical officer at the Ministry of Health in London. He was studying two strains of bacteria that were very similar, trying to figure out why one caused severe pneumonia in humans and the other did not. There was only one obvious difference: the infectious, smooth (S) form of the bacterium built a capsule around itself, while the rough form (R) did not. Griffith inoculated mice with a mixture of dead S-type and live R-type bacteria. He expected that the mice would stay healthy and the bacteria would die, because he had not injected the animals with any live infectious cells. But when he drew blood he found S-type bacteria that were alive.

Either the S type had somehow been brought back to life or something had changed the R bacteria into the S type. If the latter was the case, it meant that R bacteria were acquiring new hereditary information. Griffith began a new round of experiments to try to find out what this transforming substance was made of. One possibility was that fragments of proteins from the S bacteria were somehow being absorbed into R bacteria and were being used to build capsules, but Griffith had another idea. Rather than receiving building materials, the bacteria might be receiving the instructions to make the capsules. In other words, R bacteria had developed the capacity to make a new protein.

Griffith's investigations ended with his tragic death when London was bombed by the Nazis in 1941. But his work had attracted the interest of another scientist. Oswald Avery (1877–1955), a physician and researcher at the Rockefeller Institute in New York, was trying to develop a vaccine for pneumonia.

That work became unnecessary through the discovery of antibiotics, which very effectively killed the pneumonia bacteria. But the project gave Avery what he needed to follow up on Griffith's experiments, which looked like the most promising way to find bacteria's hereditary material. Members of his lab purified molecules from the S type and showed that DNA alone was able to transform the R type into infectious pneumonia bacteria. Avery cautiously proposed that in bacteria, DNA was the hereditary material, and that perhaps this was true of other forms of life as well. Yet other researchers remained skeptical.

One person who believed him was Erwin Chargaff (1905–2002), an Austrian working nearby at Columbia University in New York. He wrote, "Avery gave us the first text of a new language, or rather he showed us where to look for it. I resolved to search for this text." If DNA was truly the language of heredity, it could not be the same in every species, so Chargaff began trying to find differences in DNA.

He started out by simply comparing how much of each of the four bases could be found in yeast cells and the tuberculosis bacterium. By chance, he had chosen two organisms with major differences in composition of their DNA. Yeast had high amounts of A and T but much lower amounts of G and C, exactly the opposite of the bacterium. Chargaff tried the same thing with other organisms and found that each had its own particular recipe of DNA. In humans, for example, about 30.5 percent of DNA was A, 31.8 percent was T, 17.2 percent was C, and 18.4 percent was G. The tuberculosis bacterium gave a much different picture: 15 percent A, 13.6 percent T, 34 percent C, and 37.4 percent G.

The fact that each organism had its own recipe of bases meant that DNA might be the molecule of heredity. Chargaff noticed another curious fact: In any given organism, A and T were found in almost identical amounts; the same was true of G and C. Although he did not realize it, these numbers provided one of the most important clues as to how the DNA molecule was put together. It would not be explained until James Watson and Francis Crick understood DNA's structure.

PHYSICS, CHEMISTRY, AND GENETICS

Alongside Virchow's presentation of the cell theory and the first announcement of evolution, the year 1858 saw an important breakthrough in chemistry. Archibald Scott Couper (1831–92) and Auguste Kekulé (1829–96) drew the first blueprints of molecules: diagrams showing the positions of atoms and their relationships to each other. DNA, RNA, and proteins are fundamental units of life, but the atoms that make them up are even more basic. As any engineer knows, the function of a machine depends on the way its parts are assembled, and the same is true for molecules. By the mid-20th century it had become clear that understanding genes would require learning about their chemical makeup and physical structure.

Chemists knew that DNA consisted of a sugar called deoxyribose, plenty of phosphate atoms, and the four nucleotide bases. Each component has a particular shape and chemistry that determine how it snaps onto the others. With very simple molecules, it is sometimes possible to guess how the parts fit together just by looking at the chemistry of the subunits, but in this case there were too many ways that the pieces might fit.

The details of DNA and other molecules such as proteins were too small to be seen through even the most powerful electron microscopes, so chemists were trying to understand DNA's structure by watching how other molecules changed it—a bit like ramming cars into each other to study their engines. *Crystallography* took another approach, turning molecules into crystals and exposing them to X-rays. This method, developed by physicists, had provided some important information about the shapes of proteins; perhaps the same thing would work with DNA.

When an X-ray beam passes through an object, some of the waves collide with atoms' electrons and are diffracted (they scatter off in a new direction). William Astbury (1898–1961) shined X-rays through molecules and captured the scattering patterns on photographic plates. Usually the resulting image was an unreadable smear. But a molecule whose atoms were arranged in precise, repeated groups scattered the waves in the same directions over and over again, creating a symmetrical pattern that hinted at the shapes of molecules. Astbury had been trying this with proteins that had formed crystals. In crystals, molecules are often arranged in precise lattices that repeat over and over again, billions or trillions of times. This creates the regular structures necessary to obtain a clear diffraction pattern.

Very pure DNA could either be made into crystals or pulled into fibers that also provided regular diffraction patterns. When Astbury examined DNA fibers with X-rays, he obtained some basic information about the size and architecture of the molecule. His interpretation was that the bases fit together into flat disks, squeezed very tightly together like dinner plates stacked in a column. He could measure the diameter of the disks and the height of each plate. However, many of the details remained blurred; without knowing it, he was working with two different forms of DNA. In his images they were superimposed.

The problem interested the great American chemist Linus Pauling (1901–94), who carried out similar experiments in his laboratory at the California Institute of Technology. He proposed a structure for DNA showing the molecule as a braid of three strands organized in a helix, like a spiral staircase with three handrails. It was one of the few times Pauling was wrong. Considered to be one of the greatest chemists of the 20th century, he had already used crystallography to study the composition of proteins; this work earned him a Nobel Prize in chemistry in 1954. Eight years later he became only the second person in history to win a second prize in a different category (the other was Marie Curie). This time it was the 1962 Nobel Peace Prize, for his efforts to stop the testing of aboveground nuclear weapons.

He paid a price for his political activities. In 1952, he had been denounced as a communist before the House Committee on Un-American Activities. When Pauling wanted to attend a scientific meeting of the Royal Society in London, he was refused a visa. One of his colleagues, Robert Corey, went instead. During the trip, Corey met with a young researcher named Rosalind Franklin (1920–58), who was also using X-rays to investigate DNA. It is hard to tell what might have happened had Pauling attended the meeting; he might have obtained data that would have led him to an accurate model of DNA. Franklin's work was about to play a crucial role in figuring out the molecule's structure.

Another incorrect model had just been proposed by the British scientist Francis Crick and his young American partner, James Watson, working in Cambridge, England. Watson had obtained his Ph.D. at the age of 22, working on viruses that infected bacteria at the University of Indiana, and had come to Cambridge determined to solve the riddle of DNA's structure. He was now 23 and Crick was 35, but the two men quickly recognized each other as two of the brightest people on campus and hit it off. They had a lot of catching up to do when it came to DNA; neither was an expert in chemistry. Their first diagram of the molecule was so wrong that it embarrassed their boss, Sir Lawrence Bragg, and he ordered them to stop working on it.

Meanwhile Franklin, an hour away by train in the laboratory of Maurice Wilkins in London, had solved a major problem regarding the X-ray images of DNA. She had figured out that DNA came in two forms: a dry and a wet form. Under humid conditions, more hydrogen atoms were packed into the molecule and that changed its shape. The preparations of the molecule made by Astbury and Pauling held both forms and caused blurring. Using only the B form, Franklin now obtained the sharpest-ever images of DNA and began trying to interpret what they meant about its structure, but interrupted the work to go on vacation. While she was gone, Wilkins showed some of her X-ray images to Watson. One look convinced him that DNA formed a double helix.

The problem that now faced Watson and Crick was like one of those wooden puzzles in which oddly shaped pieces have to be fit together to form a tight, geometric shape. In this case, the shape that had to be built was a helix, and the pieces were sugars, phosphates, and bases. Watson made cardboard cutouts in the shape of the four bases and began working on the puzzle. No matter how he tried to attach them to each other, something always bulged outside the helix. He was stuck until his office mate—ironically a former student of Pauling's—told him that bases existed in two different chemical forms, with slightly different shapes. Watson had been using a form with an extra oxygen atom, so now he remade the shapes without the oxygen. He was idly fitting them together when he had a sudden revelation: When A snapped onto T, it had almost exactly the same size as G fit to C. Fit together, their size matched the dimensions of the helix in Rosalind Franklin's X-ray photographs.

Watson showed Crick what he had discovered. They immediately realized that the steps of the DNA spiral staircase were the bases, rather than the sugars. Each step held either an A combined with T, or a G with a C. The steps were connected by winding rails of deoxyribose sugars (the backbone). Between each of the steps, there was a slight twist, making the whole structure into a helix rather than a straight, ladder-like column. They quickly wrote a paper called "A Structure for Deoxyribose Nucleic Acid" and submitted it to the journal *Nature.* It was published three weeks later—an amazingly short time, given the fact that it first had to be read and commented on by experts.

This brief article would revolutionize biology because the molecule's building plan provided immediate insights into its behavior. It explained Chargaff's discovery that A and T occur in identical amounts in an organism, as do G and C. The pairing of the bases revealed how DNA might copy itself. If the two strands of DNA were split apart, each base would attract and link up to just the right partner nucleotide, creating a second strand. The article even suggested a way that mutations could occur, in spite of the fact that bases formed regular pairs. In rare cases, hydrogen atoms might bind differently to a base, slightly changing its shape. As one strand was copied, it might then attach to the wrong base.

A crucial point was that any sequence—any possible spelling of the four bases—formed the same shape. A long strand made up only of As joined to Ts would create the same double helix as a sequence consisting only of G-C pairs. Each organism could have its own DNA sequence; a language with four letters was rich enough to create all the diversity of life on Earth. Evolution was built on a single scaffold.

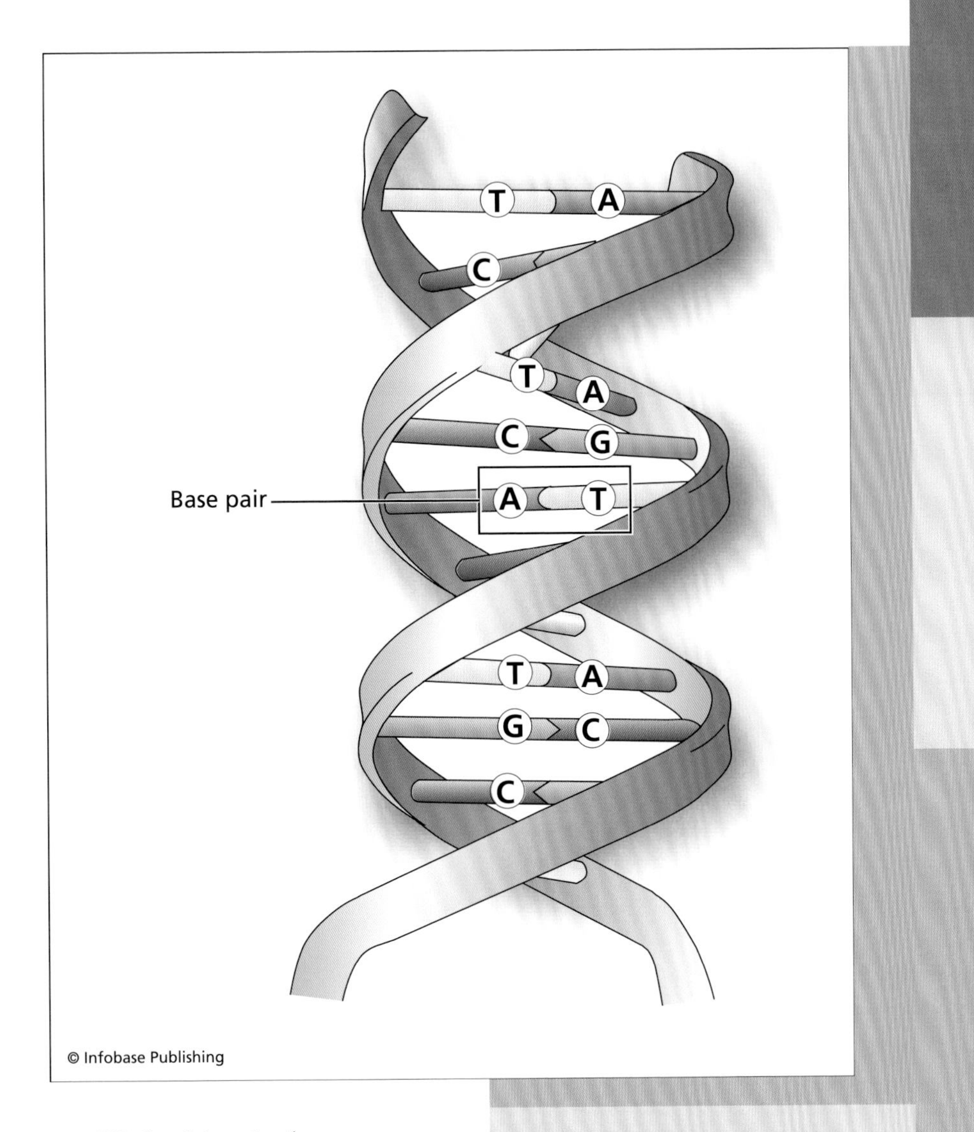

Watson and Crick's double helix model of DNA, based on data from physics and chemistry experiments, showed that bases bind in complementary pairs. This demonstrated that genes were made of DNA, explained how the molecule could be copied, and also suggested a reason for mutations.

With this single, powerful image, some of the most important questions about genes, cell replication, and evolution were resolved, all at once. Nine years later, Watson, Crick, and Wilkins were awarded the Nobel Prize in

An image of the Brookhaven National Laboratory campus in Upton, New York, as it will appear after the construction of the National Synchrotron Light Source II (NSLS II), the circular structure in the background. In the foreground is the current NSLS. *(U.S. Department of Energy)*

medicine or physiology for their discoveries. Franklin had died of cancer in 1958 and was therefore ineligible.

X-rays have continued to play a vital role in molecular biology ever since. Enormous *synchrotrons* (particle accelerators), built for physics experiments, have been harnessed to provide high-energy X-rays. Most of the projects aim to obtain high-resolution structures of proteins in crystal form. Knowledge of protein shapes has become essential in discovering their functions in health, disease, and the activity of drugs.

GENETICS, EMBRYOLOGY, AND EVOLUTION

An organism does not inherit features (such as blue eyes) fully formed from its parents. Instead, it inherits a genome that tells a single cell (the fertilized egg) how to specialize and build tissues and organs. During the nine months between fertilization and birth, they arrange themselves into tissues and complex organs such as the eye. The goal of embryology (now known as developmental biology) has been to understand the processes by which genetic information is transformed into a body.

Researchers have tried to accomplish this in two main ways: working from the developed body back down to the level of cells and molecules, and working from genes upward by studying their functions in cells and tissues. Only in the last few decades have these approaches truly found common ground with the identification of genes that help shape the embryo's body as it grows.

The earliest embryologists were physicians whose main method was to dissect fetuses that had miscarried at various stages of development. The similarity between the bodies of humans and other animals meant that a great deal could be learned from dissections of animal embryos as well. Already at the beginning of the 19th century, comparative anatomy was used to study adult animals, revealing surprising similarities between body parts such as the bone structure of human arms and hands, the wings of bats, and the flippers of whales. Karl Ernst von Baer (1792–1876) extended this work to embryos and made the discovery that animals that looked quite different as adults often went through embryonic phases in which they looked remarkably alike.

Evolution offered a possible explanation—organisms had inherited similar features (called *homologues*) from their common ancestors. This is still an important concept in evolutionary theory. Homologues appear at every level of biological organization. Related genes produce similar body structures in a huge range of species.

One of Darwin's most enthusiastic followers, the German researcher Ernst Haeckel, became known for an interesting attempt to unify embryology and evolution. Haeckel was born in Potsdam, near Berlin, and received a degree in medicine before deciding that he was cut out more for a life of research than one of dealing with sick patients. Haeckel followed in the footsteps of von Baer, armed with better microscopes and the new theory as a framework for his observations.

As he compared embryos of many species, he developed a radical new hypothesis called *recapitulation.* He believed that as an individual organism undergoes development (*ontogeny*), it retraces the evolutionary history of its species (*phylogeny*). All life began as a single cell; so does an individual. The earliest multicellular life-forms were probably ball-shaped, with just a few different types of cells; a human embryo goes through a similar phase. Only in later stages of development do animal embryos start to look markedly different from each other. For Haeckel this reflected the fact that most of today's species arose recently in evolutionary history.

Haeckel found fascinating evidence for his claims. At one stage, a human embryo develops structures like gill slits that then disappear again. This only made sense, he said, in light of the fact that the distant ancestors of mammals were fish. Haeckel drew images of embryos at various phases to show how similar their body plans were. This work has been criticized because his drawings tended to emphasize the similarities rather than the differences among embryos. Haeckel's defenders point out that such a critique is easy to make in the days of photography, where objective images can be made of samples. Drawing is always subjective; an artist must make choices about which features to emphasize after looking at many specimens, and there is always a danger of wishful thinking creeping into the process.

The recapitulation hypothesis was in many ways logical and appealing. Knowing nothing of genes or DNA or their roles in shaping organisms, researchers were struggling to understand how one species might be transformed into another. It was easy to imagine that this could happen when a species added on developmental stages or its development slowed down. But Haeckel took the idea much further, claiming that "ontogeny recapitulates phylogeny" was a law. A human embryo did not simply resemble that of a fish; he believed that it actually became a fish—an adult fish—on its way to becoming an adult human. The hypothesis claimed that evolution worked by adding new developmental stages to the end of an animal's life.

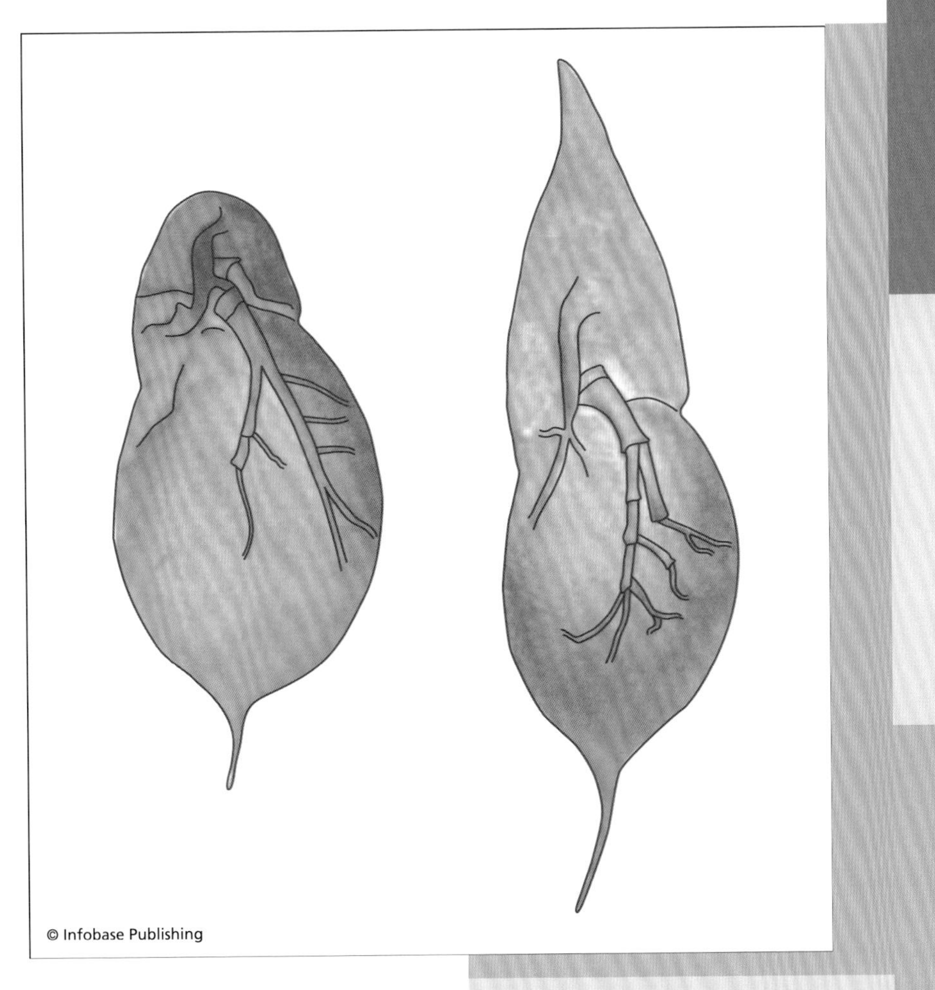

Evolutionary adaptations can take place at any stage of development. A water flea larva (left) develops differently when predators are nearby, growing a larger, helmetlike head (right) that makes it harder to swallow. This probably happens because it senses the predator's molecules in the water.

This would soon be revealed as a serious flaw in the hypothesis. Natural selection works at every stage of an animal's life to shape it, which means that the embryonic phases of an organism's growth also can be shaped in ways that are unique to a species. Fritz Müller (1821–97), a German expatriate living in Brazil, studied crustaceans to prove that evolution shaped larvae as well as adults. As free-swimming organisms, the larvae of each species would have to cope with predators and other features of the environment, so natural selection would shape them just as it affected adults.

In the meantime, numerous examples of such embryonic or larval adaptations have been found. Species such as water fleas, frogs, and carp develop differently when predators are nearby. Water flea larvae grow larger, helmet-shaped heads that make them harder to swallow. Tadpoles grow stronger tails that allow them to swim more quickly and make faster turns.

Despite its flaws, the recapitulation hypothesis encouraged scientists to begin thinking of development in evolutionary terms and to focus on the processes by which hereditary information directed the growth of body structures, rather than only end results such as fully formed limbs. If the bones in a dolphin fin could be matched one-to-one with the hand of a primate, the processes that created the bones should also be homologous. This could be followed all the way back to the earliest stage of embryonic differentiation: *gastrulation,* in which embryos of nearly all animal species develop three specialized layers. But without an understanding of the genetic code and its relationship to the molecules in cells, scientists were stuck there.

A few decades later, Walter Garstang (1868–1949) and Gavin de Beer (1899–1972) pointed out the importance of timing when comparing embryos of different species. New species did not usually arise by adding on developmental stages, as Haeckel had proposed. Instead, each organ and bodily system should be looked at as an independent module. The development of one part might speed up compared to the others, a bit like the way engineers make changes in computers. They might develop a new graphics or sound card while the rest of the machine stays the same. Of course, this may well put pressure on other parts of the machine to change—new games might be made to take advantage of the features of the graphics card and for the games to run well, they might require more RAM or changes in the keyboard. The fact that one change often prompts others could explain why a new species had longer limbs than its ancestors or why humans and chimps do not have tails.

Recapitulation is undergoing a sort of limited revival in the molecular view of evolution. Today it might be phrased in this way: "Organisms resemble each other at many stages of development because the genes they inherited from common ancestors work in a similar way to create the same kinds of body structures." This happens even when the starting points and ending points of development are different—the eggs of a chicken and human are quite different, and they are very different as adults, but particular phases of embryonic development are similar.

A few species have indeed evolved the way Haeckel believed, a process that evolutionary researcher Stephen Jay Gould (1941–2002) called terminal addition. In this process, a new species adds developmental stages beyond those of its ancestors (like adding boxcars to a train). In other cases, evolution has sent species like caterpillars off on a path that is completely different from other kinds of larvae, like trains leaving a station in different directions. And a few types of organisms underwent the opposite of what Haeckel proposed, becoming stuck at an early phase of development because of changes in the genes that were supposed to trigger the next step. An example is the axolotl, a rare salamander found only in Lake Xochimilco in Mexico. This creature remains in its larval form its whole life long and can even reproduce without becoming an adult.

A bottom-up approach to development has only become possible recently, thanks largely to the work of the German researcher Christiane Nüsslein-Volhard (1942–) and the Americans Eric Wieschaus (1947–) and Edward Lewis (1918–2004). Their work on the fruit fly in the late 1970s and 1980s created a new kind of developmental biology that was strongly tied to molecular biology and led to their sharing the 1995 Nobel Prize in physiology or medicine. This was so important because embryology had not yet truly come into the molecular age.

Thomas Hunt Morgan's lab, where classical genetics was born, had focused almost entirely on the appearance of traits in adult flies, hoping to identify the genes that were responsible. It was a good approach at the time, given that scientists knew almost nothing of what genes were made of and how they

While investigating genes that help establish body structures in fly embryos, the German researcher Christiane Nüsslein-Volhard and her colleagues changed the relationship between developmental biology and genetics. In the intervening years, the zebrafish has become one of biology's most important model organisms due to Nüsslein-Volhard's efforts. *(Association of German Foundations)*

functioned. Morgan himself was not particularly interested in the chemical nature of genes; at the time, biochemistry was not far along enough to answer the important questions. The discovery of DNA's structure and the birth of molecular biology completely changed this situation: The biochemistry of the cell was now the central theme of biology. Work with flies was regarded as old-fashioned.

This explains the skepticism that greeted Nüsslein-Volhard and Wieschaus when they arrived at the European Molecular Biology Laboratory in Germany in 1979. Their research plan was to try to identify genes from the mother fly that influenced the development of embryos. To create mutations they fed male flies sugar water containing substances that damaged DNA, then allowed them to mate with females. This often produced malformed embryos, a starting point for discovering which genes had which effects. The work required Nüsslein-Volhard and

Wieschaus to spend several months peering at embryos under a microscope with two sets of eyepieces, looking for developmental defects.

The project paid off quickly. The fly embryo turned out to be ideal for the new approach: Its body is divided into stripelike segments that disappear or become rearranged through mutations. Since each segment gives rise to particular body structures that develop later, the mutations served as a guidebook to genes crucial to the development of flies and most other animals. These genes, some of which are now known as *HOX genes,* play an important role in establishing both the overall building plan for a body and the structure of specific parts, like arms and legs. They are ancient and so important that they have been preserved throughout the animal kingdom.

THE RISE OF GENETIC ENGINEERING

As Nüsslein-Volhard, Wieschaus, Lewis, and a handful of other geneticists were planting the seeds of a new type of developmental biology, a much louder revolution was beginning in California. Barely a century after Mendel's discovery of genes, scientists had learned to read the genetic code. This accomplishment was also a triumph of blending chemistry with biology. During the 1970s, Frederick Sanger (1918–), a British biochemist, developed a method to sequence DNA. The accomplishment earned Sanger the 1980 Nobel Prize in chemistry. It was the second time he had won the prize. The first time, in 1958, came for a method of determining the amino acid sequences of proteins.

Reading the genetic code set the stage for learning to write in it through genetic engineering: a set of powerful new tools to study and manipulate organisms' genes. Genetic engineering allows scientists to alter the DNA of a cell, plant, or animal in deliberate ways for research purposes, so that they can observe how changes in genes affect an organism. This is called *reverse genetics* because it is the opposite of the classical method of starting with a phenotype and looking for the gene that is responsible (*forward genetics*). By the end of the 20th century, it had become routine to make targeted changes in plants, animals, and human cell lines. Genetic engineering also led to the development of applications such as new foods and the use of microbes and animals to produce molecules such as insulin, used in medicine.

DNA naturally undergoes mutations; starting in the 1920s, scientists began learning methods to alter the molecule artificially. Hermann Muller (1890–1967), a former assistant in Morgan's lab, showed that radiation could increase the rate at which mutations occurred. This was soon followed by the discovery of other *mutagens* (such as chemicals) that could accomplish the same thing. As useful as these techniques were, they all had a drawback: The changes they caused in genes were random and unpredictable. Sometimes it was impossible to connect a change to a specific gene. Researchers dreamed of a day when they could pick a gene, *knock it out,* and then study its effects over the course of an organism's development and lifetime.

A few key discoveries set the stage for genetic engineering, which would give them that opportunity. In the late 1950s, a Swiss scientist named Werner Arber (1929–   ) was investigating how bacteria become resistant to attacks by viruses called phages. Salvador Luria (1912–91), working at the Massachusetts Institute of Technology, had discovered that bacteria had proteins called restriction enzymes that helped protect them from the virus. Arber and Hamilton Smith (1931–   ), of Johns Hopkins University, showed that the proteins formed part of a bacterial defense system that chops foreign DNA into small pieces.

Genetic engineering requires a pair of molecular scissors (to remove a gene from one place) and a sort of glue (to paste it in somewhere else). Restriction enzymes provided the scissors. Bacteria contained another type of molecule, called a ligase, which could spot broken ends of DNA and mend the cuts. Organisms need such enzymes because DNA sometimes breaks by mistake. Ligases can scout the molecule and make repairs by matching the broken ends to matching sequences in a chromosome and gluing them into place. Sometimes they insert the fragment in the wrong place. If that happens to be in the middle of a gene, the information gets scrambled. So ligases provided a tool to interfere with existing genes. They can also be used to insert a new gene into an organism's DNA.

In 1972, Janet Mertz and Ron Davis of Stanford University combined restriction enzymes and ligases in a technique now known as DNA recombination. A year later, Herbert Boyer of University of California, San Francisco (UCSF) and two colleagues at Stanford University, Stanley Cohen and Annie Chang, put the method to work to move a gene from one species to another. They combined genetic material from a virus and a bacteria and inserted it into another bacteria. This artificial gene was taken up by the cell and used to create a foreign protein.

Modern techniques allow scientists to remove a gene, substitute another one for it, or add a new molecule to an organism's genome. These methods are now used routinely in medicine, agriculture, and all kinds of biological research. One application, for example, has been to use bacteria or other species of animals to make human insulin needed to treat diabetes. People with type 1 diabetes are unable to produce insulin, which is needed to regulate the body's uptake of glucose. Healthy people cannot serve as donors, because insulin cannot be obtained in large amounts from the body. Doctors used to administer insulin extracted from pigs or cows, but their bodies produce a slightly different form of the molecule that sometimes causes rejection by the immune system or long-term health problems. Changing the recipe of the animals' insulin genes through genetic engineering causes them to make a more human version of the molecule that can be safely used, without such side effects.

Genetic engineering has many other practical uses. Bacteria are being put to work for *bioremediation*—clearing the environment of contaminants such as pollution. In many cases, this happens naturally. Certain species of microorganisms are able to digest toxic substances such as oil. Researchers discovered that after an oil spill a thin layer of bacteria may form a biofilm, a sort of floating mat on the surface of the ocean, which breaks down the contaminants into less harmful substances. Scientists have been able to identify some of the organisms that are responsible and hope to learn to engineer the genes that permit them to do so.

Today, a rising percentage of corn, tomatoes, soybeans, rice, and dozens of other crops produced across the world have been manipulated through genetic engineering. As well as attempting to improve the size, taste, shelf life, or nutritional value of crops, scientists have transplanted genes that help protect plants from insects, fungi, and other parasites. These changes might help farmers ward off pests without the dangerous side effects of pesticides. On the other hand, growing numbers of ecologists in the environmental movement protest that genetic engineering might upset delicate balances in nature. These issues are discussed in more detail in the next chapter.

COMPUTERS AND BIOLOGY

The revolution in biology that started in the second half of the 20th century has depended on an equally amazing revolution in computers in a significant way. It is not surprising that the two fields have found a great deal of overlap. First, many of today's experiments produce huge amounts of data that cannot be captured or stored without computers. Analyzing it may involve comparing billions of pieces of information to billions of others—also impossible without the help of machines. Another use of computers is to solve extremely complex biological puzzles, such as looking at the amino acids that make up a protein and predicting the shape they will form, which is an important clue to the molecule's functions.

Now, during the first decade of the 21st century, researchers are using computers to model and simulate biological processes. Limitations in technology restricted early molecular biologists to observing a few components of a biological process at a time; today, the same events in cells are seen as taking place against a background of networks of molecules that shift from one state to the next in very complex ways. It takes a computer model to keep track of the components, let alone predict how the entire system will change when something happens. For all of these reasons, computers have become indispensable to biological research.

Previous sections of this chapter have introduced some of the early uses of mathematics and models in genetics. Gregor Mendel's talents at statistics were essential in allowing him to analyze inheritance in peas and other plants. The method exposed the dominant or recessive nature of alleles and other aspects of the behavior of genes. Thomas Hunt Morgan's laboratory used the rates at which genes were inherited together to map their locations on specific chromosomes. When the first computers were built, they were put to work on biological questions: Millions of calculations were needed, for example, to interpret the diffraction patterns produced by X-ray experiments, aiming to uncover the structure of protein crystals.

With the arrival of DNA sequencing, computers became essential partners in everyday biology. In 1982, Greg Hamm and Graham Cameron, two researchers at the European Molecular Biology Laboratory (EMBL) in Heidelberg, Germany, proposed collecting DNA sequence data in a universal public database. The project they created is now known as EMBL Bank and is hosted by the European Bioinformatics Institute (EBI), a bioinformatics center run by the EMBL near Cambridge, England. At nearly the same time, an American group created a database called GenBank, originally hosted at the Los Alamos National Laboratory, now moved to the National Center for Biotechnology Information in Bethesda, Maryland. The two groups and another project in Japan worked out a system for exchanging data to offer a free, worldwide, up-to-date service that researchers across the world can access online. As of February 2009, the databases contained 100 gigabases of information about DNA, RNA, and protein sequences.

These resources were in place in the late 1980s as high-throughput DNA sequencing became common, and the United States and other countries decided to embark on genome projects. Biocomputing methods were required to collect and assemble the information. Early in the Human Genome Project, for example, it was decided to take a shotgun approach to obtaining DNA sequences: Rather than starting at the beginning of the first chromosome and analyzing the DNA to the end in a linear way, the sequence was broken down into fragments of several hundred nucleotides. This sped up the project because it allowed parallel processing of many regions of the genome, but it required an additional step of assembling the fragments into a linear text. Each sequence had a bit at the end that overlapped the ends of sequences from other fragments; by matching up the ends, the computer could join the small bits of sequence into a whole. Accomplishing this required several days of constant processing on a farm of 100 Pentium III computers dedicated solely to the task.

Then came the next challenge: to search the sequence for genes. Protein-coding regions make up less than 2 percent of the human sequence. There are three main ways to identify them. If an RNA or protein has been found in an experiment on human cells and researchers have obtained its sequence, they can work backward from the protein's amino acids or the RNA's nucleotides and predict what the sequence of the gene must be. Then they simply scan the genome for the matching code. But some molecules are only produced for short periods of time in specific tissues and have never been detected in experiments. Here a second method comes into play: Years of work have revealed some standard features of genes that can be detected by computer programs. In bacteria, for example, most genes begin with one of a few types of promoter sequences that are usually similar to each other and relatively easy to find. Eukaryotes such as humans, animals, and plants have evolved more complex promoters and genes that have to be spliced—meaning that protein-encoding information is hidden between long stretches of nonsense called *introns*. This makes human genes much harder to find.

A third approach is based on evolution and uses sequence information from one species to detect genes in another. Experiments in flies, worms, mice, and dozens of other laboratory organisms have revealed tens of thousands of genes. Most of them have homologues in other organisms. Despite the fact that they have undergone mutations and other types of changes, a clever computer program can still detect the similarities. Thus, every new gene found in another species is compared to the human genetic code.

One of the most interesting uses of bioinformatics has been to answer questions about the history of life. Darwin's principle of common descent dictates that if two species share a feature, this usually means that their common ancestor also had the feature. The same is true at the level of genes: If two species have nearly identical DNA sequences, it is because they inherited the gene from a common ancestor. This method can be used to reconstruct the ancestral gene. Each living species may have a different version of it, but by comparing the spellings of enough versions, it is often possible to reconstruct the original sequence.

In 2005, Detlev Arendt's laboratory at EMBL used this method to compare the evolution of gene structure in animals. The scientists showed that the genome of the fruit fly is not only quite a bit smaller than the common ancestor of insects and vertebrates—its genes are also less complex. The average human gene contains 8.6 introns; in the fruit fly, the average is only 2.3 per gene. Originally scientists believed this meant that genes have been getting more complex over the course of evolution—humans are more complex than flies by most ways of measuring things and thus are often thought of as more evolved.

Arendt points out that humans are evolving at a much slower pace than insects, partly because of the speed at which they reproduce. Fruit flies reach sexual maturity two weeks after birth, compared to 14 or 15 years in humans (who usually wait longer than that to reproduce). Thus in just 250 years, flies have given birth to about the same number of generations as humans have produced in 100,000 years. Since every new generation is an opportunity for mutations and other changes in DNA to creep into the genome, flies are evolving faster.

Arendt's study compared the introns of humans, flies, and a worm that resembles their last common ancestor. His conclusion was that ancient genes were more like those of humans than flies—they had more introns. Here, evolution has not been making things more complex; it has been simplifying things by cutting out introns in fly genes.

Sequencing gave researchers a way to carry out a complete inventory of a cell's DNA sequences; it has been much more difficult to conduct a similar census of proteins. In the 1990s, scientists figured out a way to begin, using a method called mass spectrometry, adapted from chemistry. Proteins are extracted from a cell and are cut into small fragments by enzymes. These are shot past a magnet and fall into a detector, which essentially weighs them. The amino acids that make up the fragments carry different electrostatic charges, so the magnet bends them in different ways depending on their composition. The size and charge of a fragment determines where it falls. By analyzing this position, a computer program can figure out the amino acid recipe. The computer can compare this to the human genome sequence and identify the genes that match the protein sequence. The process is a bit like rearranging fragments of sentences into a text. It is dependent on the computer's ability to assemble the data in a meaningful way.

Computers also play a central role in the design of new drugs. When researchers have identified a target molecule—such as a protein or RNA that plays a central role in a disease process—they begin looking for a substance that can alter its activity, usually by docking onto it. This is normally done by screening a library of substances (known drugs and other small molecules) against the purified protein in the test tube, or cells that contain it. Sometimes computer "docking" programs are used to preselect likely candidates, through a puzzle-building-like effort that involves matching the surfaces of molecules on the screen. The real work for computers begins when a substance is found that alters the protein's activity in the desired way. It usually needs to be rebuilt to fit the target more precisely and to have stronger effects. Programs analyze its binding sites and recommend small structural changes that should make it more effective.

The most sophisticated use of bioinformatics methods is to model complex, dynamic systems in cells and organisms. In 2000, Eric Karsenti, a cell biologist at EMBL, began using models to try to understand the *microtubule* system in cells. Microtubules are fibers made of single protein subunits called tubulin. The

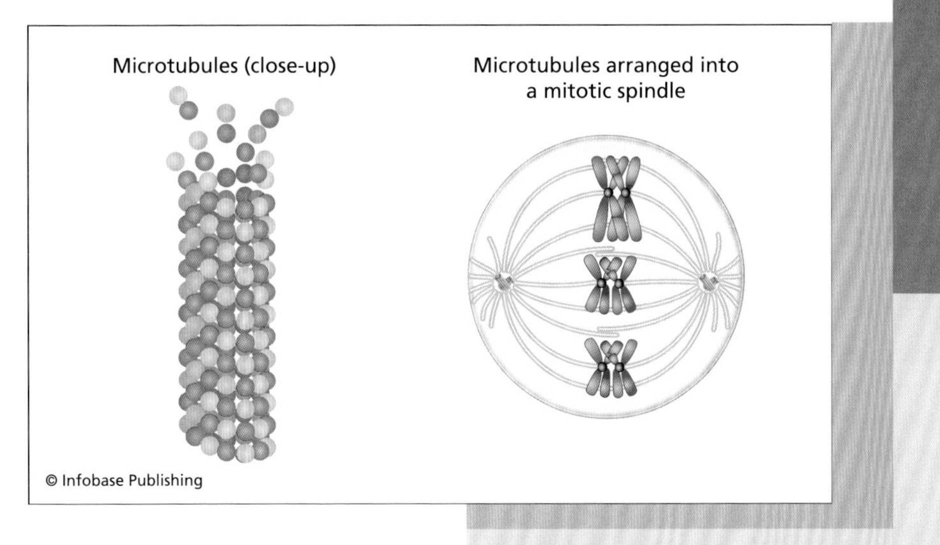

The left image shows the structure of a single microtubule, built of two types of protein subunits called tubulins (red and green). For most of the cell's lifetime they form a sprawling network of fibers aiming outward from the border of the nucleus. During cell division (right), they form a spindle-shaped structure that pulls chromosomes in opposite directions to create two new daughters.

cell builds them by stacking tubulin in long columns—like stacking styrofoam cups—and then gluing several stacks together to form a tower. At the top, proteins are continually added and removed. Other molecules determine how quickly this happens and how high the stack becomes.

Normally, microtubules sprawl through the cell and are used as a sort of scaffold that provides structural support and gives the cell an overall shape. They also serve as traffic ways along which molecules are delivered through the cell. But during cell division, the entire system is completely broken down and rebuilt into a mitotic spindle. Microtubules form a double set of towing lines that stretch from poles on opposite sides of the cell to the center, a bit like two people standing on opposite sides of a field, holding thousands of strings attached to a fleet of kites flying in the air between them. The kites are chromosomes, and when the two fliers reel in their lines they separate DNA into two equal sets. Each set will become the nucleus of a new daughter cell.

Cells have neither a brain nor a master architect to direct this rebuilding project, and Karsenti wanted to understand how the components of the microtubule system were capable of self-organizing into such dramatically different structures. One discovery was that motor proteins play a crucial role. These are molecules that drive along the surface of microtubules, dragging along cargos on a flexible tether. Motors usually have two feet—regions that bind to the microtubule. Each time a foot lands, it undergoes a chemical transformation that causes it to let go again and then move along to the next foothold. Sometimes the feet bind to more than one microtubule, which pulls the fibers into alignment.

In collaboration with Stan Leibler at Rockefeller University, Karsenti and his colleagues have modeled this behavior in the computer to show that complex cellular structures such as the spindle can be generated by a few molecules, following a few simple rules generated by their physical structure. In making the model, postdoctoral fellow François Nédélec and the rest of the team had to know the following:

- the probability that a motor will bind to a microtubule or detach
- concentrations of the molecules that are involved
- the rate at which a microtubule grows and the amount of flexibility it has
- the probability that a motor will fall off the end of a microtubule
- and the speed at which a motor travels and its direction

Careful studies of cells gave the researchers the data they needed to establish some of these parameters, and their simulation of virtual mitotic spindles on the computer screen eerily mimics what they see under the microscope. The lab can now use the model as a way to test hypotheses about the behavior of microtubules and motors. Slightly changing one parameter on the screen—for example, telling a motor protein to move more slowly—may cause the spindle to break down and microtubules to assume a completely different shape. If the scientists notice a similar rearrangement in cells, they can look for a protein or something else that is slowing the motor down.

Such blends of computational and experimental science are increasingly being used by researchers in order to understand what happens in the cell and the body. A new term has been invented to describe this type of science—*systems biology.* At the moment, this phrase is used in so many contexts to describe so many types of work that it is hard to provide a simple definition. The reason it is being used so widely is that many feel molecular biology has entered a new phase, one in which mathematical models and computer programs are essential partners in the investigation of complex biological phenomena. This is a natural evolution of the trends described in this chapter. From their origins as separate fields in the mid-19th century, evolution, genetics, cell biology, chemistry, physics, medicine, and mathematics have become interwoven in a unified science of life. The next chapters show where this evolution is likely to lead and the ways that society is trying to cope.

2

Literature, Culture, and Social Perceptions of Science

Discoveries in biology since the 19th century have dramatically changed people's views of life, human nature, and the environment. Humans are now seen as both a product and part of the natural world. The body is viewed as an assembly of cells that grows in a systematic way from a fertilized egg, rather than as a mass of tissue formed all at once through a special act of a creator. The genetic code preserves a record of the entire history of evolution, revealing a link between people and every other living organism on Earth.

When these principles were first discovered, they ran so counter to what nearly everyone had previously believed that they provoked strong reactions in society. Religious institutions, which claimed to be the authority in matters regarding human nature, were concerned about the implications of a materialist view of life—and were afraid of losing their power. Other fears soon followed. Discoveries in the other sciences—especially physics—led to the development of terrible weapons. New types of technology such as television, personal computers, the Internet, and cell phones began changing people's daily lives, often so quickly that there was little time to adapt. Biologists and doctors predict that data and tools from molecular biology will have an equally profound effect on society.

Writers and artists have a unique opportunity to explore these themes by imagining future worlds or describing fictional situations in which science has changed society. Their visions have had a deep impact on how people think of science. This chapter demonstrates how, as scientists have constructed a new view of life, writers and thinkers have explored its implications.

FRANKENSTEIN: OR, THE MODERN PROMETHEUS

In 1816, Mary Godwin (1797–1851) traveled with her future husband, the poet Percy Bysshe Shelley (1792–1822), from England to Lake Geneva, in Switzerland. The couple was recovering from the death of a prematurely born daughter. They brought along Mary's stepsister, Claire, planning to spend the summer with George Gordon Byron, or Lord Byron (1788–1824)—famous for his poetry and infamous for a number of scandalous romantic affairs. The young group was brought together by their love of literature as well as their social ideas: All were concerned by the disturbing political and industrial changes taking place in their home country, England. It was not a good time for social radicals in Great Britain. The French Revolution, which had begun in a liberal spirit, had toppled its nation's monarchy through murder, pushed Napoleon to power, and embroiled Europe in war. England was in a conservative mood, and the group sought refuge on the continent.

The summer was gloomy and rainy, and they were often confined to the house for days at a time. They wrote, talked about science and politics, and shared ghost stories by candlelight. Byron suggested that each of them write a story about the supernatural. His own contribution was a vampire tale, based on stories he had heard during his travels through eastern Europe. Another guest, John Polidori, reworked the fragment into a story called "The Vampyre," which inspired a long romantic tradition of tales about the creatures.

Among the topics of discussion were the works of Erasmus Darwin (1731–1802), poet, physician, and natural scientist (grandfather of Charles, at the time a young boy). The elder Darwin had speculated on the possibility of bringing the dead back to life. The discussions stayed with Mary, and a few days later, still haunted by the death of her daughter, she had a vivid dream:

> I saw the pale student of unhallowed arts kneeling beside the thing he had put together. I saw the hideous phantasm of a man stretched out, and then, on the working of some powerful engine, show signs of life, and stir with an uneasy, half vital motion. Frightful must it be; for supremely frightful would be the effect of any human endeavour to mock the stupendous mechanism of the Creator of the world.

This dream formed the seed of the novel *Frankenstein* published in 1818, destined to become one of the most widely read books in the world and the inspiration for dozens of films and an entire genre of literature. Mary Shelley's book would establish one of the main themes of science fiction—the scientist driven by ambition or personal motives to create something without regard for ethical concerns. Then, whatever he creates escapes his control and wreaks havoc and terror on the public. In *Frankenstein,* the creation is a human being; in other novels it would be robots, microbes, or machines transformed into terrible weapons. Researchers are constantly seen as fall-

Mary Shelley, author of *Frankenstein,* in a portrait by Richard Rothwell *(National Portrait Gallery, London)*

ing prey to hubris, overweening pride, or reaching beyond one's grasp with no concern for the consequences. It has been a central theme of literature since the ancient Greeks.

Frankenstein tells the story of Victor Frankenstein, a wealthy young man fascinated with odd scientific theories. "It was the secrets of heaven and earth that I desired to learn; and whether it was the outward substance of things or the inner spirit of nature and the mysterious soul of man that occupied me, still my inquiries were directed to the metaphysical, or in its highest sense, the physical secrets of the world," Frankenstein relates. He has a friend named Henry Clerval who concerns himself with moral themes, "The busy stage of life, the virtues of heroes, and the actions of men."

Frankenstein has little time for such considerations. He turns to the writings of medieval alchemists, hoping to find cures for diseases but also searching for incantations able to raise ghosts or devils. Eventually he becomes fascinated by the phenomenon of life, stating, "To examine the causes of life, we must first have recourse to death. I became acquainted with the science of anatomy, but this was not sufficient; I must also observe the natural decay and corruption of the human body." After days and nights of scouring graveyards and examining decomposing corpses, Frankenstein suddenly grasps a secret: "I succeeded in discovering the cause of generation and life; nay, more, I became myself capable of bestowing animation upon lifeless matter."

He begins to assemble a huge, man-shaped being—eight feet (2.4 m) tall—that he brings to life by charging it with some sort of energy. Shelley is vague about what type, but in the introduction to a late edition of the book she mentions the experiments of Luigi Galvani (1737–98), an Italian physician who discovered that an electrical current could make the muscles of dead frogs twitch.

Instead of a having created a beauty, the Adam of a new species, Frankenstein discovers that his creation is shockingly ugly. Horrified, he flees the city and to the home of his friend Clerval, where he falls into a fever and has to be nursed for many months. Upon his recovery, he receives the news that his young brother has been murdered. Frankenstein returns home and catches a glimpse of the monster near the scene of the crime. He is convinced that his creation is responsible for the murder, and shortly afterward the two meet in a forest.

The monster forces his creator to hear his tale of hiding in the woods and watching a peasant family for months, learning to read, teaching himself to speak. His observations of people and books give him a sense of humanity and of virtue. He longs for someone to speak to. But every time he approaches people, they either retreat in terror or try to harm him. Finally he comes across a small boy in the forest. "Suddenly, as I gazed on him, an idea seized me that this little creature was unprejudiced and had lived too short a time to have imbibed a horror of deformity," the monster relates. "If, therefore, I could seize him and educate him as my companion and friend, I should not be so desolate in this peopled earth." However, the boy is equally terrified and threatens the monster—if anything happens, the boy says, his relative, Frankenstein, will exact revenge.

Hearing the name, the giant realizes that he is confronting the brother of the man who created him and murders the boy. He demands that Frankenstein construct a second creature, a bride, so that he will not be alone in his miserable existence. He has turned to violence, he claims, only because he has been abandoned by the one person who should feel a moral responsibility for him:

"I am malicious because I am miserable. Am I not shunned and hated by all mankind? . . . Shall I respect man when he condemns me? Let him live with me in the interchange of kindness, and instead of injury I would bestow every benefit upon him with tears of gratitude at his acceptance. But that cannot be; the human senses are insurmountable barriers to our union. . . . I will revenge my injuries; if I cannot inspire love, I will cause fear . . ." Out of pity and a troubled conscience, Frankenstein agrees to make the monster a wife. But later he changes his mind, fearing that he will unleash two monsters on the world. In revenge, the creature murders Henry Clerval, then Frankenstein's own young bride, and flees. Frankenstein pursues him to the North, finally reaching the Arctic, where both he and the monster die.

Frankenstein is a moral tale centered around the scientific idea that one day humans might push their power over nature too far and create organisms of their own. This idea was not entirely new. Ancient Jewish mystics and the medieval alchemists believed that life arose when inanimate material was activated by a mysterious force. This philosophy was known as *vitalism,* and during Shelley's day it was a matter of intense debate. The alternative was *materialism*—the idea that living organisms obey the same chemical and physical laws as inanimate objects. Frankenstein's vitalist method (which he refuses to tell his friends, in fear someone else might repeat his experiment) remains a secret, and the book remains a fable about what happens when people use science—or magic—to overstep certain moral limits. It echoes themes of *Faust* by Johann Wolfgang von Goethe (1749–1832), in which a scholar trades his eternal soul to the devil in exchange for ultimate wisdom and power on Earth.

This stereotype of the researcher has become fixed in people's minds and has often been used to attack scientific progress. With the rise of genetic engineering, for example, people were quick to attach names such as frankenfood and frankenfish to describe genetically modified plants and animals. The names implied that species were likely to escape control despite the best intentions of their creators. The terms are used deliberately to play on public fears. There are reasons to be concerned about where genetic science is leading, but the debate should focus on the issue at hand rather than inflammatory stereotypes. Calling a thing a monster does not necessarily make it so.

EVOLUTION, RELIGION, AND SCHOOLS

The announcement of the theory of evolution in 1858 triggered a profound change in people's thinking. Prior to that time, there was no solid scientific theory that could explain the spread of life across the Earth, the relationships between species, or their origins. While many scientists suspected that life had arisen through natural processes, the laws had yet to be discovered.

Until that happened, explanations based purely on religious beliefs seemed just as valid as any other. The work of Charles Darwin and Alfred Wallace changed that, almost immediately provoking a strong backlash from theologians and many others. While a number of religious leaders had no problem with the theory—claiming that evolution might simply have been the mechanism by which a creator put life on Earth—more fundamentalist religious thinkers took it as a direct assault on religious beliefs. The controversy continues to this day.

Initially, neither Darwin nor Wallace played much of a role in defending the theory in public. Wallace was in Southeast Asia; Darwin was plagued with ill health and occupied with family matters. His infant son died just as the first paper on evolution was to be read in public. But as critics became more vocal, some of their scientific colleagues jumped to their defense. One of the most forceful was Thomas Henry Huxley (1825–95), who later called himself Darwin's bulldog in reference to his fierce support of evolution.

Strong opposition came from figures such as Richard Owen (1804–92), the first director of the Museum of Natural History in London, and Adam Sedgwick, Darwin's former geology teacher. Sedgwick followed the hard line of the movement known as natural theology, or intelligent design, which claimed that science's only purpose was to collect evidence for the existence of a creator and bring people closer to God.

On the other hand, some religious thinkers did not see the theory as a threat, as long as it left room for a creator to have set natural laws in motion that could produce the Earth and its life. Technically, evolution made no direct statements about the origins of life: Natural selection began with the first organism that had offspring. But a number of scientists believed that even the first cell may have been produced through natural processes—some type of chemical evolution—and indeed many considered the theory eliminated the need for any supernatural explanations of nature at all. As a result, many clergymen were forcefully opposed, and they condemned Darwin in sermons, stating that the theory directly contradicted the Bible and called into question the existence of an immortal soul.

Although evolution was strictly a scientific theory, like the theory that Earth orbited the Sun, everyone attacked it on any grounds whatsoever. The first major confrontation came in a huge public meeting in Oxford in 1860 that was attended by about 700 people. Once again Darwin could not attend; he had suffered a severe relapse of a stomach illness. After delivering a long speech, Samuel Wilberforce (1805–73), bishop of Oxford and the figurehead of the antievolutionists, turned to Huxley and asked ironically whether he descended from an ape on his grandfather or grandmother's side.

Huxley could not resist making a sharp reply. Later he was quoted as saying, "If I would rather have a miserable ape for a grandfather or a man highly endowed by nature and possessed of great means and influence, and yet who employs those faculties for the mere purpose of introducing ridicule into a grave scientific discussion—I unhesitatingly affirm my preference for the ape." These probably were not his exact words, but whatever he said, the battle lines were now drawn for a major confrontation. Science and religion would be divided over the issue, and the debate would spill over into politics, education, and social philosophy.

Over the past 150 years, evolution has been attacked time and time again by religious fundamentalists, most of whom defend the belief that the Bible holds a literal account of creation and that nature bears evidence of intelligent design. This trend has been strongest in the United States. A recent tactic has been to have evolution declared only a theory, implying that it is something like a religious hypothesis that cannot be proven and thus should not be given preference over any other opinion about life. The biggest battleground has been schools; state senates have repeatedly attempted to pass laws forbidding the teaching of evolution or at least promoting a religious account of creation alongside it. Usually these movements have been motivated by Christian fundamentalist groups who have a particular interpretation of the Bible that they hope to get into the textbooks in a clear violation of the principle of separation of church and state. The laws have repeatedly been struck down by the U.S. Supreme Court.

The most famous confrontation over evolution in schools was one of the earliest: the 1925 Scopes trial ("monkey trial"), in which a substitute high school teacher was arrested and tried for breaking a state law banning the teaching of evolution in school. The trial was later dramatized in the 1955 play *Inherit the Wind,* by Jerome Lawrence and Robert Edwin Lee, and became a film several times, most notably in 1960 with Spencer Tracy and Fredric March. The play incorporates historical facts and testimony from the trial, but takes liberties with the real story.

Initially, evolution was not much of an issue in American schools—likely because most of them simply did not teach it. But at the beginning of the 20th century, things began to change, probably because of a rising general level of education and the growth of the U.S. university system. In 1900, few American pupils obtained more than a primary school education—schools had less opportunity to shape young people's ideas and opinions. By the 1930s, many more children were moving to higher grades. By that time, most scientists were strongly convinced that evolution had taken place and were including it in textbooks. This created a conflict between the research community, which expected evolution to be taught in schools like other accepted scientific principles, and fundamentalist religious groups, who saw the theory as a threat to the values they wanted their children to learn.

In the mid-1920s, legislators in several states tried to pass laws forbidding the teaching of evolution in schools. Their first success came in Tennessee in 1925, with a bill making it a crime to teach evolution or "any theory that denies the story of the Divine Creation of man as taught in the Bible."

The American Civil Liberties Union (ACLU) was alarmed because the law required science teachers to promote a particular religious view in classrooms, a violation of constitutional principles and an affront to the beliefs of the many American children who were not Christians. The ACLU wanted to challenge the law in court, but this could only be done based on a specific case. The organization placed an advertisement in a Tennessee newspaper, offering to pay all the legal expenses of a teacher willing to challenge the law. They found a volunteer in John Scopes (1901–70), a young man who had been a substitute biology teacher in Dayton, Tennessee. He was convinced by the scientific case for evolution and felt that pupils had the right to learn about it, particularly because the school's own biology book had a chapter on Darwin and natural selection. How could it be against the law to teach what was in the school's own textbook?

Dayton took on a carnival atmosphere as the town prepared for the trial. Journalists and celebrities arrived. The prosecution found a spokesman in the famous orator, fundamentalist Christian, and former presidential candidate, William Jennings Bryan (1860–1925). America's most famous trial lawyer, Clarence Darrow (1857–1938), took on the defense without charging a fee. Both felt that a great deal was at stake. The ACLU regarded the trial as a major test of individual rights versus an attempt by a religious majority to force its opinions on everyone. On one hand, democratic principles seemed to imply that the majority should be allowed to decide what was taught in schools. On the other, new scientific discoveries arose all the time, and it was felt that teachers should teach theories accepted by most scientists—even those that contradicted religious beliefs.

Darrow hoped to use the trial as a public education campaign for evolution by putting experts on biology, evolution, and even religion on the stand. The judge agreed with the prosecution, however, that the legal issue was not whether evolution was correct, but only whether Scopes had violated the law. Darrow's only option was to call Bryan to the stand as an expert on the Bible. Despite conceding that the Sun did not revolve around the Earth, Bryan claimed that the Bible should be regarded as the sole authority on matters like creation and did not need to be interpreted.

Scopes was convicted of having broken the law and was ordered to pay a fine. The conviction was later reversed on a technicality, leaving no opportunity to pursue the issue in the courts. Soon several other states in the South passed antievolution laws.

The legal issue was not addressed again until 1968, when the U.S. Supreme Court took on the case *Epperson v. Arkansas.* In their ruling, the Supreme Court stated that all antievolution laws were unconstitutional because they violated teachers' rights and represented an attempt to promote religion in public schools. Since then, fundamentalist religious groups have made many attempts to subvert ruling after ruling of the courts. Common tactics are to try to redefine religious doctrines as some type of science (as in the case of intelligent design), to portray evolution as some sort of subjective religious belief system, or to demand equal time for religious views, as if scientific theories were political campaigns.

In the meantime, the vast majority of religious thinkers throughout the world have come to terms with evolution. For example, the Catholic Church has stated that evolution and the Bible need not be incompatible—no more than the fact that the Earth goes around the Sun should challenge people's faith. In a 2006 book called *Creation and Evolution,* Pope Benedict XVI wrote that rejecting evolution in favor of faith and rejecting God in favor of science were equally absurd. His book promotes a theology of theistic evolutionism—in other words, evolution was the process by which God created life—and calls for people to stop making evolution a polarizing issue.

EUGENICS AND *BRAVE NEW WORLD*

The theory of evolution reveals that human biology is constantly changing. Humans and every other species are shaped by natural selection. In the distant future, people are likely to look, think, and behave much differently than they do today. In a few million years, the changes may be as significant as the features that distinguish humans from chimpanzees. Until the discovery of evolution, most people's thoughts about the future of their species were focused on progress. But evolution is nondirectional; it does not guarantee that people will become healthier, smarter, or better in any other way. This was disturbing to the many 19th-century thinkers who believed that human society—partly through scientific and technical progress—was improving.

Was the future of humanity completely up to chance? The arrival of genetic science suggested a way of choosing between desirable and undesirable characteristics by controlling which plants or animals breed with each other. The same thing ought to be true of humans. The genetic makeup of future generations depends on who breeds in this one. So pushing human evolution down a particular path would require a program to ensure that the right people breed with each other.

This philosophy spurred a movement called eugenics, which had two forms. Positive eugenics encouraged people with desirable characteristics to marry each other and start families; negative eugenics aimed to eliminate undesirable traits by preventing people from having children. This began with the involuntary sterilization of people in prisons and mental institutions and ended in one of the most horrifying events of history—the Holocaust, carried out against Jews and other undesirables by the Nazis. The eugenics movements that arose at the beginning of the 20th century were not responsible for the Holocaust, but they gave it a philosophy that it could drape itself in. Eugenics provided a pseudoscientific justification for racism that turned out to be naïve and based on false assumptions about how heredity works.

Ironically, two grandsons of Thomas Huxley would play a role in the rise of eugenics and the way it was regarded by society. The first was Julian Huxley (1887–1975), one of the scientist-architects who had helped bring genetics and evolution together. His brother, Aldous Huxley (1894–1963), wrote a book called *Brave New World,* which explored the social consequences of attempts to improve the human race through breeding.

Before World War II, like many scientists of his day, Julian Huxley felt that taking control of evolution would be a positive thing, comparing it to the practice of agriculture. "No one doubts the wisdom of managing the germ-plasm of agricultural stocks, so why not apply the same concept to human stocks?" He believed that social status was a sign of evolutionary fitness and that people belonged to lower classes because they were genetically inferior. From this point of view, population studies revealed a worrying trend. People of the lower classes were reproducing more than prominent members of society. If this continued, their "undesirable genes" would soon overwhelm the species. In a book called *The Uniqueness of Man,* published in 1941, Julian Huxley wrote, "The lowest strata are reproducing too fast. Therefore . . . they must not have too easy access to relief or hospital treatment lest the removal of the last check on natural selection should make it too easy for children to be produced or to survive; long unemployment should be a ground for sterilisation."

At the time, Huxley was serving as vice president of the British Eugenics Society, whose membership included a number of prominent British scientists and thinkers. The society had been founded by Charles Darwin's cousin Francis Galton (1822–1911) and headed by Darwin's son Leonard. While the emphasis of the society was mostly on positive eugenics, others had few moral qualms about the negative direction things were taking. The measures Huxley proposed in Great Britain were actually being carried out—not only in Nazi Germany, but also in the United States.

The U.S. eugenics movement arose from a long tradition of prejudice against the poor and the mentally impaired; concretely, it can be traced back to 1874, when Elisha Harris (1824–84), a physician and political reformer, became secretary of the New York Prison Association. He was a talented statistician and began to investigate a pattern he had noticed in the family names of criminals in country prisons. He traced an incredible number of "convicts, paupers, criminals, beggars, and vagrants" back to a family that had lived in Ulster County, New York, in the late 1700s. At the time of his study, six generations later, they had 623 descendants, many of whom became criminals. "In a single generation there were 17 children," Harris wrote. "Of these only three died before maturity. Of the 14 surviving, nine served an aggregate term of 50 years in the state's prisons for high crimes and the other five were frequently in jails and almshouses." In the absence of a theory of heredity, heredity was presumed to be responsible.

Another young statistician named Richard Dugsdale (1841–83) conducted a very thorough follow-up study, hoping to expose the real causes of violence. After examining records and interviewing family members, employers, police officers and many others, he concluded that the environment was more to blame in the family's tragedy than biology. What was being inherited, he wrote, was a pattern of neglect, abuse, poverty, other social factors, and physiological issues: alcoholism during pregnancies and sexually transmitted diseases that affected unborn children. He believed that the only way for the family to escape its fate would be through extensive social reforms that improved their lives, probably over two or three generations, by providing a secure environment, early education for children, and foster homes for orphans and children born out of wedlock. His conclusions went unheard. Dugsdale died at an early age, and soon after his death his work was being misused to promote the idea that criminals breed criminals. The stage was set for negative eugenics—a public campaign to improve society by ridding it of unfit members. Some of America's leading thinkers supported the movement.

David Starr Jordan (1851–1931), the first president of Stanford University, soon became one of the most outspoken figures in America's negative eugenics movement. Unlike Dugsdale, but like many scientists of the late 19th century, Jordan was convinced that heredity was far more important than the environment in shaping human behavior. He began to believe that crime and poverty were spreading like a hereditary disease, and Jordan set about ridding humanity of its degenerates. The idea took an even more negative turn when he began to associate evolutionary fitness with race and nationality. In 1907, he drew these themes together in a book called *The Human Harvest: A Study of the Decay of Races through the Survival of the Unfit.*

Jordan obtained a chairmanship within the American Breeders Association and helped change its constitution to include a platform of eugenics. The Association sponsored research into genetic studies of insanity and other mental diseases, to determine whether they could be inherited. A eugenics records office was established in the town of Cold Spring Harbor on

Long Island, New York, working closely with a nearby biology laboratory headed by Charles Benedict Davenport (1866–1944). These groups began to promote the compulsory sterilization of unfit people.

It was nothing new. In the 1890s, some physicians had begun to remove the ovaries of women with a history of psychological problems, believing that this could improve their conditions. Castration or vasectomies were performed on males as a punishment for crimes or cures for mental problems. These practices, the eugenicists said, were more humane than the death penalty, and they would have the added value of protecting society by ridding it of future criminals. Many doctors were appalled, but the practices went on.

Dr. Henry Clay Sharp (1869–1940), a prison physician in Indiana, began promoting sterilization as a solution to insanity and hereditary crime around 1900. He petitioned the governor and the state legislature to pass a mandatory sterilization law over the protests of groups of physicians who claimed the practices violated patients' rights. The Indiana legislature passed the first compulsory sterilization bill and the state's governor signed it into law in 1907. By 1930, similar laws had been passed in 30 states. By the time Sharp died in 1940, more than 35,000 people had been sterilized involuntarily in the United States.

Things started to turn around in the 1930s. Scientists like Hermann Joseph Muller (1890–1967) showed that many of the so-called studies of human heredity of Sharp and Harry Laughlin (1880–1943), a former high school principal appointed to direct a new Eugenics Record Office in New York, Laughlin ignored environmental factors such as the inequality of women and huge differences in education and health among different social classes. Muller wrote, "There is no scientific basis for the conclusion that socially lower classes, or technically less advanced races, really have a genetically inferior intellectual equipment, since the differences . . . are to be accounted for fully by the known effects of the environment." Unfortunately, these voices of reason were not heard everywhere, and as the movement declined in the United States it was on the rise in Germany and elsewhere. Only with the end of World War II and the exposure of the Holocaust did the scientific community—as well as politicians and many others—reject eugenics as a reasonable way to influence the future evolution of the species.

The British Eugenics Society had been much more hesitant to turn negative eugenics into a government sterilization program, but as the writings of Julian Huxley show, some of the members of the group were thinking along these lines before World War II. Ironically, at the same time, Julian's brother Aldous was coming to completely different conclusions about the type of society that eugenics might produce. Huxley was a novelist, and in 1932 he published his vision of the future in a book called *Brave New World.* The novel is a dark, anti-utopian dissection of a society in which the government controls not only people's biology, but also their behavior and thoughts through the use of drugs and sleep-learning.

Brave New World takes place in the year 2540—although the calendar in the future has been recalculated and in the book the date is 632 after Ford (a reference to the automaker and father of the assembly line). In this future, most of the planet is ruled by the World State, whose aim is to ensure a life of comfort and peace for its citizens. In the aftermath of World War I, H. G. Wells (1866–1946) and other authors were promoting a utopian image of the future in which a global state would create a happy, healthy society run on socialist principles. Huxley was skeptical about these rosy predictions, particularly after a trip to the United States. On the boat on the way over, he found a copy of a book by Henry Ford and began thinking about where industrialism and mass production were leading society. American society struck him as materialistic, overwhelmed by commercialism, young, sexually liberal, and focused on entertainment and pleasure. Carried to an extreme, these would be the trademarks of Huxley's nightmarish vision of the future.

The book begins in London with a tour of a reproductive factory where human embryos are being grown. Sexual reproduction has been done away with; embryos are produced by artificial insemination and raised in an artificial womb. Even a happy society needs workers to do unpleasant jobs, and science has developed a way to obtain them—a reproductive caste system. Embryos from the higher castes are unique individuals who are allowed to develop naturally. The lower castes are created by a sort of cloning, in which about 100 individuals are produced from the same egg. As they develop, chemicals are used to limit their intelligence and growth. Children are raised by the state, and various kinds of psychological conditioning are used to give them its core values. The family is replaced by a feeling of belonging to everyone.

Individualism and the desire to be alone are seen as antisocial tendencies. The masses are given a pleasure-inducing drug called soma and devote themselves to entertainment and social activities. A few areas of the world are set aside as reservations where people can live without interference from the state and have children naturally. The reservations serve as vacation spots, and sometimes nonfunctioning members of society are banished there.

The main conflict of the book involves a disillusioned psychologist from the World State named Bernard—although he is one of its few misfits, he desperately wants to fit in—and John Savage, who lives on a reservation. John is also an outcast because his mother comes from the industrial world, stranded on a reservation during a vacation. She would like to go home, and John is curious about the rest of the world. When Bernard arrives on vacation at the reservation, they accept his invitation to return to London. There, John is treated like a sideshow attraction at a circus—at first adored by all—and Bernard as a celebrity for having discovered him. But John becomes increasingly skeptical of the materialist society he has been thrust into, searching in vain for value and meaning in a world devoted to consumerism and pleasure. When his mother dies, he is torn by grief and stunned by everyone else's lack of feeling. He retreats to an isolated lighthouse to try to live alone and find peace, but crowds follow him there, hoping to be entertained. Finally, unable to cope, he commits suicide.

Huxley's novel is not a direct condemnation of genetics or biology. In 1932, DNA's role in heredity was still unknown and genetic engineering did not yet exist. But controlled breeding

The Shape of Humans to Come—the Transhumans of Patricia Piccinini

Patricia Piccinini (1965–) is a gifted Australian artist whose work includes sculptures of transhumans—mixtures of human and animal forms that are reflections on genetic engineering and the idea that people may soon have the possibility of drastically intervening in human nature. Her figures are so lifelike and thought-provoking—and sometimes so disturbing—that at least one of them was the source of a major Internet hoax.

In 2005, Arabic newspapers began printing the tale of a girl who threw a copy of the Qu'ran at her mother and was then cursed, leading to her transformation into a strange creature that was half human, half dog. The image was widely circulated in the Arabic press and became a topic of discussion in Islamic circles before it was discovered that it was a sculpture from an exhibition of Piccinini's work entitled *We Are Family*. The journalist Nizar Usman from Sudan reported, "I heard about the story from my daughter (10 years of age), she heard it in her school. Then I read it in a notice board in front of the main gate of a mosque where there was a mammoth gathering. Then I read it in *Al Hayat* daily newspaper. . . . Actually what confused people here is that there are Quranic verses saying that Allah—long ago—transferred some guilty people into monkeys and pigs."

In the catalog to one of Piccinini's exhibitions, the art critic Donna Haraway wrote, "Her visual and sculptural art is about . . . worlds full of unsettling but oddly familiar critters who turn out to be simultaneously near kin and alien colonists. Piccinini's worlds require curiosity,

(continues)

(continued)

emotional engagement, and investigation; and they do not yield to clean judgments or bottom lines—especially not about what is living or nonliving, organic or technological, promising or threatening." Haraway's essay draws connections between the history of Australia and the future as Piccinini sees it. The continent has experienced enormous shocks with the arrival of Europeans and a long, oppressive period of colonization that has seen the import of foreign species and cultures. In the future, genetic science and technology may create a similar situation across the entire world. Piccinini wants people to see what it will look like. She confronts viewers with an amazingly realistic world filled with sculptures so lifelike that they fooled newspaper editors. Her figures are disturbing hybrids of humans with wombats, humans with machines, and visions of new, unidentifiable species. Just as the original residents of Australia have had to learn to cope with change, humans will have to cope with any offspring that they decide to create.

For a 2003 exhibit in Vienna, Austria, Linda Michael described Piccinini's sculpture *The Young Family* this way: "The mother of this family lies on her side like a big sow with a litter of suckling pups. . . . Despite her status as a

practices were being used to alter crops and livestock, and there was a strong possibility that the same methods would one day be used on humans. His future society uses reproductive cloning and artificial breeding techniques, manipulating the environment that embryos develop in to influence their development. Control over society is maintained mostly through behavioral conditioning and drugs. The ultimate message seems to be that new mother, she is old. Her counterpart in the real world might be the 62-year-old woman who carried her daughter to term after a pregnancy created through IVF by Italian fertility doctor Severino Antinori. Her expression is tired, world-weary and patient, and somehow profoundly sad. She has eyes and skin with moles and hairs and veins just like us—but also the hairy back, muscular arms and hands of a primate, and a snout, long floppy ears and a tail stub. . . . It is a highly defined representation or surrogate of something. But of what?"

The species she envisions might be failed experiments from Aldous Huxley's *Brave New World*. On the other hand, a society of the future might consider them to be successful experiments. By making the figures so real, Piccinini forces viewers to meet potential artificial offspring. The works of science fiction writers and filmmakers prompt people to consider how their actions will influence the future. Piccinini's work does the same thing in a way that is somehow more subtle and intense. Of the mother figure in *The Young Family,* Michael writes, "If we see her as monstrous, is it because she threatens the continuity of our species? Does she signal the untrammeled potential of creation unleashed by transgenics? Or expose a horror of bestiality? Perhaps, in contrast, she just exposes the inevitable failure of our expectations and desires."

even if it becomes possible to manipulate human bodies and control minds toward some ideal of perfection, it will be the basic values and structure of society that will determine whether people live fulfilled lives. A society based on pleasure may not be the best future. If *Brave New World* had a single moral, it might be the old adage, Be careful what you wish for, because you just might get it.

SILENT SPRING AND THE RISE OF ENVIRONMENTAL MOVEMENTS

The 19th century ended in a wave of enthusiasm for progress. However, after witnessing chemical warfare in World War I, the atomic bomb in World War II, and the horrifying arms race that followed during the cold war, many people stopped looking at science as a solution for the world's problems and saw instead that it seemed to be contributing to them. This change in attitude was not restricted to physics. At the beginning of the 20th century, breeders and geneticists were regarded as heroes, people who might one day solve the problems of hunger and disease. Today, they are more likely to be associated with Frankenstein. This change has been influenced by the dangers evident in other branches of science—particularly concerns about the effects of chemical pollution on the environment. These burst into public consciousness with the publication of the marine biologist and nature writer Rachel Carson's *Silent Spring* in 1962, a book warning of the dangers of the long-term effects of pesticides.

One of the pesticides Carson was most concerned about was DDT, a chemical that had initially been regarded as a miracle pesticide and solution to disease. Paul Hermann Müller (1899–1965), a chemist at Geigy Pharmaceutical, created the compound. The widespread use of DDT saved millions of lives that would have been lost to typhus and malaria—a disease that was completely eliminated in many parts of the world—because it killed the insects that carried the diseases. Müller was awarded the 1948 Nobel Prize in physiology or medicine for his work.

But DDT quickly acquired a completely different reputation and along the way played a central role in the development of a large environmental movement in the United States. The reason for the birth of the environmental movement and the eventual banning of DDT was *Silent Spring,* which has been called both one of the "25 greatest science books of all time" by *Discover* magazine and received many votes as one of the "ten most harmful books of the 19th and 20th centuries" by the conservative magazine *Human Events.*

Rachel Carson's book *Silent Spring* gave a scientific account of the effects of DDT on the environment, woke up the American public to the dangers of pollution, and helped spawn the modern environmentalist movements. *(NOAA/Department of Commerce)*

Carson's book provided scientific evidence that DDT could cause cancer and other types of environmental damage. DDT had been "sold" to the public in extensive publicity and advertising campaigns in which its virtues were emphasized—had potential dangers been covered up? A lone woman was saying that it was absorbed by crops and entered the diets of animals and humans. Her data suggested that it had a negative effect on birds, fish, trees, and many other forms of life.

Carson immediately found herself in the middle of a public relations war in which researchers ridiculed her science and the chemical industry attacked her personally; she was labeled hysterical, and industry spokesman Robert White-Stevens said, "If man were to follow the teachings of Miss Carson, we would return to the Dark Ages, and the insects and diseases and vermin would once again inherit the earth." Most of the attacks were grossly unfair. Carson had never proposed banning all pesticides, and she was well aware of their importance in fighting disease. In *Silent Spring* she wrote: "No responsible person contends that insect-borne disease should be ignored. The question that has now urgently presented itself is whether it is either wise or responsible to attack the problem by methods that are rapidly making it worse." She pointed out that the overuse of pesticides had produced insects that were resistant to them: "The list of resistant species now includes practically all of the insect groups of medical importance. . . . Malaria programmes are threatened by resistance among mosquitoes. . . . Practical advice should be 'Spray as little as you possibly can' rather than 'Spray to the limit of your capacity.' . . . Pressure on the pest population should always be as slight as possible."

This was in good keeping with evolutionary science: A challenge such as a pesticide can suddenly put enormous pressure on a species. If chance provided some mosquitoes, for example, with genes that offered them partial resistance to a pesticide, they would likely undergo rapid positive selection, and within a short time the genes would spread through the population. Soon farmers would be back at the same place they started. It was an important lesson that would need to be learned over and over again. For example, it took doctors decades to learn that the overprescription of antibiotics was promoting the evolution of highly resistant bacteria. Such cells would have arisen anyway—bacteria reproduce so quickly and undergo so many mutations that they are amazingly adaptable—but the overuse of drugs has cleared a path for resistance genes to spread at an alarming rate. It increases the chances that a person who desperately needs antibiotics could be infected with a strain that does not respond to them.

Silent Spring was well written and controversial, which made it so popular that many people became alarmed. Its publication put pressure on the government, the chemical industry, and scientists. A committee was appointed by President John F. Kennedy to look into the matter. Its findings supported those of Carson and led to changes in the way pesticides were regulated by the government. This planted the seeds for a strong American environmental movement that turned its scrutiny on pollution and chemical waste products, many of which turned out to cause cancer.

From the beginning, the issue had a polarizing effect. Environmentalists felt that they could not entirely trust scientists, the chemical industry (who might be paying them), or government regulations to protect citizens from the products of research. The result was a conflict that continues to the present day. It can be seen in the multitudes of environmental partners as well as platforms of the green parties in the United States and many other countries. Environmental concerns blend themes of consumer advocacy and quality control, skepticism of industrialization and global business practices, concern for the environment and endangered species, and fears that meddling with nature on the part of scientists will lead to disaster. Thus when genetic engineering came on the scene, the stage was set for a confrontation.

GENETICALLY MODIFIED CROPS AND THE MARKETING OF SCIENCE

Genetic engineering arose in the late 1970s and early 1980s against a backdrop of new public concerns about technology and the environment. By the time the first genetically modified foods were put on the market in the 1990s, nearly every day's news brought reports of some new, cancer-causing substance. In other words, it was dangerous to release the products of science into the environment. Industrial chemicals and pollution had caused a huge hole in the ozone layer over Antarctica, allowing dangerous solar radiation to penetrate the atmosphere and increasing people's risk of skin cancer. Atmospheric studies warned of a greenhouse effect that could dramatically change the global climate. Asthma, allergies, and cancer were on the rise. Governments began to pass laws to minimize environmental pollution and to ban cancer-causing substances.

From the beginning, scientists were interested in using genetic engineering to improve crops. One reason was the concern that traditional methods of food production could not keep up with the number of mouths to feed. At the World Food Summit in Rome in 1996, experts stated that the world would have to double its production of food within the next 30 years just to keep pace with population growth. It has been estimated that 800 million people on the planet currently suffer from malnutrition and starvation. A solution might be to create *genetically modified organisms* (GMO) by directly modifying the genes of crops. As well as improving their size, taste, shelf life, or nutritional value, foreign genes can offer protection from insects, fungi, and other parasites without the dangerous side effects of pesticides. Members of the growing ecological and environmental movements protested that genetic engineering might upset delicate balances in nature.

One issue that makes the debate so complex is that so many of the participants have political, financial, or ideological agendas. Companies interested in creating new foods—citing humanitarian reasons and the hope of making profits—claimed that farming had always produced highly artificial crops and the products of genetic engineering would be no different in any significant way. Critics said that artificially modified organisms would automatically have an unnatural advantage over plants bred by other methods and that introducing them into the environment in huge quantities would change the way genes normally move through a population.

The first food brought to market was the Flavr Savr tomato, by a California-based biotech company called Calgene, in 1994. Research had shown that a protein called polygalacturonase played an important role in how tomatoes rot because it softened cell walls as the fruit ripens. Inserting a second gene that interfered with the protein yielded tomatoes that could be stored longer without losing their taste.

The U.S. Food and Drug Administration (FDA) examined the plant, deciding that it did not pose a health hazard to people and could be put onto the market without special labeling. Although customers in the United States and Europe were initially enthusiastic, in the long run Flavr Savr was not competitive. It lost out to other long-lasting, non-GMO brands that customers preferred—Calgene had not used the best strain of tomato to begin with, and the company had little experience in growing and marketing foods. But the GMO era had begun, and the following year Calgene was bought by the company Monsanto, which has become a major producer of many types of genetically modified foods.

In Europe, public acceptance of GMOs quickly plummeted as consumers became concerned that there might be unknown risks. Protesters demanded strict governmental controls (such as bans on imports, or at least clear labels marking food as a product of genetic engineering). The change in attitude was partly due to outbreaks of a deadly disease called bovine spongiform encephalopathy, or mad cow disease, which was caused by cows, normally herbivores, being fed the remains of other cattle. The disease was then passed along to humans who ate the animals. Once again, political and business agendas came to bear as governments initially shrugged off the threat. By the time their attitudes changed, many Europeans had lost confidence in governments' ability to regulate foods.

Tomatoes were quickly followed by genetically modified soybeans, cotton, and maize. Some of the new varieties improved the nutritional value of foods that are the core of people's diets in many parts of the world. Corn and rice lack vitamin A, which is essential to the development of the eye. Adding genes to these staple crops has helped reduce blindness and other symptoms of malnutrition that have plagued children throughout the world. Plants have also been made resistant to herbicides so that weeds can be killed without damaging crops. Tomatoes, cotton, corn, and many other crops fall prey to caterpillars; researchers have added a natural toxin, a protein called Bt from the bacteria *Bacillus thuringienses,* that kills the insects. Sweet potatoes in Africa have been made immune to viruses. Changes in species of rice have produced strains that can survive floods, and other plants have been modified to tolerate high levels of salts or acids in the soil.

The number of GMO crops continues to increase dramatically, particularly in the United States, Argentina, Canada, and China. Recently it has been estimated that about 75 percent of foods on the shelves of stores in the United States contain at least one modified ingredient. In other countries, the trend has grown at a somewhat slower pace, but overall GMO crops are winning an increasing share of the world food market. By 2005, approximately 60 percent of the world's soybean fields, 28 percent of the cotton, and 14 percent of the maize were devoted to these crops.

Decisions to develop and grow GMO foods are based on the profit they are expected to bring, as well as other motives. Businesses have sometimes engaged in questionable practices to gain an advantage over their competitors to the detriment of farmers and economies in developing nations. The practices have also raised new legal issues such as questions of ownership. The creation of a new crop requires a huge investment in basic research, laboratory experiments, costs of growing, and risk assessments. Companies need to recapture these expenses through profits, which are best ensured by maintaining ownership of their crops.

There has been a growing interest in the production of genetically modified animals for foods as well, but the efforts have met with technical, ethical, and legal challenges. It is much more difficult to develop these animals than plants. Often, a new plant can be grown from an existing one simply by taking a single cell. In animals, new genetic material can be introduced into the very early embryo so that the animal's egg or sperm cells contain the gene. The methods are not perfect, and many generations may be needed to obtain a strain with the gene.

Other efforts are underway to create pigs that produce leaner meat and to use animals as factories for drugs like insulin. The same strategy has been used to make another hormone called erythropoietin, which stimulates the development of red blood cells, and is used as a treatment in anemia and some forms of kidney disease.

Another use of genetic engineering is to create drugs that might be delivered to people through foods: chickens whose eggs contain antibodies or bananas containing vaccines. Human genes have been added to animals so that they produce the human forms of proteins, sometimes in their milk. The goal is not to produce molecular "cocktails" to deliver therapeutic genes in a drink—but to produce them in a form that can easily be extracted from an animal (by milking it) and then purified.

A major concern for the public has been the fear that genetically modified plants and animals might have an unforeseeable impact on the environment or their bodies. Here too some precedents had shaped public attitudes about introducing new species into the ecosphere. In the 1960s, scientists brought a fish called the Nile perch into Lake Victoria in Central Africa, the source of the Nile River. The perch was so well suited to its new home and multiplied so quickly that native species of fish have been virtually wiped out. The story is told in a 2005 documentary called *Darwin's Nightmare,* written and directed by the Austrian filmmaker Hubert Sauper. The film explores the human and economic impact of the fish, now such an important source of food that it is bartered to Russian buyers in exchange for weapons. Other cases of transplantation have been unintentional, such as the transport of pests in food containers or small animals such as snails that are often carried along in the ballast water of ships. When ships travel across the world and illegally empty their tanks, the result may be infestations or the disruption of local food chains. (This has been going on as long as people have traveled the globe, carrying along seeds, animals, and hitchhiking parasites—so today's normal environment must also be seen as the product of steady contamination.)

These situations are not directly related to genetic engineering, but they have set the backdrop for people's responses. While some of the arguments against genetically modified foods or other organisms are based on scare tactics, rather than realistic estimations of risk, scientists admit that the effects of these foods or other organisms are impossible to predict with absolute certainty. Every organism lives in a complex network of interactions with every other, from bacteria in the soil to other plants and animals. Testing a new strain's effects on all of them would be impossible.

Legally, however, the question became whether genetically modified species should have to meet far stricter standards of safety than any other new product brought onto the market. Many people thought that they should because they felt that genetic engineering was tampering with nature. Defenders of GMOs point out that farming also alters organisms through selection; it works with the changes in plants and animals that arise through mutations and other natural processes. These changes occur in random genes and are completely unpredictable; they may have equally strong effects on the environment, but these effects may never be studied under controlled conditions in the laboratory.

In the United States, the FDA is responsible for approving modified foods intended for market after extensive testing in laboratories. The Flavr Savr tomato was approved after the FDA determined that it did not pose a health hazard. Producers were not required to give it special labeling. In 2003, a survey conducted by ABC News showed that 92 percent of Americans believed that "the federal government should . . . require labels on food saying whether or not it has been genetically modified or bio-engineered." The percentage had steadily risen since similar surveys in 1998 (82 percent in favor of labeling) and 2000 (86 percent).

Some common fears, such as the idea that modified genes from a plant might enter a person's body and cause health problems, are simple misunderstandings about how genes work. Scientists have never discovered a case where a gene from a plant has been taken up by the human genome by eating; all food contains foreign DNA, and it is destroyed during digestion. (Some of the concerns may have arisen because of mad cow disease, but there the cause is a protein fragment; genes would not behave the same way.)

While most scientists admit that it is impossible to calculate all the risks involved in the creation and spread of genetically modified organisms, they are concerned that the debate has not been balanced and sufficiently informed by facts. Many feel that science fiction movies, negative publicity, and misunderstandings have given the public a false idea of what GMOs are and how risky their use might be. Just as the media present much more bad news than good, problems with GMOs receive far more attention than the positive effects they have had on millions of people's lives. GMOs have been designed to create new foods, curb hunger and starvation, and prevent disease. Not using science to try to solve some of these very grave problems—when there is no evidence that these crops pose a greater threat than the products of traditional agriculture—would be ethically very questionable. Discussions should not be entirely focused on risks; they must also weigh the potential benefits and give equal consideration to the consequences of not taking action at all.

Genetically modified corn growing in a field in Europe. In recent years, Europeans have shown a growing resistance to the planting of these crops; in 2009, a popular brand from the company Monsanto was outlawed in Germany.

Some studies have shown that GMOs do occasionally have unintended consequences on the environment. Cotton bearing the bacterial toxin Bt was put onto the market in 1996 by the company Monsanto, a major developer of GM crops, in order to ward off a type of moth larva called the bollworm. This pest puts such a huge dent into farmers' yields that growers in China, India, and the United States plan to start planting Monsanto cotton, despite the fact that seeds cost three times as much as other plant strains. In the meantime, more than one-third of the cotton grown across the world has the Bt gene. In the United States, more than 70 percent of cotton crops have been genetically modified.

Initially, this saved farmers money because they could cut back on the amount of costly pesticides needed to protect the crops—their use dropped by more than 70 percent. But a study completed in 2006 by researchers from Cornell showed that within seven years, Chinese farmers were using just as many pesticides as before. Not because the bollworm had evolved resistance—but because in the absence of the larva, other pests such as leaf bugs were moving in. Those, too, have to be combatted using pesticides.

These facts have rightly been cited as examples of the unintended problems that can arise through the use of GMOs. Yet it is important to remember that bollworms are not really natural pests that are being chased out by unnatural ones. The vast cotton fields that they infect are not natural; the plant itself is the product of millenia of artificial breeding practices. Farmers in the Indus Valley (located in today's Pakistan) began cultivating it from a wild plant more than 6,000 years ago. Within 1,000 years, another form of the plant was being cultivated in Mexico. Today the largest fields are found in West Texas. They do not grow there naturally, which means that the bollworms that infect them are not more natural than secondary pests such as leaf bugs. Growing cotton also requires a great deal of water that has to be diverted from other sources, so another aim of breeding and genetic engineering is to reduce the amount of water that the plants need to thrive.

Cotton is a good example of the complexity of this issue. Inserting the Bt gene is not the only way researchers are modifying the crop. Cottonseeds cannot be eaten because the plant produces a toxic substance called gossypol—which also acts as a natural pesticide, because very few organisms can digest it. However, scientists at Texas A&M University have used genetic engineering to produce a strain of the plant whose seeds do not contain gossypol. This may make it possible to turn cottonseed into a source of food for livestock and even humans. That would be of immense importance because cotton is already one of the most-cultivated plants in the world.

These factors point out some major social issues concerning GMOs that has little to do with science—they are also commercial products. Businesses, governments, and others have invested heavily in their development and obviously have an interest in turning crops into profit. Like any application of science, whether they ultimately have a positive or negative influence on society will depend on how wisely they are used by those who develop them, those who sell them, and ultimately the consumers who decide to buy them. The same types of issues surround other potential uses of genetic engineering that are explored in chapter 4.

CLONED DINOSAURS AND INVADERS FROM SPACE

Frankenstein, Brave New World, and many other science fiction novels established one of the most prominent themes of the genre: Bad things happen when scientists overextend their reach, trying to manipulate things that they do not completely control, using technology that they do not fully understand. These were prominent themes in the fiction of Michael Crichton (1942–2008), a trained physician who became one of the most popular authors of the late 20th and early 21st centuries. After writing a number of thrillers, many of which somehow integrated technology or scientific themes, Crichton achieved his breakthrough with a 1969 novel called *The Andromeda Strain.* The plot of *The Andromeda Strain* centers around a team of biologists fighting a deadly microorganism that seems to have come from outer space, brought to Earth aboard a satellite. Researchers retrieve the satellite from the small town of Piedmont, Arizona, near its crash site. It has been opened by curious citizens, unleashing a plague that wipes out everyone in the town except for an old man and a small baby. The team takes the satellite and the survivors to an underground laboratory in Nevada, where they hope to isolate the microbe and find a cure. The facility is equipped with an ultimate fail-safe measure—a nuclear weapon that will destroy it (and everyone inside) in case the organism escapes.

Throughout the book, scientists are confronted with situations that arise because they assume they are in control of things. But nature is messy, chaotic, and complex, and unforeseen events nearly lead to disaster. Problems range from the trivial—a piece of paper gets stuck in a teletype machine, preventing the transmission of a vital message—to mistaken assumptions that have

A predominant theme in many of Michael Crichton's science fiction novels is how the complexity of living systems brings near disaster to scientists who think they have nature under control.

nearly fatal consequences. Having no previous experience with extraterrestrial life, the researchers assume that the radiation from an atomic blast will kill it. By the time they discover that this would have the opposite effect—causing the organism to absorb energy, mutate rapidly, and become infinitely more dangerous—it is almost too late. Ironically, the primitive technological failure and the huge scientific mistake cancel each other out. At the climax of the novel, the bomb must be stopped, but a design flaw in the facility makes this almost impossible. The world is saved not by brilliance, but by a combination of ingenuity and good luck.

Most of Crichton's later books are built around stories with similar elements. In his 1990 novel *Jurassic Park,* a wealthy biotech billionaire named John Hammond has found a way to create living dinosaurs, using DNA found in insects that drank dinosaur blood and then became preserved in amber. The DNA is not intact, but researchers have filled in the gaps with genes from modern relatives, such as birds and amphibians. The information has been used to create complete, artificial genomes of several dinosaur species. They are inserted into cells, transplanted into eggs, and cloned. The animals are raised and kept on an island called Isla Nublar, which Hammond intends to turn into a new entertainment park.

Because of the obvious dangers of dinosaurs, various safeguards have been put into place to maintain control. To prevent the dinosaurs from reproducing, they have all been engineered to be female. They are meticulously tracked, and every animal is accounted for. But the book opens with signs that things are not going as smoothly as planned. Tourists have been attacked by an unidentified beast in nearby Costa Rica. The billionaire brings a team of experts to the island to investigate, hoping for a clean bill of health. The team soon discovers that nature has found a way to overcome man's limitations; the dinosaurs are having offspring. One by one, the island's security systems fail. As the group's mathematician has predicted from the very beginning, complex systems—like organisms—can never be completely understood and controlled.

Jurassic Park is a good example of how scientists are portrayed in Crichton's works. They are both heroes and antiheroes, human beings who suffer from the same weaknesses as everyone else. The crisis in the novel is partly the fault of researchers and their attempts to use gene technology to an end that is far too ambitious. The problem is compounded by human error, the greed of individuals and companies who are trying to profit from scientists' work, mistakes in planning, and flaws in technology. The situation leads to a rapid breakdown of control on the island, and the humans become prey of a dinosaur population that is completely out of control. As in *The Andromeda Strain,* however, things do not end nearly as badly as they might. A few people manage to escape the island, through a mix of knowledge and luck. Supposedly the dinosaurs are destroyed—only to return in a 1995 sequel called *The Lost World.*

Crichton's 2004 *State of Fear* takes the relationship between fiction and science further, taking aim at the politics of global warming. The antagonists are a group of ecoterrorists who are staging a number of natural disasters (actually of their own making) that will cost a huge number of lives; their aim is to convince the world that the environment is rapidly being destroyed. As he wrote the novel, Crichton believed that there was not enough scientific evidence to support some of the dire predictions that many scientists were making, and that there was little evidence that the solutions proposed would actually have an effect on the problem.

State of Fear was heavily criticized by a number of experts on global warming, journalists, and environmentalists—including many whose papers Crichton had cited in the footnotes. They claimed that Crichton had misinterpreted and misused their findings. In an article in the January 20, 2005, issue of *Nature,* Myles Allen of the University of Oxford's Climate Dynamics Group wrote, "Although this is a work of fiction, Crichton's use of footnotes and appendices is clearly intended to give an impression of scientific authority."

The history of science fiction shows that visions of the future can be useful and help people consider the ethical implications of research in fiction, before they have to be confronted in the real world. But as Allen and many others point out, the distinction needs to be preserved. A hallmark of the early 21st century is a rapid blurring of lines between science and fiction, science and politics, entertainment, news, and many other types of reality and fiction. Novelists have always taken liberties with scientific facts in an attempt to create art, promote personal points of view, and sway people's opinions. This has taken on a new dimension at a time when the technology of the entertainment industry is able to create compelling images that often seem more real than the real world. It is easy and dangerous for people to become confused and mistake fictional risks for real ones. If people can no longer recognize the difference between real science and entertainment, their opinions will be at the mercy of those who plan next week's television schedule and who are best at using the media in support of ideological agendas. And then the sort of brave new world imagined by Huxley and other anti-utopians will likely not be far behind.

3

Studying Life in the Post-Genome Era

Modern molecular biology began with the discovery of the structure of DNA in the early 1950s. Over the two decades that followed, scientists worked out the way that cells use the information in the genome: by building *messenger RNA* molecules (mRNAs) based on the sequences of genes, and then using the mRNAs to make proteins. Once the roles and relationships of these molecules had been clarified, the main aim became to understand the function of each gene in the life of a cell and organism.

Everything that happens in the cell involves complex networks of dozens or hundreds of molecules. Until the 1990s, however, technological limitations meant that scientists could usually investigate the functions one by one; at best, they could observe the behavior of a few molecules at a time. Now this has changed thanks to rapid developments in biotechnology. At the dawn of the 21st century, which many researchers call the beginning of the post-genome era, the strands of biology, physics, medicine, and modern disciplines such as biocomputing have come together to produce a rich, modern view of life. This chapter introduces some of the most important, cutting-edge methods in biology and describes how they have begun to change our view of life.

GENOTYPES AND PHENOTYPES

A huge amount has been learned about life since the birth of molecular biology in the mid-20th century, but some of the most interesting questions have yet to be answered. A main goal of current research is to understand the relationship between an organism's *genotype*—the complete set of hereditary information in its genome—and its *phenotype,* which means the structure of its body and anything else about it that can be observed. The phenotype includes tiny features such as the molecules present in its cells, as well as much larger phenomena such as details of the stages of an organism's development or its behavior. The genotype is invisible—except in the sense that it can be read by obtaining a DNA sequence—and some of the information it contains may never appear in the phenotype. For example, a woman may have inherited a form of a gene that causes color blindness and pass it along to her sons without herself becoming color-blind.

Ideally, scientists would like to be able to look at a genome sequence and make detailed predictions about the phenotype it can produce. This is already possible to a certain extent, in situations such as the following.

- Matching a fragmentary DNA sequence to a particular species or its place in the living world. All organisms on Earth have related genes that have undergone mutations as species evolve. Each species inherits the "spelling" of its genes from its direct ancestor; subsequently they undergo new changes. This chain of events can be read from DNA sequences, so an organic sample can be used to identify the species from which it comes. If the species is unknown, it can usually be classified into a *clade* (a group of organisms that belong to the same evolutionary branch). The early summer of 2005 saw an amusing application of this principle when a number of people reported seeing a large, unidentified creature near the town of Teslin, in Yukon, Canada. Media reports suggested the animal might be bigfoot—an unidentified, apelike creature—that some people believe inhabit sparsely populated areas of North America and other regions of the world. Strands of unusual hair were found in the vicinity of one of the sightings. The hair was given to

David Coltman, a geneticist at the University of Alberta, for analysis. Coltman obtained DNA from the sample, sequenced it, and compared it to other known species. The DNA turned out to have come from an American bison. Coltman told a bigfoot enthusiast that he believed the sample had come from a rug made of bison hide and had probably been deliberately planted.

- Predicting that a plant or animal will have specific features based on an analysis of its alleles. Mendel showed that greenness is dominant in peas, so a plant will have green seeds even if it also carries a recessive allele for yellowness. Therefore, if scientists discover the greenness allele in a pea plant, they can predict the color of the seeds it will produce. This is also true of other *monogenic traits* where one form of a gene is dominant. But only a few traits in humans are truly monogenic. One is having a cleft chin; another is the ability to roll the tongue into a U-shape. Most characteristics, such as skin color, are the result of contributions from many genes. In those cases, it is much more difficult (and often impossible) to make an accurate prediction of the characteristics a person will develop based on an analysis of his or her DNA.
- Predicting that an organism will develop, at some point, certain diseases. Researchers have connected thousands of alleles to monogenic diseases. For example, a person born with a particular form of a gene called huntingtin is virtually certain to develop Huntington's disease. The defect causes the loss of cells called medium spiny neurons, which play an important role in coordinating movement. The problem is usually discovered when a person begins to experience uncontrollable, jerky movements that come from this loss of coordination. Other diseases are thought to be caused by combinations of specific alleles. These are much more difficult to identify. Many seem to be more susceptible to environmental influences than monogenic conditions, so it may be impossible to predict with accuracy whether a particular individual will develop the disease. There are often

big differences in the degree to which people who have inherited such a multifactorial trait develop symptoms. Even interpreting the results of a test for a monogenic disease has to be done by experts with care, because in some cases a second gene might be able to reverse the effects of a defective one.

- Reconstructing ancient forms of genes and features of organisms that no longer exist. Two species usually have similar features because they inherited the characteristics from their common ancestor. There are exceptions. Fangs evolved many times—snakes and cats did not inherit them from the same ancestor. Wings evolved separately in birds and bats. But in general the principle holds, and it has been used to make hypotheses about shared ancestors. For example, scientists currently estimate that chimpanzees and humans descend from a primate that lived between 4.5 and 6.5 million years ago. Fossils of this ancestor have not yet been found, but a comparison of the two species gives some good hints about what it must have been like. Researchers possess more information about its genome than its appearance or behavior. Overall, 96 percent of the two genomes are identical, and many neighborhoods within the genome, containing genes and other information, are about 99 percent identical. This allows scientists to reconstruct nearly the entire genome of the ancestor. If they can learn more about how genes interact to build a body, they may be able to construct a very accurate model of its biology and appearance.

All of these types of work would be given a boost by a detailed understanding of the connection between genotypes and phenotypes, especially the early diagnosis of genetic diseases. In 1990, Mary-Claire King (1946–   ), professor of genetics and epidemiology at the University of California, Berkeley, discovered that some alleles of a gene called BRCA1 are strongly correlated with a much-increased risk that a woman will develop breast cancer. At the time, the idea that there might be a genetic basis for cancer was very controversial. Further research has shown that she was right and has also offered a partial explanation. The function of BRCA1 is normally to help repair damage to DNA, which can occur as a result of exposure to radiation, other types of environmental contamination, or through mistakes in cell division. A healthy form of BRCA1 can step in and repair the damage, which might otherwise trigger cancer. For this reason BRCA1 and similar molecules are called *tumor suppressor genes.*

Mary-Claire King, professor of genetics and epidemiology at the University of California, Berkeley, discovered that some forms of the BRCA1 gene inherited by women are strongly correlated with a high risk that they will develop breast cancer. *(Peter and Patricia Gruber Foundation)*

If BRCA1 breaks down, it no longer does its job, and cancer-causing defects can slip through. However, not all women who have inherited the defect develop tumors. Cancer often requires other random, unpredictable mutations in genes—either through natural mistakes as DNA is copied or influences from the environment—and a woman might be lucky enough to avoid them. Finding a way to calculate risks more accurately would be very helpful both to patients and doctors as they face difficult decisions about therapies.

GENOMIC TECHNOLOGIES

In the post-genome era, the molecule-by-molecule approach to studying life has given way to technologies that can observe the activity of the entire set of genes in a cell and throughout an organism's body. One aim of these experiments is to get a better idea of how molecules participate in cellular processes. There are also more ambitious goals, such as discovering the causes of cancer and other diseases, finding better ways to diagnose them, and learning why different people respond to the same medication in different ways. This information will be an important step on the road to personalized medicine, in which details of a person's individual genome are taken into account while diagnosing tendencies to diseases and designing treatments.

These methods will be the most powerful when researchers and doctors have access to the complete genetic codes of individuals. (The human genome sequence that has been obtained now is a reference version, obtained by combining the DNA of several anonymous individuals; no one person has exactly this sequence.) Realistically, this will only happen if the cost of obtaining a complete sequence drops to a reasonable level. The U.S. National Institutes of Health (NIH) has launched a program whose target is to bring the price to $1,000 dollars by the year 2014. If that target can be met—in 2014 or the near future—it will be feasible to sequence individual genomes on a large scale, and it will signal the beginning of an era of personalized genomics. How far the information affects a person's medical care will depend on what has been learned about the genetic causes of disease.

Currently, researchers have linked about 6,000 single genes to diseases. Multiple genes are thought to contribute to many more illnesses, but finding them requires studies of very large families. If a parent has a disease caused by two genes, only a fourth of his or her children will inherit both of them. If the disease is recessive, then both parents have to have both genes—and only one in every 16 children may display symptoms. Not many families have that many children, so the disease might not appear for several generations. If it does, it may not be recognized as a hereditary problem. A similar situation confronts researchers who wish to understand very complex interactions between organisms and the environment, such as the links between diet and disease. There are so many variables in the environment that isolating the important ones might be impossible without collecting huge amounts of data.

Genomic technologies will provide a shortcut in diagnosing known genetic problems. In early 2009, nearly 2,000 types of genetic tests were available, but most target a specific disease (or just a few). With access to a person's complete genome, on the other hand, the sequence can be scanned by computer, for all known disease markers.

Most of the technologies that monitor the activity of the genome do so by detecting RNAs, which are produced when a gene is switched on, or proteins, which are made using the information in messenger RNAs. The main methods are described below.

- *DNA microarrays* (DNA chips) use probes made of DNA attached to glass or another material to detect RNA molecules. DNA and RNA are made of the same basic subunits—nucleotides—that bind to each other if they have complementary sequences. One common type of probe contains sequences that are complementary to RNAs for every human gene. RNAs are extracted from cells, and if the sample contains an RNA made from a particular gene, it binds to the probe and emits a fluorescent signal. Microarray experiments usually compare two types of cells, such as healthy human cells and those taken from a tumor. The readout of the experiment shows which genes are more active in one cell than the other and which behave the same way. If a gene is more active in one type of cell, this may show that it plays an important role in the process that is being studied—for example, the development of a tumor. The method was developed in 1994 by Pat Brown, a biochemist at Stanford University and the California-based company Affymetrix, and is now used almost universally in laboratories throughout the world.
- *Tiling arrays* are specialized micro-arrays that are often used to look at the behavior of regions of the genome that do not contain genes—in humans, this material accounts for at least 98 percent of the sequence. Until recently, the function of most of this DNA was a complete

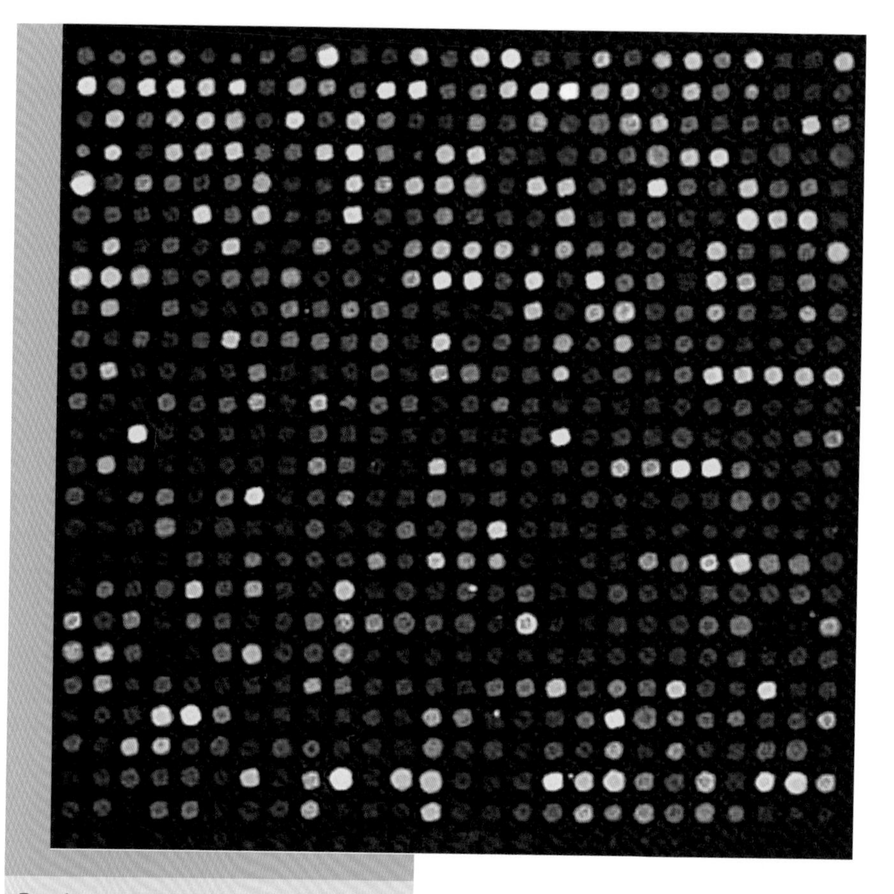

Readouts of DNA chips like this one show researchers how gene activity varies in different types of cells. Comparing a cancer cell to a similar but healthy one, for example, reveals genes that play a role in the development of tumors. *(WormBook)*

mystery—it was often called junk. Many researchers supposed that it was simply extra material that had accumulated over the course of evolution, such as artifacts of ancient genes that had undergone mutations and become nonfunctional. Now scientists have discovered that cells transcribe an enormous amount of this material. Often the products are small RNA molecules called *microRNAs.* They are not used to make proteins; instead, many of them dock onto other RNA molecules with complementary sequences and prevent their translation into proteins. Tiling arrays usually consist of very short probes to which small RNA molecules bind. In 2004, Jason Johnson and Eric Schadt used this method to survey the complete gene activity of two human chromosomes. Making the array was a huge task involving the construction of more than 3,700,000 individual probes. The results were stunning, revealing that cells made about 3,000 RNAs unconnected to any known gene—just on the two chromosomes. Schadt and his colleagues concluded that about one-fourth of these might represent real genes that had gone undetected. In 2005, Jill Cheng, Philipp Kapranov, and their colleagues at the company Affymetrix put together a tiling array for 10 human chromosomes, containing more than 74 million probes, covering about 30 percent of the human genome. They discovered that on the average, about 10 percent of each chromosome is transcribed into RNA. This is between five and 10 times the amount of RNA known to encode proteins.

- *Chromatin immunoprecipitation* (ChIP) surveys the genome to discover where particular proteins bind to DNA. Usually the activation of a gene begins when a protein called a transcription factor binds to a sequence. But a single transcription factor can usually activate many genes, and it has been difficult to discover all the targets of these molecules. The first step in the method is to fix proteins to DNA. Then enzymes are used to chop combinations of DNA and proteins into fragments. Antibodies are used to extract specific proteins and the DNA that they are bound to. ChIP is often used in combination with microarrays to identify the target DNA sequences.
- Mass spectrometry, described in chapter 1, identifies the proteins that are present in a sample taken from a cell. One of the most interesting uses of the method is to analyze the composition of protein machines. Most proteins carry out their jobs in complexes ranging

from a few up to 100 molecules—machines which are continually dismantled and rebuilt to perform different functions. This activity is central to understanding the cell, but it has been extremely difficult to observe. In 2002, researchers in Heidelberg, Germany, discovered how truly dynamic this situation is. Anne-Claude Gavin and Giulio Superti-Furga of the young biotech company Cellzome worked with scientists at the European Molecular Biology Laboratory to capture the first complete view of the machines at work in a yeast cell. They used a new method to extract whole machines from cells and analyzed their components with mass spectrometry. They found that 17,000 proteins form at least 232 machines. Many of them work in a snap-on way; they have a core of preassembled pieces and when it comes time to do a certain job a few more are added on. A machine may remain inactive until that happens. This gives the cell a way to control its activity. To be switched on, it may need to borrow the missing pieces from other protein complexes, or components may have to be made anew.

- Live cell arrays are slides made of glass, silicon, or another material on which living cells are grown in different compartments. Molecules or substances are introduced into the compartments to study their effects on the cells. One common use is to watch how cells respond to a drug or a toxin. Another type of experiment exposes the cells to various microRNAs, which block the production of specific proteins. This often reveals the functions of the molecules. If the loss of a protein disrupts the cell cycle and makes cells divide too often, it may be a sign that the molecule plays a role in cancer.

Each of these methods reveals a slightly different aspect of the complete molecular activity that takes place in a cell. In combination, they are giving researchers a new look at how information in the genome guides the life of an organism.

BIODIVERSITY AND METAGENOMICS

How many types of viruses and bacteria live in a person's body or the soil of a farm? What effects do pesticides, genetically modified crops, or the transplantation of organisms to new regions have on the environment? How great is the effect of global warming and human overpopulation on food chains across the globe? Answering these questions will depend on our ability to measure *biodiversity*—a survey of all the organisms in a particular environment or on the Earth as a whole. Even without considering most microbes, scientists have already identified and named about 1.6 million species (more than half of which are insects), but they estimate that the Earth holds many times that number. Some researchers estimate the number of insect species alone at 10 to 30 million. Few scientists are willing to venture a guess as to the number of types of bacteria and viruses that exist; it is sure to be many, many times more. Recently, the arrival of rapid DNA sequencing methods and databases of known sequences have given researchers their first deep look at this invisible world, which humans, plants, animals, and fungi are heavily dependent on.

The new approach, called *metagenomics,* was conceived by Norman Pace, a molecular biologist now at the University of Colorado at Boulder, in 1985. Up to that time, DNA sequencing efforts had focused on humans, important laboratory organisms such as flies and mice, and microbes that had been cultured in the laboratory. The idea was to start with one species, obtain its complete sequence, then move on to the next. Pace wondered what would be found if he simply sequenced all the DNA in a sample of water or soil—more like the genome of a global positioning system (GPS) coordinate.

This was particularly important, he felt, because most microbes could not be raised in laboratory cultures. In nature, they usually live in complex communities in which thousands of different types depend on each other for survival. These living networks fulfill vital functions for humans and the ecosphere. They lie at the base of every food chain, and they play a crucial role in regulating the chemistry of the atmosphere and the water supply. Every liter of ocean water, for example, holds billions of cells that help plants remove carbon dioxide (CO_2) from the atmosphere. But scientists have likely only seen a small fraction of them and have little idea of their roles in supporting other kinds of life, including humans.

In 2002, Mya Breitbart and Forest Rohwer of San Diego State University began taking an in-depth look at the ocean using a metagenomics approach. They discovered that 200 liters of seawater contain DNA that comes from more than 5,000 species of viruses, and one kilogram of marine sediments may contain up to one million species. Samples of human feces contain more than 1,000 species. Hardly any of these had been seen before. Along with viruses were signs of huge numbers of species of bacteria and other organisms.

The approach has also been taken up by Craig Venter (1946–), a biologist and pharmacologist who founded The Institute for Genomic Research (TIGR) in Rockville, Maryland, and later the company Celera Genomics. (Both organizations played an important role in the Human Genome Project, completing a second version of the genome at the same time as the international public version.) Since leaving Celera at the completion of the project, Venter has been sailing the world in a 95-foot (29-m) yacht called *Sorcerer II,* sampling the world's oceans and investigating other environments—including the human body. In a lecture given in 2007, Venter summed up the dis-

Craig Venter, founder of The Institute for Genomic Research and Celera Genomics, has dedicated his recent efforts to metagenomic studies of the world's oceans. *(Craig Venter and Liza Gross)*

coveries by stating, "Earlier this year . . . we published a single scientific paper describing more than 6 million new genes. This one study more than doubled the number of genes known to the scientific community and the number is likely to double again in the next year."

Metagenomic studies also provide a deep look at the effects of natural selection. Different environments are unique. Each one is more advantageous to some types of biochemical processes than others. A 2002 study of farm soil carried out by bioinformaticist Peer Bork and his colleagues at EMBL revealed a wide range of genes involved in the way different species respond to each other; it also turned up dozens of molecules involved in breaking down plant material. None of these were found in whale bones, taken from the ocean floor, or samples from the Sargasso Sea. The latter region yielded hundreds of genes similar to bacteriorhodopsin, a pigment which responds to light and allows cells to snatch energy from the environment.

A typical metagenomics study reveals thousands of types of genes that have never been seen in laboratory species. Some represent biological processes that do not occur in human cells. Investigating what these molecules do in exotic species of microorganisms will keep scientists busy for years to come. Researchers expect that many will have applications in medicine and industry.

This type of research is necessary to establish a baseline by which to measure the effects of global warming, human activity, or other types of changes on the environment. Researchers know that the growth of human populations, deforestation, and chemical and biological pollutants have caused the extinction of a large number of species, and the rate at which this is happening is increasing. The International Union for Conservation of Nature (IUCN) regularly evaluates the risk of extinction faced by thousands of species. Of the 40,117 species that were being monitored in 2006, more than 40 percent were listed as threatened. Only metagenomic approaches can determine whether the same phenomenon is occurring at the level of microorganisms.

Metagenomics, Extremophiles, and the Search for Extraterrestrial Life

For years, the evidence had been accumulating. Finally, on June 20, 2008, NASA confirmed the discovery: The Phoenix Mars Lander had dug a small hole and exposed a bright patch just below the surface that turned out to be water. This raised hopes that the planet might also hold living organisms. Many researchers believe that if water is present, there is a good chance that life will evolve, and in the past the surface of Mars held enormous quantities of water. An array of instruments would be used to analyze a small sample of soil, looking for signs of organic activity. But extraterrestrial microbes would undoubtedly have a completely different biology than Earth organisms. If alien life existed on Mars, would machines built by humans detect it? Not knowing exactly what to look for, scientists have been using the results of metagenomics studies and research into organisms called extremophiles that live in environments other forms of life are unable to cope with in hopes of obtaining hints about alien biology.

A major difficulty in searching for life beyond the Earth is that every living organism on this planet—from the simplest bacterium to human beings—shares a core set of molecules and biological processes. For example, all species store hereditary information in the form of DNA, which they transcribe into RNA molecules, which are then used as templates to make proteins. Cells have to be able to copy their DNA, convert the raw materials in food into useable molecules, and respond to changes in the environment. The universality of these processes on Earth makes it hard to imagine a form of life based on another chemical system.

An alien biology might not be totally different, however. Some researchers believe that extraterrestrial organ-

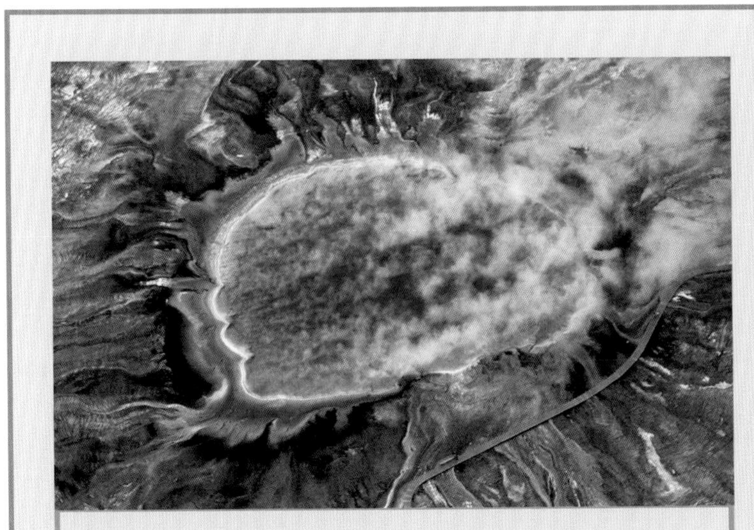

Metagenomic studies of extreme environments like hot springs and the organisms that live there may show scientists what to look for in the search for extraterrestrial life. This is the Grand Prismatic Spring, a hot spring in Yellowstone National Park, which is home to archaeal cells. *(Jim Peaco, National Park Service)*

isms might share some common elements with life-forms on this planet. Experiments attempting to reproduce the environment of the very early Earth, before life arose, have shown that some of the building blocks of proteins can arise from inorganic conditions. Even when the conditions have been changed, amino acids are almost always detected; they have even been found in clouds of gas in deep space. Amino acids are not alive, and the laboratory experiments have not produced entire proteins, DNA, or RNA, which are much more complex. That is not surprising, because it may have taken hundreds of millions of years for these molecules to arise in the vast laboratory of the early Earth's oceans. On the other hand,

(continues)

(continued)

the production of the first RNA and DNA molecules may have been a unique event that happened only on Earth. (It might also have taken place in space, and the molecules arrived on Earth as hitchhikers on meteorites, seeding the planet.)

It is also possible that extraterrestrial life might be based on an entirely different chemistry, particularly in environments that are quite different from the Earth. The surface of Mars, for example, is composed of 14 percent iron—nearly three times the amount in the Earth's crust. Organisms here make use of iron, taking advantage of its high chemical activity; at the same time, they have to control it carefully because even a slight overdose will disrupt cell chemistry. Life on Mars would need to have mechanisms to cope with this.

It might be possible to get a glimpse of what it would be like by studying organisms that live in iron-rich environments on Earth. Extremophiles such as bacteria or archaea that live in the boiling waters of hot springs or soil with high amounts of salt need special mechanisms to cope with such conditions. Extraterrestrial organisms may have adopted similar chemical strategies, which would give NASA an idea of what to look for.

Before they can hope to understand alien biology, researchers may need to get a better grasp of what is hap-

Scientists universally agree that a drop in biodiversity could have extremely serious effects on humanity. Over half the pharmaceutical compounds that have been developed for use in the United States are derived from compounds found in plants, animals, and microbes. Insects—a huge number of which are also threatened—play a crucial role in pollinating plants. It would be impossible to replicate that activity artificially.

pening on Earth. Metagenomic studies are revealing a range of processes that have never been observed in the laboratory because so few organisms can survive there. They are also revealing chemical signatures of different types of environments such as the farm soil and ocean floor. Living beings on a moon of Jupiter may have started with a much different chemistry than the cells born in the oceans of the early Earth, but natural selection may have pushed them to adapt in ways similar to organisms living in comparable environments here.

Until the first extraterrestrial is found, there is no way to really guess what it will be like, so the instruments needed to detect it are being designed in a very flexible way. The Phoenix Mars Lander is equipped with an instrument called a thermal and evolved gas analyzer (TEGA), constructed by the University of Arizona and the University of Texas at Dallas. The TEGA contains eight tiny ovens that will slowly cook small samples of soil, looking for changes in energy and chemistry. In the end, the temperature is so hot that the material turns into a gas, which is analyzed in a mass spectrometer. The hope is that the chemical experiments will reveal some sort of respiratory process—a conversion of substances that could only be carried out by a living organism—and that the mass spectrometer will reveal complex organic molecules. The latter may be detectable even if life vanished from Mars long ago.

The human body itself is host to an entire universe of microorganisms; scientists estimate that there are 10 times as many types of microbes on the skin alone as there are cells in the body. According to a 2007 report from the National Research Council (NRC), somewhere between 10 and 100 trillion microbes live in the intestine, where they "perform functions that humans have not had to evolve, including the extraction of calories from otherwise indigestible components of our diet and the synthesis of essential vitamins and amino acids. The complex communities of microbes that dwell in the human gut shape key aspects of postnatal life, such as the development of the immune system, and influence important aspects of adult physiology, including energy balance. Gut microbes serve their host by functioning as a key interface with the environment; for example, they defend us from encroachment by pathogens that cause infectious diarrhea, and they detoxify potentially harmful chemicals that we ingest."

MANIPULATING GENES

The tools of genetic engineering introduced in chapter 1 have given rise to a wide palette of methods by which researchers can manipulate genes. The previous chapters have described some of the most important applications: creating crops resistant to pests or herbicides and using bacteria or animals to produce proteins for medical use. Yeast cells have been altered to yield better beer, and researchers are using bacteria to clean up environmental contamination through bioremediation. The late 1990s and early 21st century have seen the development of several new methods to give researchers a much more precise control of genes. At the moment, these techniques are mainly being used to discover gene functions, but one day they may produce new types of medical therapies.

Some of the techniques include the following:

- knock outs, which delete a gene
- *knock ins,* which add a gene to a cell or organism that does not normally have it
- overexpression studies that raise the amount of RNA and/or proteins produced from a given gene

Hermann Muller was the first researcher to deliberately introduce mutations in animal genes, using X-rays that caused random changes in DNA bases. Later, scientists began to use chemical mutagens. These measures often lead to knock outs by creating a gene sequence that encodes a defective protein. Another effect may be *constitutional activation,* in which a gene is switched on even when it should not be. If the molecule is involved in stimulating cell division, it should normally be turned off; an overactive version may lead to a tumor.

Modern reverse genetic techniques allow scientists to make precise, targeted changes in specific genes and watch what happens to cells, plants, or animals. The first knockout methods were all or nothing, completely removing a gene and eliminating its functions in all of an organism's cells throughout its lifetime. If the molecule plays an important role in an animal's embryonic development, this likely leads to such severe defects that the fetus dies before birth. Obviously, that makes it impossible to study a gene's functions during later phases of its life. It is often the case that the same gene may be needed at different times to do different things in various types of cells. For example, a protein called PS1 seems to act as a switch for different types of functions: It is needed to pass important signals that tell some types of cells to grow and develop. At other times and places in the body it is involved in *apoptosis,* a type of cell suicide that is necessary as tissues form. If the gene for PSI is removed, organisms lose control of these processes.

It is not surprising that proteins have multiple functions or even tasks that may seem contradictory. Human genes evolved from much smaller genomes in ancient ancestors with much simpler bodies and fewer genes, and those ancestors can all be traced back to a single cell. The development of new cell types and tissues did not necessarily require that new genes arise; often it occurred because cells began using the existing set of molecules in new ways. Just as a variety of electronic devices have some of the same components, multiple systems in the body rely on common proteins that have adapted to different tasks. So yeast, which is a single cell, contains proteins that now help build brains, eyes, and other highly complex organs in animals.

Getting a handle on fundamentally important genes required a way to shut down genes in specific types of cells at specific times. In the mid-1990s, the geneticists Klaus Rajewsky, Frieder

Schwenk, and their colleagues at the University of Cologne in Germany figured out how to do this with the invention of *conditional mutagenesis.* Their method relies on the fact that genes are accompanied by control elements—sequences that proteins dock onto in order to control when they are activated. Rajewsky's lab built genes with artificial switches that gave the scientists control over when and where a gene was shut down in an organism.

As with many other methods in genetic engineering, the technique is based on molecules from bacteria. An enzyme in bacteria called Cre recognizes patterns in DNA called loxP sequences. If Cre finds two of these sequences in DNA, it binds to the sites and draws them together, making a loop of the DNA that lies between them. This looped sequence is cut out, destroying a gene or any other information that it contains (such as regions that control a nearby gene). The cell then repairs the break by gluing the cut ends together. Thus, the first step in creating a conditional mutant is to build an artificial gene centered between loxP sequences.

The DNA is only knocked out in cells that produce both Cre and loxP sequences. If they are active in all of an organism's cells, the effect is like a complete knock out. Since the whole idea behind conditional mutagenesis is to avoid this, Rajewsky and his colleagues had to find a way to activate Cre only in particular types of cells. The solution was to find other genes that were only switched on in specific tissues or cell types, such as a molecule which is only produced in the brain. By combining Cre with the control regions of such genes, scientists could ensure that it too would only become active in the brain. The same technique could be used to study genes in any other tissue, providing a unique control region could be found.

Further refinements now allow scientists to determine the time as well as the type of tissue in which Cre becomes active. This is accomplished by attaching a switch, such as a receptor protein called LBD, to the Cre gene. LBD only becomes active in the presence of a specific hormone. Since animals do not naturally produce this hormone or obtain it through their diets, Cre

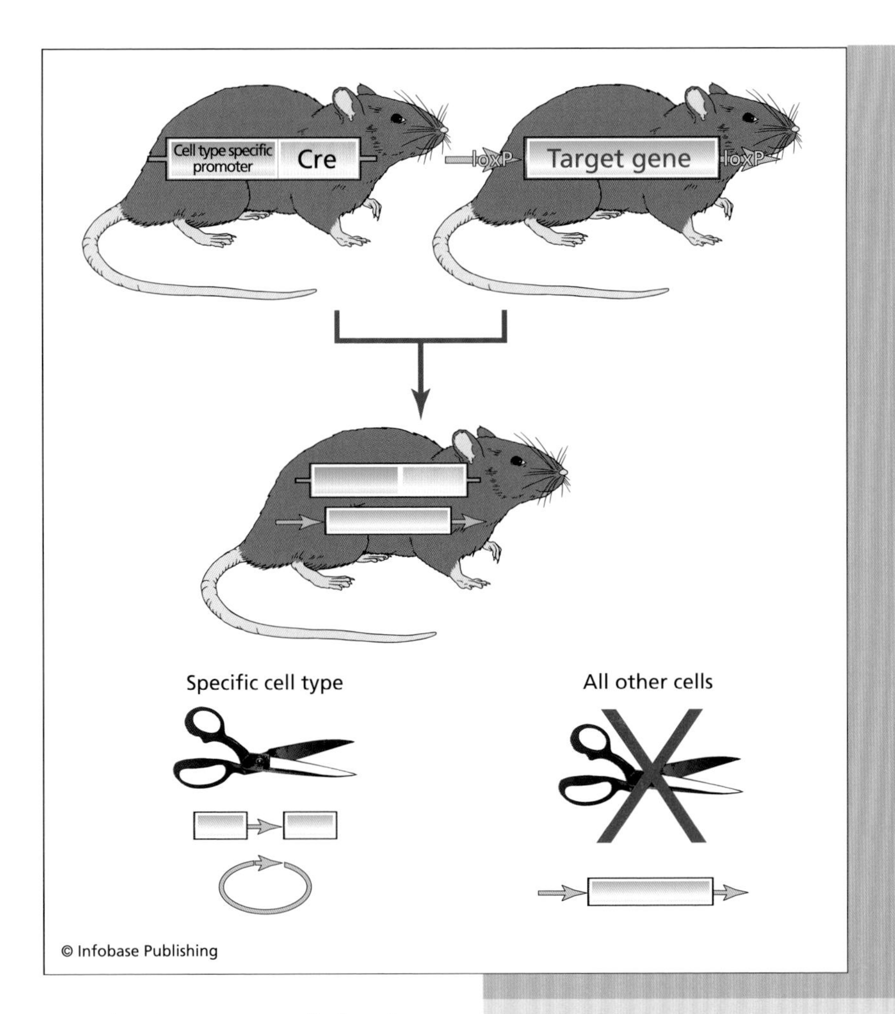

Conditional knock outs using Cre. This method requires developing one strain of mouse with the Cre gene—which acts like a pair of scissors—and a second strain with the target gene (blue). When the two mice are mated, they produce a mouse with the scissors and the gene that it will remove. By placing the gene for Cre behind a promoter that activates it only in a certain kind of cell (green), scientists can knock out the target gene only in that type. The rest of the animal's cells will keep the target gene.

remains inactive until the desired time. Then the hormone is administered in an animal's food or by injection.

The Cre and loxP marked genes have to be introduced into separate strains of mice, which are then mated. Some of their offspring will have cells with both Cre and the targets. While this means waiting at least two generations for a mouse that has both elements, it also permits scientists to mix and match strains with Cre under the control of different tissues with those with different genes marked by loxP. For example, with the same Cre mouse, researchers can investigate the functions of different molecules in the brain by mating it with animals that have loxP attached to different target genes. And the reverse is also true. If the same protein is needed in the brain and the kidney, for example, and its gene has been tagged with loxP, scientists can mate the mouse with one Cre animal to test its functions in the brain and another to see what it does in the kidney.

Ideally, researchers would like to have a strain of mouse in which each gene is surrounded by loxP elements and other strains that express Cre in each tissue and cell type. Theoretically, this would allow researchers to test the function of every gene in every kind of tissue. It would be an enormous amount of work—mice have at least 13,000 genes and at least several hundred different cell types. Yet the usefulness of the mouse in creating human disease models has convinced many researchers that doing so could be worth the effort. This has encouraged scientists to start creating Cre zoos—collections of animals expressing Cre in different tissues. These animals are commonly shared by different labs, saving time and reducing the number of animals used in research. Centralized collections of mouse strains have been established at Jackson Laboratories in Maine, the European Mutant Mouse Archive near Rome, Italy, and elsewhere.

Studying these animals will not solve all questions about the functions of genes in humans or even in mice, because everything that happens in cells and organisms requires the collaborative efforts of many genes. One day, scientists dream of developing research animals with switches on every gene; this would give them control of many genes at the same time and allow them to investigate complex patterns.

In the meantime, researchers have discovered other ways to shut down genes—for example, by introducing small RNA molecules that block the production of proteins from other RNA molecules (see section titled Genomic Technologies on page 91). If researchers can find a way to introduce these molecules

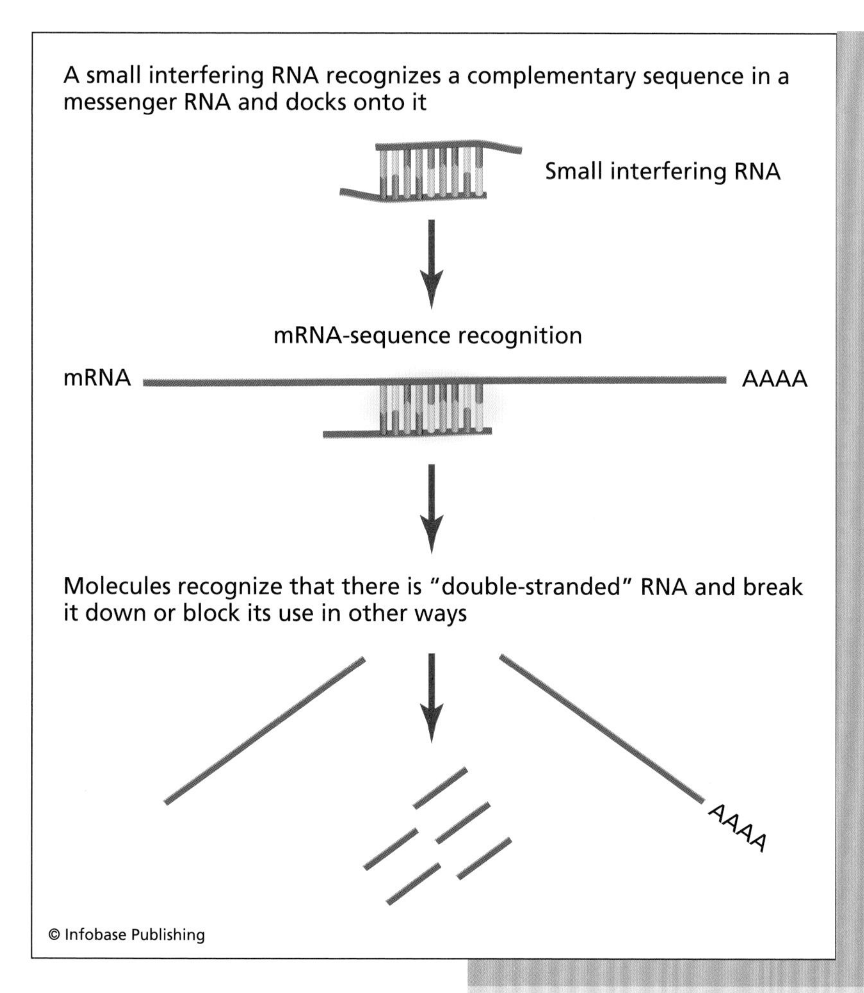

Small interfering RNAs (siRNAs) provide a new method of knocking out genes. When siRNA are introduced into cells, they dock onto a messenger RNA with a complementary sequence and prevent it from being used to make the protein it encodes.

into specific cells in an organism, it would give them a useful new knockout technology that might even be turned into medical therapies.

The first successful clinical trial of such *small interfering RNA* (siRNA) molecules was carried out in 2005 by Sirna Therapeutics, a pharmaceutical company based in Boulder, Colorado, on patients suffering from macular degeneration. People with this disease lose their eyesight because of the death of vision cells called rods and cones or because blood and proteins leak into the inner lining of the eye. Doctors treated the patients with siRNAs to block the production of a protein that plays a key role in the disease. The experiment was considered safe because the eye is relatively closed off from surrounding tissues, which meant that there was little danger that the molecules would escape to other parts of the body. All of the patients showed improvements over the course of 157 days with no signs of side effects. More clinical trials are planned to treat other diseases.

IMAGING MOLECULES WITHIN LIVING CELLS

Google Earth allows an Internet user to zoom in on any region of the world, close enough to pick out houses, cars, and people. But the resolution is not high enough to identify a single person. Similarly, the most powerful electron microscopes can sometimes spot large single molecules and complexes containing many of them, but the image is not sharp enough to directly identify a specific protein. That would be helpful in understanding the functions of single molecules and the processes that go on in cells.

The resolution of light microscopes is about 1,000 times poorer than that of an electron microscope. Even so, light microscopy has been undergoing a renaissance because of the discovery of fluorescent proteins that can be used in new ways to study living processes.

In the early 1960s, Osamu Shimomura (1928–) and his colleagues at Princeton University isolated two luminescent proteins from *Aequorea victoria,* a species of jellyfish. Doing so required that Shimomura undertake a seven-day drive from New Jersey to Friday Harbor Laboratories of the University of Washington to capture and then dissect 10,000 jellyfish in just a few months every summer. Shimomura and his colleagues managed with the help of schoolchildren. By the mid-1970s, the researchers had perfected their routine and were collecting more than 3,000 jellyfish every day. In 1986, laboratories in the

United States and Japan simultaneously isolated the DNA sequence encoding one of the proteins, called aequorin. The second, called *green fluorescent protein* (GFP), was isolated in 1992 by Douglas Prasher and colleagues at the Woods Hole Oceanographic Institution in Massachusetts. These were crucial steps on the way to being able to work with molecules and turn them into tools for research.

GFP absorbed blue light emitted by aequorin and flashed brilliant green. Martin Chalfie of Columbia University immediately realized that it might be possible to turn the protein into a tool for microscopy. A laser microscope could play the role of aequorin; shining the right wavelength of light on GFP might trigger it to flash. The real use of the tool would come from the fact that the light-emitting part of the protein was located in one small, compact region (or domain) of the molecule. It might be possible to attach the domain to other proteins. If so, it would serve as a beacon that would allow molecules to be tracked under the light microscope.

Several steps were necessary to turn GFP into such a tool. S. James Remington, a structural biologist at the University of Oregon, obtained crystals of GFP and used X-rays to obtain a high-resolution map of the molecule. One of Remington's collaborators, Roger Tsien, and his colleagues at the University of California, San Diego, discovered a way to alter the molecule's structure so that the wavelengths of light produced by laser microscopes could activate it. Additional changes made the module much brighter and allowed GFP to work efficiently at body temperature. Since the late 1990s, Tsien's laboratory and others have developed versions of the molecule that emit other colors, including blue, cyan, and yellow. And, using proteins from coral and other animals, more fluorescent tools have been produced, which can be attached to other genes that make proteins that are fluorescent but otherwise normal. For their work, Shimomura, Chalfie, and Tsien were awarded the 2008 Nobel Prize in chemistry.

The method has several important uses. The first is simply identifying whether a molecule is made by a particular type of cell and where it carries out its functions: in the nucleus, the membrane, or another region of the cell. Another use of fluorescent microscopy is to find out whether switching on one gene leads to the activation of another—by marking them with different colors and watching to see when they are made.

Some of the more elaborate uses of GFP and similar proteins are based on the physics of how they absorb energy. Each GFP-tagged protein gives off a particular signal with precise characteristics. The signal shifts whenever the protein's activity changes—for example, when it binds to another protein or a small substance such as a drug. Thus fluorescence microscopy has become an important part of the drug development process as well as a tool to investigate the functions of molecules. If two molecules are labeled with different fluorescent modules and they bind to each other, each absorbs a bit of the light energy given off by the other. This can be detected by measuring the light that they emit. For the first time, researchers could directly observe the binding of two proteins in a living cell—even though the molecules themselves are too small to be seen in the microscope.

Next, researchers learned to apply the methods to tissues and organisms. The light from a microscope's laser can penetrate several layers of cells and excite a fluorescent molecule that lies below the surface. This principle was used in the late 1980s in the development of confocal microscopes. These instruments use a point of laser light to scan a sample, focused on a particular depth. It excites the fluorescent proteins; then the laser is refocused on the next lower layer. An image is captured of each layer and then a computer assembles the slices into a three-dimensional image. As digital imaging technology improved and computers became faster, it was possible to do this with living samples, such as insect larvae or fish embryos suspended in a liquid. A favorite subject was the zebrafish, a tiny fish that was becoming popular for genetic experiments. The fish is virtually transparent, which makes it easy to observe internal structures over the course of development.

Before these methods can be used, a scientist has to use genetic engineering techniques to attach GFP or another fluorescent module to specific genes. This means that molecules

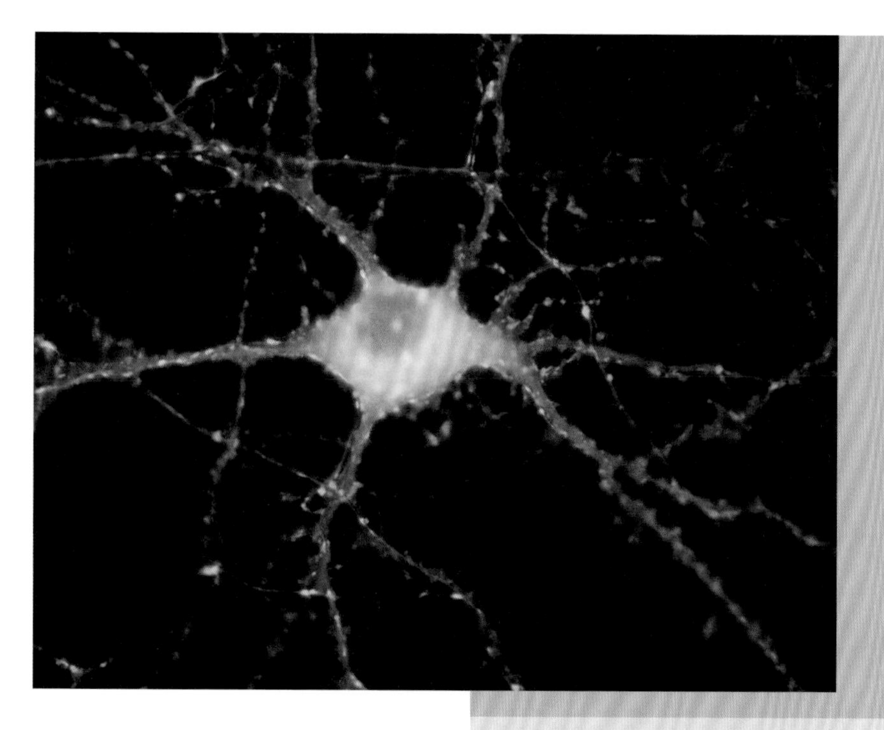

Molecules in this neuron have been labeled with a fluorescent protein, allowing researchers to observe their locations and behavior using light microscopes. *(Michael A. Colicos, Division of Physical Sciences, USCD)*

have to be investigated one by one; researchers cannot peer into a sample and survey all the proteins that are found there. A new method gets around this restriction by combining microscopy with mass spectrometry. A beam is passed through a thinly sliced sample, creating ions that are captured by the spectrometer. They are weighed, as described earlier in the chapter, and from the results it is possible to identify the proteins that are present in the tissue.

THE DIGITAL EMBRYO

The early 21st century has seen the development of a range of techniques that can be used to study the activity of molecules and other biological processes in living organisms. One of the most interesting new applications is to watch developmental processes in embryonic fish and other laboratory organisms under the microscope over long periods of time. An ongoing project at the European Molecular Biology Laboratory (EMBL) in Heidelberg, Germany, is doing this in a unique way.

In the 1990s, Ernst Stelzer and his laboratory at EMBL gained a reputation for developing innovative types of microscopes that have helped bring light microscopes into the molecular age. Alongside making improvements in confocal instruments, they have constructed new types of microscopes that can make high-resolution, three-dimensional films of living fish and other small embryos.

The recent project is a collaboration between the lab of Stelzer and the developmental biologist Jochen Wittbrodt, one of his colleagues at EMBL. Wittbrodt wondered whether it might be possible to carry the method even further and study a single organism over long periods of time, perhaps even following its entire embryonic development. The embryo could be kept alive in a small, water-filled chamber that served as a sample chamber. But first Philipp Keller and his colleagues in Stelzer's group needed to make some improvements in the microscope. One problem was blurring and shadows that prevented scientists from obtaining a sharp look at details below the surface of a sample.

The problems stemmed from the fact that the laser of the microscope and the detector that captured images were right next to each other, aimed at the sample from the same direction. This was like taking a picture of an aquarium using a digital camera with a flash mounted right next to the lens. The glass of the aquarium might pose a problem by reflecting the flash; in the same way, a fluorescent molecule at the top of a sample might interfere with seeing molecules that are underneath it. Additionally, the picture of the fish inside might be very sharp, but it would be hard to guess whether they are at the front of the aquarium or the back, because most cameras do not provide very good information about how far away something is. Fish at various distances might be in focus. Researchers were having the same problem with laser microscopes when they wanted to make three-dimensional images—the resolution from side to side was very good, but along the line of sight, things were much blurrier.

Photographers sometimes get around these problems by using a remote flash, aimed at the subject from another angle, and the solution developed by Keller and other members of Stelzer's group was similar. In their new method, called digital scanned laser light sheet microscopy (DSLM), the microscope lens examines a specimen from the front while light enters from the side. The light consists of a very thin sheet that is slowly passed through an organism, from front to back. Only objects within that sheet are illuminated, which tells the researcher exactly how far away they are. The sample is then rotated and illuminated from different directions. This produces sharp slices that can be assembled into three-dimensional images with the help of the computer.

Keller and Wittbrodt began using the method with embryos of zebrafish and another small fish called medaka. One aim was to make very detailed maps revealing the tissues that produced specific molecules at various stages of development. Then they embarked on a much more ambitious project to create a living genealogy of each of the cells needed to make up a fully formed fish.

This has been an aim of biology since the mid-19th century, when Rudolf Virchow proposed that every adult cell stems from a single, fertilized egg. The best way to understand development would be to be able to trace the complete life history of every adult cell, back through each stage of specialization and development, all the way back to the egg. Researchers had some basic techniques to do this. In the late 19th century, embryologists learned methods to stain particular types of cells in the early embryo. Each time such a cell divided, it passed along the stain to its offspring. But these methods were imprecise because they required dissecting the embryos; cells could not be tracked in a single, living organism. And even simple animals consist of so many cell types that following them all would be impossible. According to Wittbrodt, a comprehensive study of development in the small worm *C. elegans* requires tracking only 671 cells, whereas "the analysis of complex vertebrate species requires the simultaneous determination and tracking of the positions of tens of thousands of cells."

The limitations left open many questions about developmental processes. Wittbrodt hoped that some of them could be answered using the new microscope, which could observe the same embryo for several days. This posed significant technical and computational challenges. "In order to observe and follow the nuclei of, e.g., the 16,000 cells of an 18h-old zebrafish embryo," Wittbrodt writes, "a volume of 1000 × 1000 × 1000 cubic micrometers must be recorded at least once every 90 seconds." This was the maximum amount of time that could be taken to make a complete scan of the embryo. If it could be done, it would give the researchers a smooth, rolling film that would allow them to track each of the cells in the embryo. Cells divide and migrate quickly in early embryos. They can move away from their original positions in 90 seconds, but they do not move far enough to get lost. At longer intervals, the computer would lose track of them.

Additionally, the researchers wanted more than just a series of images. They wanted to teach the computer to identify single cells and keep track of them, then record each position in a database. This would provide a digital representation of each stage of the embryo's development that could be used to create hypotheses about the influences of genes on cell behavior.

The light sheet had to be moved at tiny increments through the sample, resulting in about 400 image slices that had to be combined to create each frame of the three-dimensional film. For a 24-hour recording, this added up to 400,000 high-resolution images per embryo that had to be combined in the computer.

In 2008, the machine was ready, and the data was captured. The group began sifting through the data. Each experiment recorded the history of an embryo from the first egg to a point at which it consisted of about 20,000 cells. The project shed new insight into some key events in embryonic development, such as gastrulation. This is the process by which a ball of undiffer-

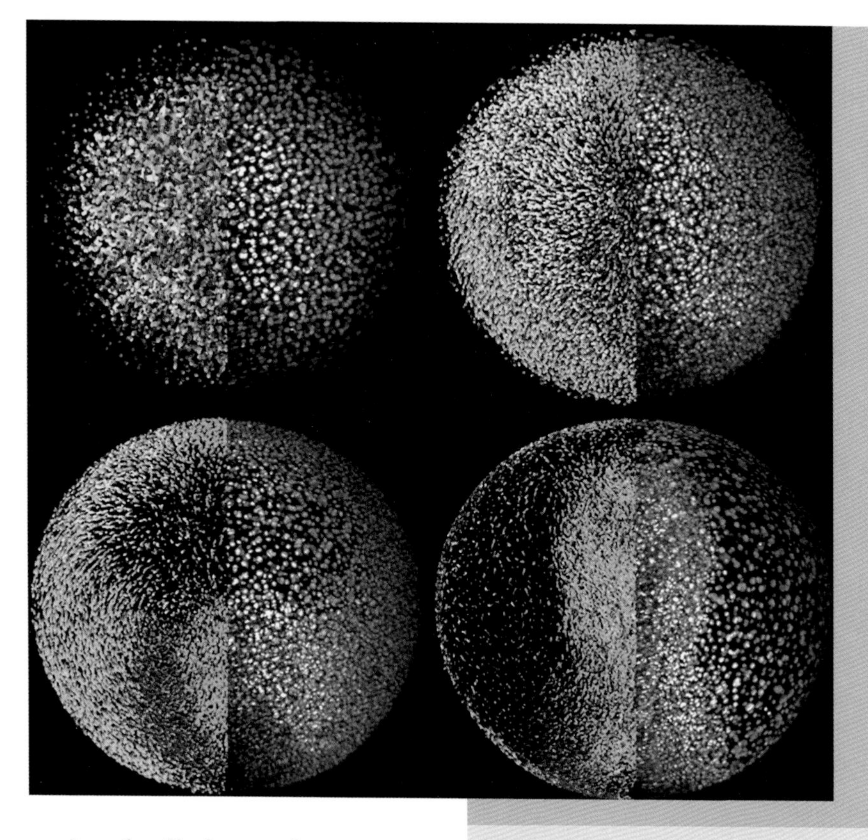

Using a new microscope to create a digital embryo. These images were made by Philipp Keller, a student in the groups of Ernst Stelzer and Jochen Wittbrodt at the European Molecular Biology Laboratory in Germany. The digital scanned laser light sheet microscope, described in the text, makes three-dimensional films of fish embryos from the earliest stages of development up to the formation of clear body structures. Each cell can be tracked in the computer. The right half of each image is the actual microscope picture; the left half is a computer reconstruction. *(Philipp Keller, EMBL)*

entiated cells forms three specialized layers, each of which goes on to form specific tissues and organs. At a very early stage, when the embryo has taken on the shape of a ball, made of a thin layer of cells, some of them migrate inward through a gap in the surface to create new layers. The study showed that this migration happens through a sudden influx of cells—like grains of sand pouring through a hole—rather than as a continuous, rolling-under movement of a sheetlike layer of cells.

Wittbrodt and his colleagues think that the method has the potential to answer a wide range of questions about animal development and disease. It will allow them to track any cell from its origins to its final form, when it is fully specialized and embedded in an adult tissue. The researchers can observe the formation of organs in real time. Some of the most interesting applications, Wittbrodt says, involve watching how development occurs in organisms with genetic defects. For example, proteins on the surfaces of neurons direct the way the cells form networks in the brain. Very subtle changes alter these connections and can cause the brain to develop in a different way. This process of hardwiring can be observed as it takes place in the embryo using the new microscope, and it may provide insights into the causes of some types of brain damage that have been linked to genes.

IMAGING BODIES

Large animals or human beings cannot be put under a microscope for observations; nor are they transparent or small enough to be studied using the DSLM instrument described in the previous section. But a number of other new techniques are available to capture images of the behavior of living cells and tissues in bodies. These methods are currently being used in medical applications such as visualizing tumors without surgery, discovering how much damage has been caused by a stroke, or simply exploring the function of the brain.

The earliest noninvasive technology that could be used in patients was X-rays. The German physicist Wilhelm Röntgen (1845–1923) discovered that when these waves were shined through a body, they were blocked or absorbed by various types of tissues in different ways. Capturing the absorption pattern on a photographic plate, behind the body, yielded an image in which bone, fat, and muscle could be clearly distinguished. X-rays gave doctors a look at broken bones and problems in other tissues, particularly the lungs.

Using X-rays in conjunction with substances that absorb them, such as barium, gave doctors a closer look at soft tissues through which the substances move. These images were two-dimensional up to the development of *computerized tomography* (CT) scans, initially based on a moveable X-ray machine attached to a scanner. The method was simultaneously announced in 1972 by the British researcher Sir Godfrey Hounsfield (1919–2004) at the Central Research Laboratories of Electronic & Musical Industries Ltd. (EMI) and Allan Cormack (1924–98) of Tufts University in Massachusetts, leading to the joint award of the 1979 Nobel Prize in physiology or medicine. (Best known for its record label, which featured albums by the Beatles, EMI was deeply involved in technological research and had helped build radar devices during World War II.)

The instrument is passed over the body, creating image slices that are assembled by a computer into a three-dimensional image. The method is limited because barium or hard parts of the sample block the passage of the waves, preventing researchers from seeing the soft tissues in much detail. The original CT machines relied on X-rays. Even though small doses were used, there was always a small risk of damage to patient tissues.

In the meantime, CT was being used with other types of imaging. *Magnetic resonance imaging* (MRI) was also the product of applying discoveries in physics to a biology problem. The principles behind the technology were discovered in the 1930s by Isidor Rabi (1898–1988), a physicist at Columbia University in New York City. He was using molecular beams to investigate the forces that hold electrons to the nuclei of atoms, work which earned him the 1944 Nobel Prize in physics. It also had practical applications in the development of radar, a project that Rabi was recruited to work on in World War II.

Two physicists who had also been recruited, Felix Bloch (1905–83) and Edward Purcell (1912–97), adapted nuclear magnetic resonance (NMR) so that it could be used to investigate liquids and solid objects, an accomplishment recognized with the award of the 1952 Nobel prize in physics. NMR slowly became an important method for the determination of structures of biological molecules thanks to the efforts of Kurt Wüthrich (1938–), a Swiss chemist who now heads laboratories at the Swiss Federal Institute of Technology in Zürich and the Scripps

Institute in La Jolla, California. Wüthrich began working with NMR when he joined Bell Laboratories in Murray Hill, New Jersey, in 1967, where he had access to one of the best instruments in existence at that time. He used it to study the structure and behavior of proteins in liquids; thanks largely to his efforts, it has become a standard tool in structural biology, drug discovery, and related fields. The 2002 Nobel Prize in chemistry was awarded to Wüthrich for this work.

Unlike X-rays or other imaging methods, MRI does not require radiation or ionization that might be harmful to living cells. Theoretically, patients can be scanned again and again with no adverse health effects; the only risk is that tissues are slightly heated and the procedure should not be used on people fitted with pacemakers or other electronic devices.

The method is based on exposing a sample or patient to strong magnetic fields. The nuclei of atoms are sensitive to these fields and absorb the energy. This effect is like placing a strong magnet next to a compass and then removing it. At close proximities, the magnetic field draws the needle of the compass. Removing the magnet again makes the needle return to its normal position. Something similar happens with atoms in an NMR experiment. The magnetic field aligns their nuclei. When the field is relaxed, they snap back to their normal state, but the way that they do so depends on what other atoms are nearby. Applied to a tissue sample or an organism, this produces an image that is particularly good at showing differences in soft tissues or liquids.

The method can also be tuned to detect the presence of particular substances or molecules. One use has been to follow the activity of the brain as it performs different tasks. MRI shows changes of blood flow to various regions, which has been associated with brain functions.

MRI has also been adapted in clever ways to other types of problems. A recent study by the laboratory of Peter Schlag, a German cancer researcher at the Charité University Medical School in Berlin, proposes a use that might help doctors diagnose the severity of breast cancer. Traditional mammography depends on X-rays to reveal abnormalities in breast tissue. By showing the size, location, and shape of a tumor, mammographies help doctors look for signs that a tumor is undergoing transformations that lead to metastases. But the method is imprecise, Schlag says, and it cannot be used to look for molecular events that might allow a doctor to predict how the tumor will develop. Magnetic resonance mammography, on the other hand, might provide more information.

Schlag's new method is based on the fact that molecules that can be detected and tracked by MRI might behave differently in tumors and noncancerous tissue. For example, rapidly growing tumors stimulate the formation of new blood vessels; tumor cells are just as dependent on nutrients delivered through the blood as healthy cells. But they grow much more quickly than other adult tissues, which means that special mechanisms are required to meet their demands. The adult body often builds new blood vessels—to repair damage caused by injuries, for example—but this happens at a slow rate, and the process is carefully regulated.

Tumors have to overcome the regulatory systems, and in doing so, they create vessels that are slightly different than those of surrounding healthy tissue. "One of these differences," Schlag said in a personal interview with the author, "is that the new blood vessels are not sealed as tightly as other vessels. Gaps between the cells permit large molecules to slip through and seep into the surrounding tissue. This led us to start looking for a substance of the right size—able to slip through tumor-related vessels but not healthy ones—that can be detected by MRI. We found such a substance in indocyanine green, a fluorescent molecule that leaks into tumors but not into healthy tissue. Most of the substance stays in the bloodstream and moves to the liver, where it is cleared from the body. After this happens, the tumor has absorbed the molecule and stands out in high contrast against the surrounding tissue. In the patients we examined, the contrast is highest in malignant tumors, those which have metastasized or will do so soon."

The same approach can show how the body absorbs other substances, such as drugs. MRI is being used to look for substances that can cross the blood-brain barrier, a defensive mechanism that has evolved in the brain to protect it from toxins, viruses, and most parasites. The blood vessels of the brain are especially tightly sealed. This means that many drugs that are successfully delivered through the bloodstream to other parts of the body do not reach brain tissue. MRI can be tuned to pick up the traces of particular substances. By tracking their presence in blood vessels and surrounding tissues, the method can be used to find new substances that penetrate the blood-brain barrier.

4

The Future of Humanity and the World

"The advance of genetic engineering makes it quite conceivable that we will begin to design our own evolutionary progress," wrote the science fiction author Isaac Asimov in a collection of essays called *The Beginning and the End.* Even though genetic engineering had barely begun when the book appeared in 1977, Asimov and others clearly understood its potential. The idea that humans will take control of their own genetic future has been around ever since, and it has aroused both curiosity and fear. Part of the fear stems from the fact that researchers do not yet fully understand the human body and mind; some believe they never will.

Scientists have been manipulating human genes since the early days of genetic engineering—transplanting them to bacteria or other species, which can be used as factories for medically important molecules such as insulin—or manipulating lines of human cells grown in the laboratory. There is great interest in learning to replace defective genes in humans, as a cure for genetic diseases, but a safe and reliable method of doing so in embryos or adults has not yet been perfected. The idea of correcting the defects in egg or sperm cells or very early embryos has been rejected for ethical and technical reasons. Within a few years, the technical issues may be resolved, and it would likely be possible to extend the techniques commonly used in animals to humans—if society were to allow it to be done. One theme of this chapter is to explore the possible consequences of taking direct control of human genes and human evolution.

People are already shaping their future evolution. The driving force that changes species is natural selection, which implies that by altering the environment, humans will indirectly change themselves. The human impact on the environment involves technological inventions, changes in diet, overpopulation, genetically modified crops, pollution that causes global warming, and behavior that reduces the world's biodiversity. Over the long term, these factors will inevitably transform the species. This chapter looks at these issues from the perspective of some of the most exciting frontiers of genetics and biology. Each of these fields is so complex that it could easily fill a book of its own. Here, each topic will be introduced through the eyes of one of the most prominent thinkers and scientists working in the field.

DESIGNER BABIES AND A POST-HUMAN FUTURE

Until the publication of Darwin *On the Origin of Species* in 1859, nearly everyone regarded humans as far superior to any other living creature, the pinnacle of nature, nearly godlike. People expected to change in technological and social ways, and they hoped to improve themselves morally, but there was no real notion that humans might one day become an entirely different species. With the discovery of evolution and the laws of heredity, mankind suddenly saw itself as one small step along a branching, open-ended pathway of life, rather than an end point. The future of the species was suddenly up in the air. Evolution could not be stopped, so humans would inevitably change. Their descendants—if they survived—would likely be as different from the people of today as *Homo sapiens* is from the early primates that wandered the African savanna millions of years ago.

There is no way to predict what direction these changes will take as long as reproduction is left to the roulette of nature, through which each new child is a chance mixture of parental alleles and a few new mutations. There is no guarantee that the people of the future will be smarter or more peaceful or that they will live in a healthier relationship to the environment. The only way to change this situation and steer evolution in a desirable direction may be through genetic engineering. But even if these methods could be used safely in humans, many ethicists and thinkers are concerned about the end result. Just as the eugenics movements of the early 20th century would not have worked—because they were based on a misunderstanding of human heredity and the nature of genes—deliberately manipulating the human genome might have entirely unwanted and unpredictable consequences.

In a 2002 book entitled *Our Posthuman Future: Consequences of the Biotechnology Revolution,* the American philosopher and political economist Francis Fukuyama (1952–) eloquently gives voice to these concerns. A decade earlier, Fukuyama gained widespread attention with his book *The End of History and the Last Man.* The title is not meant to imply that the species is doomed; instead, it raises a subtle question about whether people will need to write histories in the future. In the past, Fukuyama says, the historian's main focus was the struggle between competing political systems and ideologies. Such conflicts might soon be a thing of the past. Recent decades have shown that communism and most other types of regimes are unsustainable, he believes. With the arrival of Western liberal democracy, mankind has reached the logical end point of social evolution. "What we may be witnessing is not just the end of the cold war, or the passing of a particular period of postwar history," he wrote, "but the end of history as such: that is, the end point of mankind's ideological evolution and the universalization of Western liberal democracy as the final form of human government."

Just 10 years later, however, Fukuyama felt the need to refine his hypothesis. Discoveries in science—particularly biology and genetics—might change society or even human nature, which could well create the need for new forms of social organization. In *Our Posthuman Future,* Fukuyama points out that political systems are ultimately dependent on human nature, science, and technology. Liberal democracy is a product of human lifestyles and the mind—so what would happen if, for example, gene technology were used to change the brain?

These questions have led Fukuyama to consider the potentially dangerous and ethically questionable consequences of genetic engineering. Current uses of the technology include diagnosing serious birth defects, looking for genetic markers associated with diseases, creating genetically modified crops and animals, and developing therapies for health conditions by trying to repair or replace defective molecules. All of these methods are in their infancy, and most biomedical researchers believe that in the near future, they will be major tools in fighting diseases, such as cancer, and possibly in the repair of developmental defects caused by flaws in genes.

But the potential goes much further. If molecular medicine develops in the way most researchers expect, scientists will soon find genes linked to mental conditions such as schizophrenia and bipolar disorder. Research has also revealed connections between genes and intelligence, tendencies toward addiction, and other behavior such as attention deficit/hyperactivity disorder (ADHD). This condition is a good example of the concerns that Fukuyama and many others have about the potential uses of gene therapy—not because of ADHD itself, but because of the way the disease has been handled by the medical community and the family members of those affected.

The causes of ADHD are largely unknown. Symptoms include inattentiveness, hyperactivity, and a wide range of other behaviors that make it challenging for children to adapt to school and other social settings. Studies of twins reveal that there is likely to be a genetic component, although specific genes have not yet been identified. This means that there is no objective test to establish whether a person has the disorder. Diagnosis is difficult because it is usually hard to tell the difference between normal—although somewhat exaggerated—childlike behavior and the symptoms of ADHD. Studies of the rates at which the condition occurs in schoolchildren give widely varying results—from 2 to 14 percent. There is a large ongoing debate in the medical community about the disorder, motivated by concerns about misdiagnosis, ethical concerns about medicating children, and worries about the long-term effects of the drugs they are given. In spite of these issues, a growing number of parents are turning to medications to treat children that have been diagnosed with ADHD. While these treatments have certainly helped many children and their families, researchers worry that they might be overused. The way the drugs affect the brain is not completely understood, and there has been a lack of research into effects that they may have over the long term.

Fukuyama's concerns with genetics center around this issue of defining what is normal. He casts his mind toward the future and predicts a time when "Knowledge of genomics permits pharmaceutical companies to tailor drugs very specifically to the genetic profiles of individual patients and greatly minimize unintended side effects. Stolid people can become vivacious; introspective ones extroverted; you can adopt one personality on Wednesday and another on the weekend. There is no longer any excuse for anyone to be depressed or unhappy; even normally happy people can make themselves happier without worries of addiction, hangovers, or long-term brain damage."

This scenario, which sounds strikingly similar to the pleasure-intoxicated society of Aldous Huxley's *Brave New World,* is one of several ways that Fukuyama sees people adapting to and using the new possibilities of genetic science. Other medical discoveries, originally intended only as last-resort measures to save lives, have now been adapted for purposes like cosmetic surgery. If people have the chance to eliminate disease or improve themselves—making themselves more attractive or intelligent without serious side effects—what would keep them from doing so?

Giving parents control over their children's genes leads to other concerns; Fukuyama uses homosexuality as an example. One intensely debated topic has been whether sexual preference has a genetic basis or is determined almost entirely by environmental influences. While some studies have linked male homosexuality to a position on the X chromosome and others have reported differences between the brain structures of homosexual and heterosexual males, there is not yet a definitive scientific answer; scientists have not yet found a gay gene.

But suppose that such a molecule is found or that at least researchers discover a biological basis for homosexuality. Knowledge about the causes of things, Fukuyama writes, "will inevitably lead to a technological search for ways to manipulate that causality." In the case of homosexuality, he proposes a thought experiment: "Assume that in 20 years we come to understand the genetics of homosexuality well and devise a way for parents to sharply reduce the likelihood that they will give birth to a gay child. This does not have to presuppose the existence of genetic engineering; it could simply be a pill that provided sufficient levels of testosterone in utero to masculinize the brain of the developing fetus. Suppose the treatment is cheap, effective, produces no significant side effects, and can be prescribed in the privacy of the obstetrician's office." Even if homosexuality is completely accepted by society at this future date, Fukuyama wonders how many pregnant women would take the pill. He thinks that many would, even if they were not prejudiced against homosexuals. "They may perceive gayness to be something akin to baldness or shortness—not morally blameworthy, but nonetheless a less-than-optimal condition that, all other things being equal, one would rather have one's children avoid. . . . Wouldn't this form of private eugenics make them more distinctive and greater targets for discrimination than they were before? . . . Should we be indifferent to the fact that these eugenic choices are being made, so long as they are made by parents rather than coercive states?"

At the moment, parents cannot pick the characteristics of their children, except in the sense of prenatal screening for disease genes—a practice that is also troubling to many ethicists. The more choices parents have, the more expectations they are likely to have—that their children will be better behaved, will not become bald, or will develop in certain ways. However, human genetics is so complex that even if scientists decided to take drastic measures to alter the human genome, the children produced by genetic engineering would be unpredictable.

At the moment, people know (or quickly learn) that creating a new child is like spinning a roulette wheel: What happens is a bit of heredity, but mostly chance. The same thing will be true of engineered humans, but people may have a harder time adjusting when their expectations are dashed.

This modern form of eugenics—steered by genetic engineering—is only one of many ways that Fukuyama can imagine biotechnology influencing the future of human beings. "While it is legitimate to worry about unintended consequences and unforeseen costs, the deepest fear that people express about technology is . . . that, in the end, biotechnology will cause us in some way to lose our humanity—that is, some essential quality that has always underpinned our sense of who we are and where we are going, despite all of the evident changes that have taken place in the human condition through the course of history. Worse yet, we might make this change without recognizing that we had lost something of great value. We might thus emerge on the other side of a great divide between human and posthuman history and not even see that the watershed had been breached because we lost sight of what that essence was."

CURING BRAIN DISEASES AND THE NEW PHARMACOLOGY

At the beginning of the 21st century, the study of diseases of the mind and brain is entering a new phase. These diseases serve as a good example of the way medicine is being transformed by genetics and molecular biology. Researchers are starting to uncover the genetic and physical causes of numerous mental health problems, and this has significantly affected the way that the diseases and their victims are seen by the medical community and society as a whole. This is an enormous change from the state of things just a few decades ago, and the new perspective brought by the molecular revolution in the neurosciences will likely be the key to finding cures for Alzheimer's disease, strokes, and many other problems of the brain.

How far things have come can be seen in the fact that people with mental disabilities, particularly the residents of mental institutions, were among the most unfortunate victims of the eugenics programs of the early 20th century. They were often sterilized without their consent or even that of a relative. Under the Nazi regime, many were simply murdered. Systematic sterilization was finally halted in the United States in the 1930s because of ethical concerns and a recognition of patient rights. But at the same time, another equally questionable medical practice began making the rounds of hospitals and psychiatric clinics.

In 1935, the Portuguese physician Antônio Egas Moniz (1874–1955) began performing a type of brain surgery on human patients after learning that it successfully cured monkeys of aggressive behavior. Cutting a particular tissue in the front region of the brain—separating the white matter from the rest of the organ—rendered monkeys calm and friendly. Egas Moniz began performing the operation on humans. In many cases such *lobotomies* calmed the person, brought an end to epileptic seizures, or stopped other undesirable behaviors—because it broke connections that allowed impulses to spread from some regions of the brain to others. In the days before antipsychotic drugs such as Thorazine, the procedure was regarded as a huge breakthrough, and in 1949 Egas Moniz was awarded the Nobel Prize in physiology or medicine. In America, the neurologists Walter Freeman and James Watts streamlined the procedure so that it could be carried out routinely in psychiatric clinics, by pushing an icepick-like instrument into the brain through the patient's eye socket. Eventually, it was performed tens of thousands of times.

However, it often had terrible side effects. For example, the operation was performed on Rosemary Kennedy, the sister of President John F. Kennedy. The aim was to reduce what the family called aggressive behavior. Before the operation she was considered mildly mentally disabled; afterward, she was reduced to an infantile state. Other patients had similar problems or experienced undesirable personality changes. Long-term studies and a more careful examination of patients began to reveal many serious, undesirable side effects. Finally, the surgery fell out of favor as antipsychotic drugs became available.

Brain research and the way researchers searched for cures changed dramatically over the next decades, due to the arrival of noninvasive techniques such as magnetic resonance imaging (described in chapter 3). This shifted the focus of the work toward discovering how modules in the brain communicate with each other along electrical and chemical pathways to form thoughts, feelings, perception, and behavior. In disease, these conduits are interrupted. Imaging techniques have permitted the discovery of links between specific tissues and mental behavior and some ideas about how normal processes are disrupted in disease. Sometimes, this information can be woven into stories of how the brain functions. In her book *Mapping the Mind,* for example, the medical writer Rita Carver summarizes a current hypothesis of clinical depression: "Depression is caused by the firing of a circuit in which the amygdala feeds negative feelings to consciousness, the prefontal lobe pulls out long-term memories to match the feeling, the anterior cingulate cortex fastens on to them and prevents attention from shifting to anything more uplifting, and the thalamus keeps the whole circuit alive and firing."

Geneticists and molecular biologists, on the other hand, look at the brain from the bottom up, asking different sorts of questions. The contrast is most obvious in a case like Alzheimer's disease. More than 100 years ago, the German physician Alois Alzheimer (1864–1915) discovered that the brain of a woman who had suffered from dementia contained *amyloid plaques*: fragments of proteins that had formed tangled clumps in the space between brain cells. Eventually, the brain shrinks, and one-by-one its functions fail. Using imaging techniques, researchers can help diagnose the condition as it develops and watch the changes that take place.

Most of the work of molecular biologists has focused on the amyloid plaques, which block communication between the cells, prevent them from getting nutrients, and eventually cause their death. Only today are scientists beginning to understand how the plaques form. They begin as part of a protein called APP (amyloid precursor protein), which is made by neurons and some other types of cells. APP rests in the cell membrane, where it seems to play a role in memory and learning, but its healthy functions are not yet really understood. Quite a bit has been learned about its biochemistry, however, and some of the processes by which it contributes to disease.

Researchers have discovered that other proteins come along and compete to slice APP at various places, which leads to different types of fragments. Some of these are harmless, but others latch onto each other and form amyloid plaques that do not dissolve. Molecular biologists hope to learn why brains sometimes stop making healthy fragments and start producing the unhealthy form. The answer, they have learned, partly depends on which proteins do the cutting and the order in which they do so. Other molecules play a role by binding to the fragments and helping weave them into fibers.

It is possible that an existing drug—or one of the millions of compounds in the libraries of pharmaceutical companies—will block one of these processes, prevent the accumulation of the plaques, and stop the course of the disease. Knowing what to look for makes the search for treatments infinitely easier. Controlling the development of Alzheimer's disease will probably require learning to control the activity of these other molecules. That will be easier in some cases than others. Some families are particularly susceptible to the disease, probably because of the influence of other genes.

This is typical of the molecular age's approach to the study of other types of brain disorders and many other diseases. The first step in looking for a treatment is to understand how a problem affects cells and molecules. In a stroke, for example, the blood flow is cut off to cells in a particular region of the brain and they die. The reason for their death is not directly suffocation or starvation. Instead, the loss of the blood supply cuts off signals that cells need to survive and triggers other signals that tell them to die. In 2005, Oliver Hermann and Markus Schwaninger of the University of Heidelberg, Germany, discovered that in the wake of damage, a self-destruct signal is likely passed to cells' genes via a molecule called NF-kappa-B. Using mice that had suffered a strokelike condition, they showed that blocking the signal helps cells stay alive much longer and recover, even if the treatment comes a few hours after a stroke. Their finding

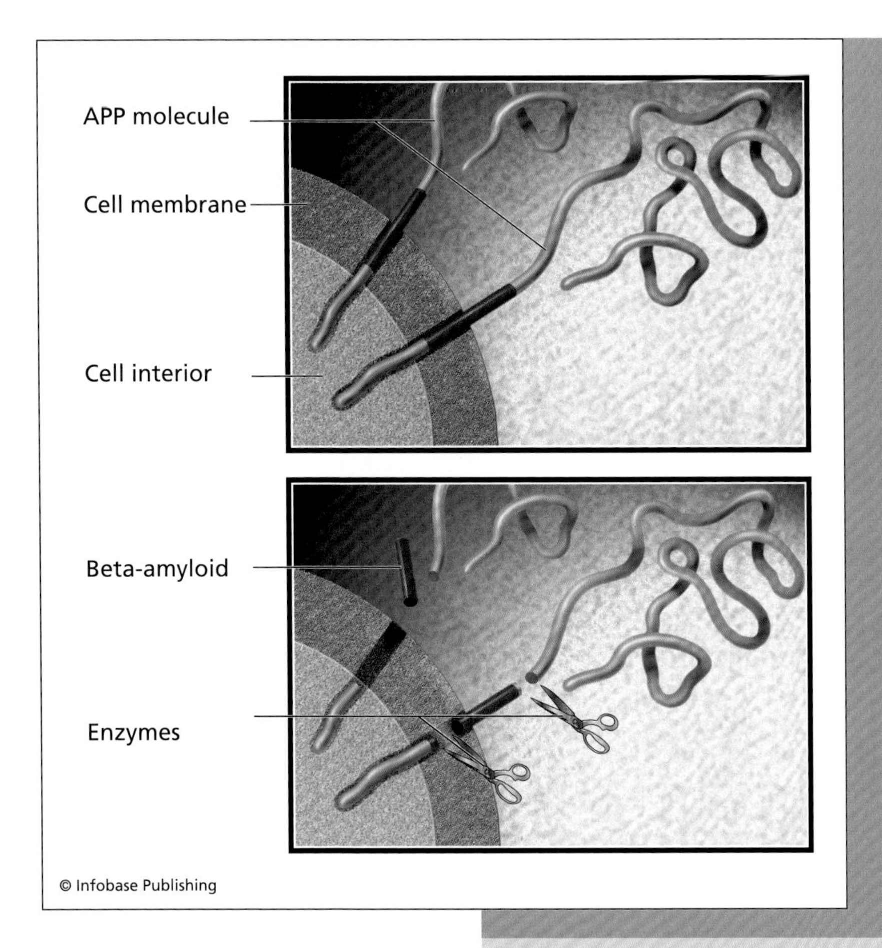

A molecule called APP is embedded in the surface of neurons, where it can be cut at various locations by enzymes. Some types of cuts (lower left) produce a fragment called Beta-amyloid that accumulates between nerve cells, causing the symptoms of Alzheimer's disease. Other ways of cutting the molecule (lower right) produce a harmless fragment.

has been used as a starting point for the development of new, experimental therapies.

In the future, even the death of cells may not cause an irreversible health problem because there may be ways to replace them. Until the late 1990s, it was almost universally believed that nearly all of a person's neurons developed early in life; any that were damaged or died could not be replaced. In the meantime, researchers have discovered that the creation of new neurons continues into adolescence and later. In 1998, Peter Eriksson and other members of the laboratory of Fred Gage, working at Sahlgrenska University Hospital in Göteborg, Sweden, discovered that cells in a region of the brain called the hippocampus could differentiate into new neurons. Eriksson has gone on to show that additional areas of the brain hold stem cells that can do so. But with age the creation of neurons likely becomes rare, and it is also limited to specific parts of the nervous system. In a spinal cord injury, for example, cells are unable to repair breaks that prevent the brain from communicating with the rest of the body. But if biologists could tap into the potential of stem cells, by learning to control the signals that guide their specialization, it might be possible to activate the body's cells to do so.

In many other brain diseases, protein fragments accumulate and cannot dissolve. In Huntington's disease they collect in the cell nucleus rather than the space between cells—eventually with the same result, killing neurons. Under the right circumstances, so many different types of proteins form amyloid clusters that many researchers believe that eventually nearly anyone who lives long enough will suffer from one of these types of neurodegeneration. And the numbers are sure to rise if cures are found for other main killers that strike earlier, such as cancer and cardiovascular diseases. As James Thorson puts it in his book *Aging in a Changing Society,* "Remember, everyone has to die of something, so if one cause goes down, something else has to go up." Truly eliminating these diseases may require solving an even greater scientific question—why cells and bodies change over time—in other words, why people age. That theme is taken up later in this chapter.

Several new approaches are being taken to address the symptoms of neurodegenerative and other systemic diseases such as cancer. Regenerating tissues using stem cells is one. Additionally, of course, researchers continue to try to find natural substances or develop artificial compounds that can be used as drugs. The way that this is done has changed dramatically over the past 20 years and is increasingly moving toward what scientists call rational design.

This begins by pinpointing the cellular processes that cause a disease and identifying the molecules involved in it. The next step is to find a target: a particular protein or gene whose manipulation would give researchers a way to control the process. In Alzheimer's disease, for example, this might be one of the enzymes that cleaves the APP protein to generate a dangerous fragment. Once such a molecule has been identified, researchers try to obtain its precise three-dimensional structure (a picture of the arrangement of its atoms, which gives details of its shape and chemistry). This may reveal a location in the molecule for a drug to dock onto and change its activity. If that is successful, a researcher can scan databases of drug compounds in hopes of finding another molecule with the right configuration and chemistry, a molecule that can snap on, a bit like looking for the right electrical adaptor to fit a cell phone. Then the molecule is screened to see whether it really influences the activity of the target, and how well it does so. If an appropriate drug or compound is not found, researchers screen the target molecule against a library of substances and hope for a lucky hit. Promising candidates can then be rebuilt by chemists to be more effective.

Over the past decades, the methods to carry out such screens have been automated and improved, making it relatively easy for scientists to test thousands or even millions of substances for pharmacological activity. There are different types of screens. First, experiments are carried out in test tubes, looking for signs of changes in a molecule's chemical activity. Next, substances are introduced into cells, also using automated methods. If a molecule passes the test, it can be tried in animals; promising results may convince a pharmaceutical company to step in and conduct human clinical trials.

This is a much different strategy than scientists used in the past. It is based on an exact knowledge of molecular cell biology and the selection of a precise target, rather than simple trial and error and experiments that were often based entirely on animals. Researchers are continually on the lookout for new compounds that might make effective drugs. Other species—even unusual, exotic ones—are one important source; another is the

Cures from Shamans and Traditional Healers

Mark Plotkin (1955–), an ethnobotanist and president of the nonprofit organization Amazon Conservation Team, has dedicated his career to traveling to remote parts of the world, living with indigenous peoples and learning about the way their healers use plants. Plotkin's story is featured in the IMAX film *Amazon*, partly based on his 1994 book, *Tales of a Shaman's Apprentice*. His work is also the subject of a 2001 film called *The Shaman's Apprentice*, directed by Miranda Smith.

"Over thousands of years, through a method of trial and error, indigenous tribes have built up a storehouse of knowledge about the native vegetation," Plotkin writes. "There exists no shortage of wonder drugs waiting to be found in the rain forests, yet we in the industrialized world are woefully ignorant about the chemical—and, therefore, medicinal—potential of most tropical plants. . . . The approximately 120 plant-based prescription drugs on the market today are derived from only 95 species. A quarter of the prescription drugs sold in the United States have plant chemicals as active ingredients. About half of those drugs contain compounds from temperate plants, while the other half have chemicals from tropical species."

The Amazon alone is home to about 25 percent of the world's plant species. Why should organisms in such exotic locations be a source of interesting pharmaceutical compounds? One reason has to do with the equally varied species of insects that inhabit the rain forest. "The fact that the forest has not been devoured by this entomological onslaught is testament to these plants' abilities as chemical warriors. Plants protect themselves by producing an astonishing array of chemicals that are toxic to insects, thereby deterring predation. When ingested by

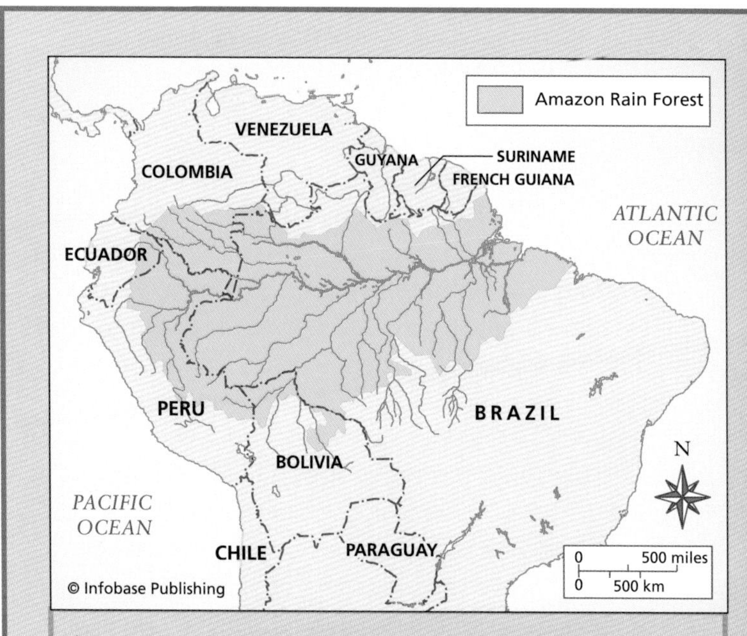

Amazon rain forests are home to a rich variety of plant and animal species, many of which are found nowhere else on the globe. While these species potentially hold substances that can be used as medicines and for other purposes, the rain forests are shrinking at an alarming pace due to human activity, and many species are becoming extinct.

humans, these same plants—and their chemical weapons—may act in a variety of ways on the body: they may be nutritious, poisonous, or even hallucinogenic. And in some cases, they are therapeutic."

This is one reason that Plotkin and many other conservationists are racing against time to save the Amazon and other endangered ecospheres. There are many more (see the section titled Saving the World, later in this chapter). Tropical plants frequently contain alkaloids, natural chemical compounds that contain nitrogen and often have a powerful effect on the body. Caffeine, nicotine,

(continues)

(continued)

morphine, cocaine, and quinine (long the most effective treatment for malaria) are alkaloids.

Another natural substance that has currently attracted great interest among biologists is epigallocatechin-gallate (EGCG), found in green tea. The tea has been used as a remedy in traditional Chinese medicine for hundreds and probably thousands of years. Recently, Erich Wanker's group at the Max Delbrück Center for Molecular Medicine in Berlin, Germany, discovered that in the test tube, EGCG reduces the clustering of proteins responsible for Huntington's disease. "If EGCG was helpful in one case," Wanker told the author in an interview, "it might be helpful in others. So we have tested proteins responsible for Alzheimer's disease and Parkinson's disease. In each case, EGCG prevents protein fragments from clustering into fibers. We followed up in a second study to see if it would also have beneficial effects on cells. In each case, the result was positive." Wanker and his colleagues are now purifying and improving the substance, hoping that a form of EGCG can be developed which will slow or stop the progress of these brain diseases.

enormous range of plants or natural substances used by traditional healers.

Alongside stem cell therapies and the development of new pharmaceutical substances, the molecular age has spawned other new ideas about treating diseases. These methods include correcting defective molecules by delivering healthy versions to cells—in essence, giving the body new genetic instructions by which it produces its own therapeutic molecules. The biggest problem has been finding a way to get the molecules into cells, which have evolved defenses that protect them from taking up foreign genes or RNA. Some viruses manage to overcome these defenses, however, so one method that is being tried is *viral therapy.* This approach starts by taking a relatively harmless virus, removing any information that might cause an infection, and replacing its genetic material with the healthy form of a human gene. The virus is altered so that it cannot infect healthy tissues or be transmitted to another person. The technique has been tested in a number of clinical trials—in some cases very successfully, but there have also been deaths and negative side effects. Thus, researchers are searching for new ways to deliver therapeutic molecules to cells.

Another approach is to extract immune system cells from a patient and train them to recognize new types of problems such as cancer or amyloid plaques. White blood cells called T cells and B cells are the major tool used by the body to defeat parasites, viruses, or toxins. They recognize these invaders because they have randomly created antibodies or receptor proteins on their surfaces that are able to dock onto foreign molecules; when this happens, they summon immune cells to break them down or destroy them. One therapeutic strategy that seems very promising is to remove T cells from a patient, grow them in the laboratory, and then equip them with receptors that can recognize unusual proteins that might be found on the surface of cancer cells, dangerous amyloid plaques, or other disease molecules. If this works, it might be possible to teach the body to confront cancer, degenerative diseases, and possibly even conditions such as aging in the same way that it fights infections.

Francis Collins (1950–), an American geneticist who heads the National Institutes of Health and led the Human Genome Project, recently discussed how he expects the relationship between genetics and medicine to evolve over the next decades. Some of his predictions, which appear on the MSNBC Web site (see the Further Resources section in the back matter), include the following:

- By 2010, tests will be developed to screen patients for genes linked to common diseases such as colon cancer, and over the next decade several types of gene therapy will be proven successful.

- By 2020, doctors will be using designer drugs to treat conditions such as diabetes and high blood pressure; therapies for cancer will have been developed based on molecules found in tumors, and it will be possible to diagnose and treat a number of genetically based mental illnesses.
- By 2030, researchers will have a list of genes involved in aging and will be carrying out clinical trials to extend people's lives; computer simulations will replace many types of laboratory experiments, and it will be common to sequence individual genomes.
- By 2040, medicine will have become individualized based on people's genetic profiles; in many cases molecular testing will warn doctors that there is a problem in advance of the appearance of disease symptoms, and gene therapy will be available for the treatment of most diseases.

Given the increasing pace of discoveries and technological developments, no one will be surprised if some of these milestones are reached earlier. On the other hand, it is entirely possible that scientists will discover new aspects of living systems that change the way they think about some neurodegenerative diseases, cancer, or other health problems. This may reveal that the diseases are much harder to treat. Yet it may also reveal ways of coping with them that are much simpler.

CONSCIOUSNESS AND THE BRAIN

"It's a scandal that science leaves out consciousness," said Christof Koch (1956–), a neuroscientist at the California Institute of Technology, in a 2006 interview with Caltech's Institute of International Studies. "Ten or twenty years ago, when we started, many scientists, probably the majority, said, 'Well, consciousness: we've got to leave that to the religious people, we've got to leave that to the philosophers, we've got to leave that to the New Age cult. That's not something scientists can study.' But that's silly. We are conscious and I believe it's the most essential aspect of my life, it's the fact that I'm a conscious being, and if I leave that out, then I will forever deprive science of one of the key aspects of the natural world."

Today's biology is a materialist science; it aims to explain things in physical terms. As Koch points out, for centuries philosophers, religious thinkers, and scientists have considered consciousness inexplicable, even off limits to materialist investigations. But no science of the human brain can be complete or satisfying unless it can explain what people usually regard as its most interesting feature, and Koch believes that neurobiology is ready to take on the theme. Over the course of 20 years, he pursued the topic with Francis Crick, codiscoverer of the structure of DNA and a founding father of molecular biology. Long before his death in 2004, Crick called consciousness the "major unsolved problem in biology," and began a quest to discover how the biology of the brain could produce this unique phenomenon. He found an excellent sounding board, critic, and partner in Koch, who continues to pursue the question in his laboratory at Caltech. The aim, he says, is ultimately to understand how a physical system like the brain can feel things—pain or pleasure, the sense of being angry, and self-awareness.

The researchers settled on a unique approach to the problem. Koch points out that while consciousness is an integral part of some human activities, others are done without conscious control. People digest food, ride bikes, and even have conversations without having to plan every sentence deliberately. Suddenly a person finds himself sitting in the car in the driveway, with no memory of a drive home from work, because his mind has been on something else. Koch calls such automatic activity zombie agents and says that life would not be possible without them. Some animals might have a sort of consciousness, he says—especially complex ones—but in other species, zombie agents might be able to manage all the activities they need to survive and reproduce. Realizing that zombie agents exist in humans permits scientists to look at how the brain manages them, and then the goal is to discover what makes them different from consciousness.

An important concept in Koch's work is the idea of a "neuronal correlate of consciousness" (NCC). He defines it as the "minimal set of mechanisms in your head that you need in order to be conscious." It is likely to be much smaller than the entire brain, because consciousness can operate when entire parts of the system are inactive. "I can close my eyes and I can visualize, so I don't need my eyes. Do I need my cerebellum? Probably not for visual consciousness or . . . any consciousness. So you can ask the question, what are the minimal set of mechanisms in your head that you need in order to be conscious? Is there a specific neurosignature, are there specific types of neurons, is there a particular type of neural activity, are there particular types of molecules, particular types of synapses, do they sit in a particular part of the brain? . . . Can you track them, can you catch a brief picture of them using some fancy imaging technique? Can you influence them?"

Most of the work of Koch's lab has focused on visual consciousness because it is easy to set up experiments to test what people see and what they do not. Magic tricks are often successful because the performer successfully directs people's attention away from what he does not want them to see—even when an object or a movement is plainly visible. The same effect can be achieved with a test subject looking at a video screen. By monitoring brain activity with an imaging technique such as MRI, the researchers look for areas of the brain that light up when a person becomes conscious of something. "It's not going to be one area, we know that," he says, "it's going to be a series of areas, probably distributed, that have different properties."

In this view, consciousness cannot be pinned down to a specific region of the brain or set of neurons. Koch says it is more like a "coalition of neurons, a little bit like in a democracy where you have coalitions that form, and that assemble and then disassemble . . . For a hundredth of a millisecond you may have this coalition of 5 million neurons. . . . They may give rise to a feeling of 'darn it, I'm late today,' and then this is suppressed because then there's this other 5 million neurons, or 10 million

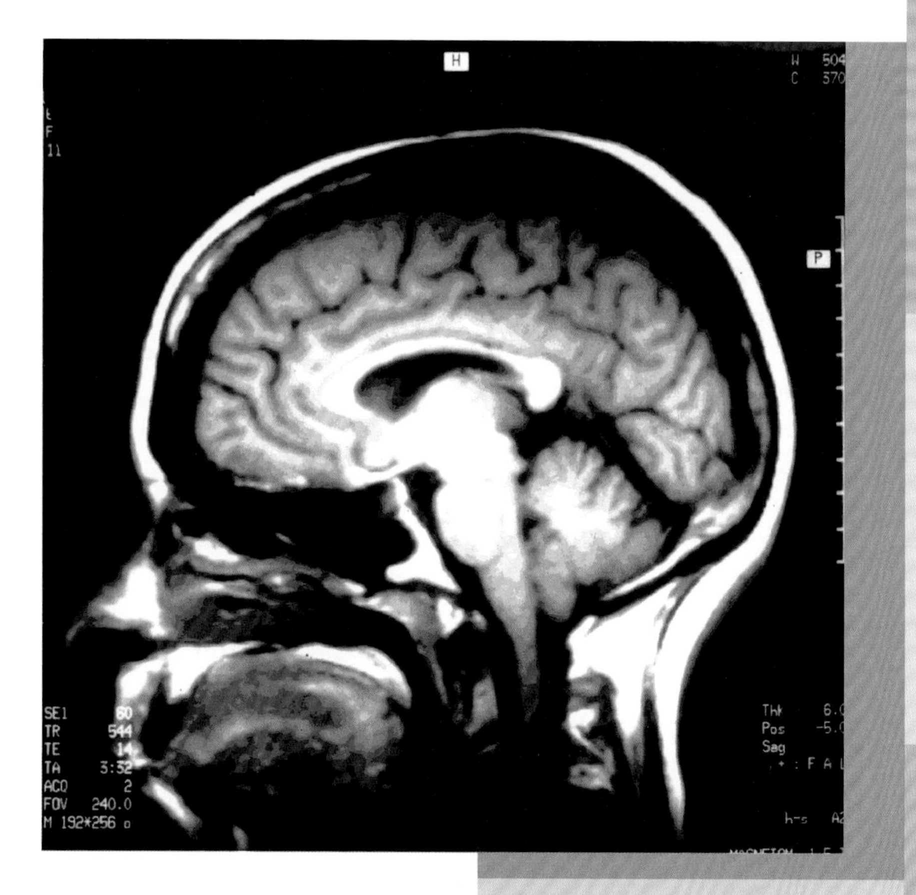

MRIs reveal the flow of blood through the brain as it performs various functions. This technique has been important in allowing researchers to determine which regions of the brain are crucial to various types of activity, as well to assess damage that has occurred through injuries, strokes, and other diseases. *(Dan S. Heffez)*

neurons, who now give rise to the 'oh, I see my daughter over there.'"

The coalitions compete for the attention of the host. Finding the NCC will involve looking at a brain pattern, tracking the formation of coalitions, and learning to recognize which ones lead to consciousness. If Koch and his colleagues can get a grip on visual processing, they hope that the same principles will apply to the way the brain manages awareness of other types of sensory information. Then it should be possible to look for telltale signs of consciousness in a fetus, a person in a coma, a dog, or a fly.

The scientists have already established that the frontal lobe—a higher part of the brain that arose in mammals, relatively recent in evolution—seems to be actively involved in directing a person's attention to specific parts of a stimulus. One amusing example that Koch has frequently used in lectures challenges the audience to watch a film of a group of six people, moving around and tossing two basketballs to each other. The task is to count the number of times the basketballs are tossed. Doing so requires such concentration that very few people in the audience notice that an actor in a gorilla suit walks leisurely by, stops and beats his chest, then moves on. He is perfectly obvious to anyone who is not busy counting. To Koch and his colleagues, this shows that the brain does not simply passively review visual information coming in through the eyes and assemble it into a story—instead, it helps direct attention and awareness through feedback loops that tell the eyes what to look for.

While Koch obviously hopes that the search for the signature of consciousness will be successful, he admits that there is no guarantee. Just as there is no chance that an ant can understand the theory of relativity, people might not have the mental capacity to understand consciousness. On the other hand, he feels that the present state of technology makes it worthwhile to pose the questions. His laboratory is now investigating whether there are specific types of neurons devoted to consciousness, to determine whether specific parts of the forebrain are required for consciousness, and to map the routes of feedback mechanisms between the frontal lobe and other parts of the brain.

Some answers to these questions may come from studies of the brains of people as they enter different conscious states—deep, unconscious sleep and dreaming; more may come from studies of people whose mental life has been disturbed because of an injury or a disease. One fan of the work of Koch and Crick is Oliver Sacks, a neurologist at the Albert Einstein College of Medicine in New York. Sacks is a brilliant observer who has written a number of insightful books on the way people adapt when things begin to go wrong in their brains. He deeply empathizes with these people and sees their conditions as gateways to understanding questions about the mind and consciousness.

In 2005, following a talk by Koch at the New York Academy of Sciences, Sacks commented: "As a clinician, I see patients with problems. They can be thought of as experiments of nature . . . Pathologies and illusions are a wonderful subject for examining the connections between the mind and the nervous system."

THE QUEST FOR ETERNAL YOUTH AND IMMORTALITY

In 1993, researchers in the laboratory of the biochemist Cynthia Kenyon at the University of California, San Francisco, discovered that a change in a single gene doubled the life span of a laboratory organism, a small worm called *Caenorhabditis elegans*. It was a startling finding. "These mutant worms still looked and acted young when they should be old," Kenyon wrote on her laboratory Web site (see the Further Resources section in the back matter). "Seeing them was like talking to someone that looks 40 and learning that they were really 80. This was a stunning finding because no one thought it was possible." Further work on the worm, mice, and human cells suggests that similar mechanisms may control the life span and the process of aging throughout the animal kingdom. The discovery has triggered a new way of thinking about aging, and serious efforts across the world to find ways to cure it.

Kenyon had long wondered whether genes influenced the process of aging—an unconventional question. Most of her colleagues considered the deterioration of the body as simply a natural process: an accumulation of errors in cells, leading to damage in DNA that might cause cancer, defective proteins, and accumulations of junk such as amyloid plaques that eventually disrupted the functions of organs. Many thought that aging was like taking a photo of a painting and then photographing the photo, over and over, losing quality each time, until the image became unrecognizable. The immune system had not evolved mechanisms to prevent aging because natural selection has only a very weak effect on organisms once they pass the age of reproduction. Still, Kenyon thought that there had to be genetic controls on the mechanisms that controlled aging and an organism's life span.

Some hints that she might be right had appeared a few decades earlier. Until 1962, most researchers had believed that human cells grown in laboratory cultures could keep reproducing forever. In that year, Leonard Hayflick, a professor of research medicine at the Wistar Institute in Philadelphia, showed that they had a limited life span. Depending on the conditions in which they were grown, they reproduced themselves from 50 to 70 times and then the whole population died. Part of the reason became clear in the 1970s with the discovery of *telomeres,* unusual regions of DNA at the end of chromosomes.

The postdoctoral fellow Elizabeth Blackburn, working at Yale University in the laboratory of the cell biologist Joseph Gall, was investigating a curious phenomenon related to cell division. DNA is copied by molecular machines, but they have a limitation: They cannot copy all the way to the ends of chromosomes. This means that each time a cell divides, they lose a bit of information at the tips. If there were genes in these regions, this process would take larger and larger bites out of them, quickly destroying key parts of the molecules. Blackburn and her colleagues found that evolution had provided a solution: long DNA sequences that did not contain genes had evolved at the tips of chromosomes. Sequences are still lost with every cell division, but a lot of junk has to be carried away before any genes are affected. This acts like a timer that allows the cell to use all of its genes until the telomere is gone.

Blackburn and Gall also discovered that the timer was sometimes turned back a bit to extend the cell's life span. Some types of cells made proteins called *telomerases* that added new DNA to the telomeres. In a speech given as Blackburn accepted the 1998 Australian Prime Minister's Science Prize, she compared the process to shoelaces: "If you don't have those little tips on both ends of your shoelace, the shoelace frays," she said. "Even worse, without telomeres, broken chromosome ends combine with any other end they find and that is not good for the health of the organism. It's as though someone ties your shoe laces together and makes you fall over." Telomerases allow stem cells, embryonic cells, and a few other types to divide more times because they produce telomerases. But that production stops in most adult cells, and the timer begins its countdown toward aging and eventual death. One way to extend life might be by fooling more types of cells into making telomerases and to have them keep doing so for a long time. On the other hand, the effects of the molecules are not always positive. Cancer cells sometimes use the same trick to overcome the cell division timer. So telomeres and telomerases are of interest to cancer researchers as well as those working on aging.

Such findings supported Cynthia Kenyon's feeling that genetic mechanisms might have something to say about life spans and aging. "After all, rats live three years and squirrels can live for twenty-five, and these animals are different because of their genes. Also, most biological processes are subject to tight control by the genes. If so, then by finding genes that control aging, and then changing the activities of the proteins they encode, one day we might be able to stay young much longer than we do now."

In the early 1990s, the worm *C. elegans* was becoming a favorite of scientists. Sydney Brenner (1927–), was using it in his laboratory at the Medical Research Council in Cambridge, Britain. Mutations in genes had immediate, obvious effects on the worm's body plan, and a series of discoveries about animal development by Brenner's lab earned him a 2002 Nobel Prize in physiology or medicine. Kenyon had worked with him as a postdoctoral fellow. When she got an independent position at UCSF in the mid-1980s, she brought the organism along, intending to use it to study aging. In 1993, Ramon Tabtiang, a student in her lab, discovered mutations that doubled the life span of *C. elegans.* Normally, the worms live about 21 days, but mutations in a gene called daf-2 produced animals that lived about 45 days—and they remained active and healthy to the end.

Interestingly, the gene has an important role in the biology of the worm: When faced with overcrowding or starvation, its larvae go into a sort of holding pattern called the Dauer state. They stop developing and aging and survive four to eight times longer than their counterparts. If food becomes available again, the worms complete their development into adults. Daf-2 plays an important role in switching this pause condition on and off. The discovery that the same gene extended the worms' life span—without triggering the Dauer state—convinced Kenyon that it was part of a more general life extension mechanism.

The group identified a second gene, called daf-16, which also contributed to keeping the worms young. Since these discoveries, Kenyon's lab and others have unraveled some of the reasons why. "We now know that these genes, daf-2 and daf-16, allow the tissues to respond to hormones that affect life span. We showed that daf-2 and daf-16 ultimately affect life span by influencing the activities of a wide variety of subordinate genes that influence the level of the body's antioxidants, the power of its immune system, its ability to repair its proteins, and many other beneficial processes. . . . This knowledge has now allowed us to extend the life span of active, youthful worms by sixfold."

One conclusion from the work has been to demonstrate that life span and aging are not necessarily tightly bound to each other. Few people would choose to double their life span if it meant living for another century with Alzheimer's disease or if the body underwent more and more severe deterioration. But as Kenyon writes, "Especially wonderful is the fact that these long-lived animals are resistant to a variety of age related diseases, including (in various animals) cancer, heart failure, and protein-aggregation disease. Thus these mutants not only look young, they are young, in the sense that they are not susceptible to age-related disease until later. . . . This link between aging and age-related disease suggests an entirely new way to combat many diseases all at once; namely, by going after their greatest risk factor: aging itself. This is an extremely exciting and important concept that could revolutionize medicine, human health and longevity, and it has just now begun to be studied in earnest, still in only a handful of labs."

Can Kenyon's results be extended to humans and possibly turned into a method of extending the length and quality of life? Many researchers are convinced that they can be, at least to some extent, especially since the discovery that the genomes of animals ranging from flies to mice to humans contain molecules related to daf-2—and they have similar functions. In the worm, daf-2 sits on the surfaces of cells, where it senses hormones. These are small molecules that carry signals through the body, helping it adapt to changes—sometimes sudden ones. In humans, the hormone insulin, for example, acts as a monitoring device that helps the body adjust to the presence or absence of food. It travels through the bloodstream, docks onto receptor proteins on cells, and triggers the activation of genes. *C. elegans* also produces an insulin-like molecule, and the molecule it docks onto is daf-2.

Where the worm has one molecule, evolution has given human beings two. The closest relatives of daf-2 in people are insulin and another hormone called the insulin-like growth factor 1 (IGF-1). To Kenyon and a number of other researchers, this suggests that the human receptors might also have played a role in aging. Interestingly, it would also provide a connection between that process and a person's diet.

Researchers across the world have been investigating this question in mice and other laboratory animals. It is one theme being pursued at the Italian station of the European Molecular Biology Laboratory near Rome, in the group of the developmental biologist Nadia Rosenthal (who wrote the Foreword to this book). A main focus of Rosenthal's work is muscle, particularly the heart, and diseases related to muscle development and deterioration over the course of a lifetime. IGF-1 drew her attention in the 1990s, while

Nadia Rosenthal, an American geneticist with colleagues in Italy and Australia, is author of each foreword in this multivolume set. Her research is providing insights into the molecular signals that prompt the regeneration of tissues by stem cells and genes involved in the process of aging. *(Nadia Rosenthal)*

she was working at Harvard University, because the hormone was thought to play an important role in triggering the formation of muscle cells. Normally, the signal is active in embryos and cases where damaged muscle needs to be repaired.

Rosenthal wondered what would happen if adult muscles could produce the growth factor themselves. In collaboration with H. Lee Sweeney's lab at the University of Pennsylvania, her group developed a strain of mouse in which particular muscle cells produced IGF-1 locally throughout their entire life spans. The researchers discovered that the factor seemed to be activating a "regenerative program" that could recruit stem cells to form new muscle very efficiently. The mice were healthy and so muscular that lab technicians gave them the nickname "Schwarzenegger mouse," and they remained amazingly fit even at the "advanced age" of 20 months—the mouse equivalent of retirement age. IGF-1 was holding off the normal deterioration of muscle and helping to rebuild it in mice that had already lost muscle mass. And it was significantly increasing the animals' health span. In the case of the Schwarzenegger mouse, the local production of IGF-1 was acting to protect the tissue environment from the deterioration of age by inducing new signals in the muscle.

This is only one example of an enormous amount of ongoing work that has established a connection between the body's hormone systems—which are closely linked to diet—and aging. Insulin and IGF-1 provoke different responses in different tissues. While both molecules are vital to growth and the way the body processes food, Kenyon and many others believe that keeping insulin levels low is generally good for animals. "What's really interesting is that you can get the life span benefits by taking away the insulin receptor in individual tissues," she says. "So it might not be overall percentage of insulin function we need to concentrate on, but a selective percentage in different tissues—like fat cells."

Kenyon has formed a company called Elixir Pharmaceuticals that aims to develop therapies based on manipulating the body's response to insulin. A cure for aging is not anywhere near on the horizon. But many scientists now believe that genes may be a key to lengthening life and improving its qual-

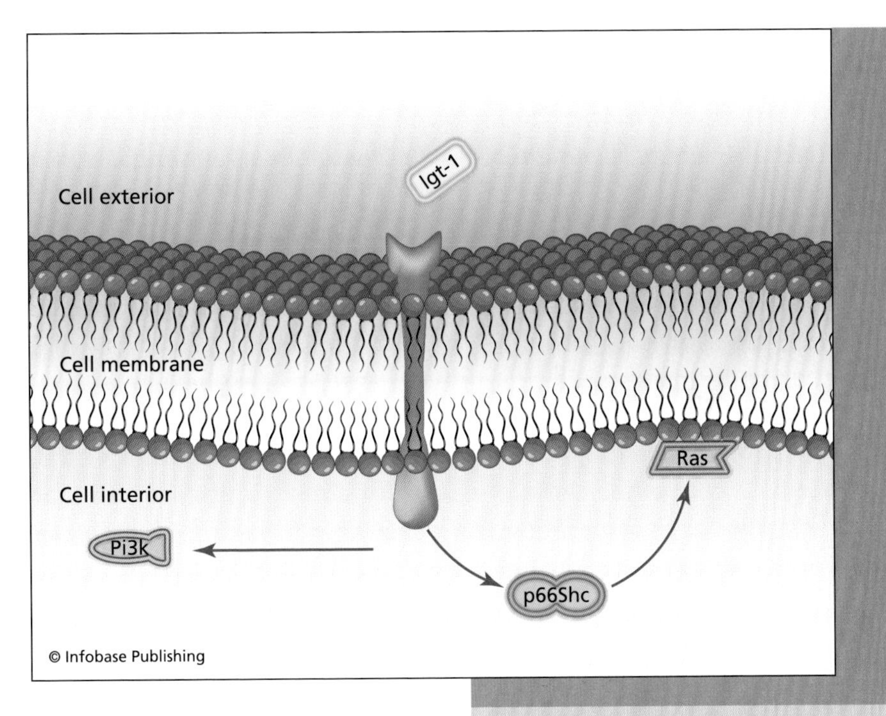

The IGF-1 receptor is embedded in the surface of cells and responds to an insulin-like hormone. Researchers have shown that its activity is related to aging.

ity in old age. In an interview for the journal *PLoS Biology,* Kenyon put the issue into an evolutionary perspective. Hundreds of millions of years ago, humans' ancestors were also small worms. "If you'd asked me . . . 'Cynthia, you have a two-week lifespan, do you think that you could [live longer]?' And if I'd told you, 'Well, I think our descendants will live 1,000 times longer,' you'd have said, 'Oh, come on!' But we do. It happened."

SAVING THE WORLD: ANTS, HUMANS, AND THE FATE OF THE EARTH

It might seem strange that a scientist whose work has mainly focused on a tiny insect spends a great deal of his time considering the fate of the entire planet. But Edward O. Wilson (1929–), a professor at Harvard and the world's foremost expert on ants, has always had a much larger perspective on life. He has been awarded two Pulitzer Prizes for his writings on ecology and the natural sciences. Part of his broader view of the world has come from witnessing the increasingly rapid destruction of habitats occupied by ants and many other species—mostly as a result of human activity. Wilson believes that the situation is dire, and in 2002 he collected his thoughts in a book called *The Future of Life*—a vision of what the world will be like if present trends such as human population growth and the destruction of ecological systems continue.

Thomas Malthus (1766–1834), a British scholar and pastor, seems to have been one of the first people to understand that overpopulation could lead to disaster. He was the first to describe a connection between the growth of populations—which puts stress on the food supply and the environment—and poverty. This relationship was so misunderstood in the late 18th century that the British government was providing subsidies that encouraged the poor to have more children. As a result, the country's population was growing at an alarming rate.

"The power of population is indefinitely greater than the power in the Earth to produce subsistence for man," Malthus wrote in the first edition of *An Essay on the Principle of Population,* published in 1798. The message was that every new human child has the potential to go on to create many new mouths to feed, whereas the surface of the Earth does not inflate as human populations expand. It has finite resources, and the amount that it can produce grows much more slowly. Malthus also saw the connection between overpopulation, wars, and epidemics. When Charles Darwin and Alfred Russel Wallace read Malthus's essay, they began thinking about how nature kept species in check—for example, why the surface of the world was not a towering pile of ants that rose miles into the air—and the result was the theory of evolution.

Evolution and ecological thinking arose hand in hand, because the theory showed how dependent species are upon one another. In some cases, these interdependencies are obvious—if one species is wiped out by a disease, those that feed on it will also suffer. But often the relationships are more subtle, based on networks of interactions between many species that are difficult to uncover. Those must be understood, Wilson believes, because only then will people be able to perceive the dangerous effects that their own behavior and lifestyle are having on the planet as a whole.

The Future of Life opens with some startling facts. For example, Wilson describes a person's "ecological footprint—the average amount of productive land and shallow sea appropriated by each person in bits and pieces from around the world for food, water, housing, energy, transportation, commerce, and waste absorption." In developing nations, an individual's footprint is currently about 2.5 acres, whereas in the United States it is nearly 10 times as large (24 acres). Bringing everyone on Earth to this level, he writes, "would require four more planet Earths. The 5 billion people of the developing countries may never wish to attain this level of profligacy. But in trying to achieve at least a decent standard of living, they have joined the industrial world in erasing the last of the natural environments."

This erasure involves the clearing of land to build new homes and cities, of course, but there are many other factors. Conservationists summarize the reasons for the decline of species with the acronym HIPPO, which stands for habitat destruction, invasive species, pollution, population, and overharvesting. For the first time, metagenomics methods (described in the previous chapter) are allowing scientists to determine just how bad the damage really is. Even if the situation proves not as serious on the microbial scale as it is for larger species, whose numbers are easier to measure, Wilson says that the worldwide situation is already extremely serious, as can be seen through the example of Hawaii, which he calls a laboratory in which it is possible to understand what is happening throughout the rest of the world.

Hawaii is so distant from the nearest major landmass that it took a long time for other species to settle there. Wilson and his colleagues estimate that on average, one new species may have arrived every 1,000 years, carried by winds or floating on bits of wood. "Extremely few made a successful landfall. Even then the pioneers faced formidable obstacles. There had to be a niche to fill immediately upon arrival—the right place to live, the right food to eat, potential mates immigrating with them, and few or no predators waiting to gobble them up." Once a species had settled in, it began to adapt into forms unique to the islands.

Settling Hawaii took a long time; killing off its species is going much more quickly. Originally, Hawaii was home to more than 125 species of birds that existed nowhere else; only 35 remain, 24 of which are considered endangered. The island's unique plants and insects face a similar situation. Humans have imported a wide range of other species—either deliberately for food or other uses or by accident—that have wiped out the native species. This sort of replacement happens naturally, of course, but with the help of humans, what used to take millions of years now happens in decades. Three-quarters of Hawaii's land has been converted into living space or fields for crops. Humans brought along pigs, some of which escaped and became wild predators. But perhaps the greatest blow to the environment, Wilson says, is one of the tiniest threats—ants. Hawaii never had them, so its species never had to adapt to them. Their arrival with humans caused a huge shock to Hawaii's system.

Once a unique species is lost, it is lost forever, barring a rapid jump forward in cloning technology and a massive effort to revive a species. There are practical reasons to be concerned about these losses: Many of today's most potent drugs to fight cancer and other diseases have come from exotic sources such as peculiar fungi found on the bark of trees. But the real issue is a much deeper one. "Earth, unlike other solar planets, is not in physical equilibrium," Wilson writes. "It depends on its living shell to create the special conditions on which life is sustainable. The soil, water, and atmosphere of its surface have

(opposite page) The Hawaiian Islands are so distant from the continental mainland that they were settled very slowly by foreign species, which have evolved in unique ways over millions of years. When humans arrived they brought along species that have been wiping out many of these unique plants and animals at a rapid pace.

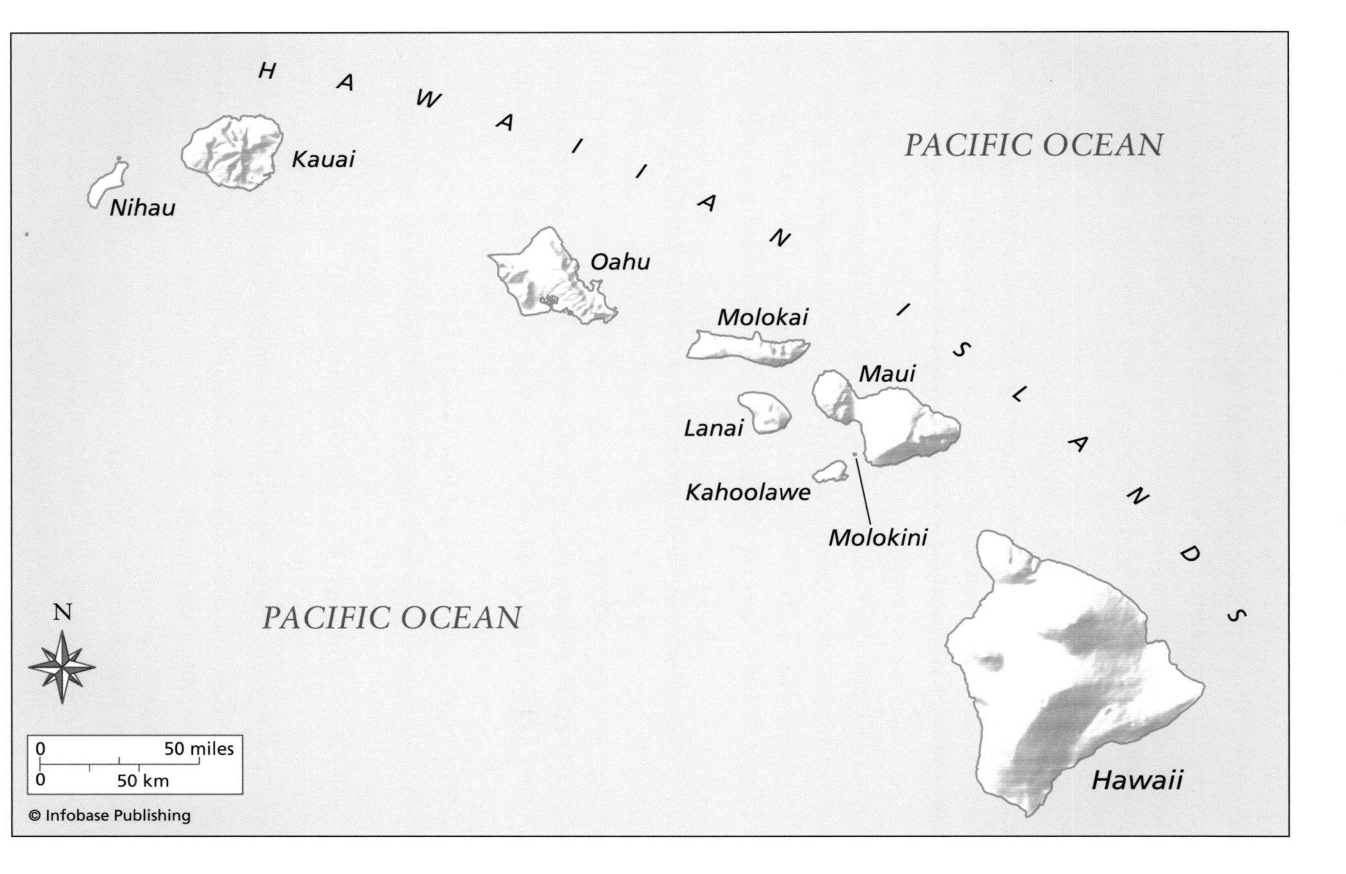
H A W A I I A N I S L A N D S
PACIFIC OCEAN
Kauai
Nihau
Oahu
Molokai
Maui
Lanai
Kahoolawe
Molokini
Hawaii
PACIFIC OCEAN
N
0 50 miles
0 50 km
© Infobase Publishing

evolved over hundreds of millions of years to their present condition by the activity of the biosphere, a stupendously complex layer of living creatures whose activities are locked together in precise but tenuous global cycles of energy and transformed organic matter. The biosphere creates our special world anew every day, every minute. . . . When we alter the biosphere in any direction, we move the environment away from the delicate dance of biology. When we destroy ecosystems and extinguish species, we degrade the greatest heritage this planet has to offer and thereby threaten our own existence."

Of the five elements of HIPPO, population has likely been the most serious and will continue to be, affecting both the natural world and the evolution of society. Just as Malthus discovered links between overpopulation, disease, and war, modern ecologists see connections between the birthrate in developing countries and huge social problems such as government instability and terrorism. In some cases, the absolute number of people living in a country may be less important than its age structure, which has an enormous influence on the economy and many other aspects of society. In 1999, in the West African country of Benin, nearly 50 percent of the population was under 15 years of age. Wilson writes, "A country poor to start with and composed largely of young children and adolescents is strained to provide even minimal health services and education for its people. Its superabundance of cheap, unskilled labor can be turned to some economic advantage but unfortunately also provides cannon fodder for ethnic strife and war. As the populations continue to explode and water and arable land grow scarcer, the industrial countries will feel their pressure in the form of many more desperate immigrants and the risk of spreading international terrorism."

There is a limit to how many people the planet can support. If everyone ate grain, the Earth might be able to provide enough for a population of 10 billion; if instead, people used the grain to feed animals to provide meat, Wilson estimates the number might only be 2.5 billion. But food production is coupled to many other things, particularly the need for water. Wilson explains: "A thousand tons of freshwater yields a ton of wheat, worth $200, but the same amount of water in industry yields $14,000." If the main interest is profit, this means that water will inevitably be diverted from food production to industry. This leads to pollution, which makes prices rise higher because the water will have to be treated and purified before it can be returned to agriculture. Industrial countries may be able to afford the exorbitant prices, but others will not.

Wilson is not a doomsayer who stops with a critique of the current situation; he proposes a number of concrete measures that could have a significant impact over the long term, including:

- Set aside and stringently protect the natural hot spots with the greatest biodiversity that are currently most at risk. Wilson estimates that 25 percent of these ecosystems take up only 1.4 percent of the Earth's land, but are home to 43.8 percent of the vascular plants and 35.6 percent of known animal species.
- Keep the rain forests of the Amazon and four other remaining frontier forests intact
- Cease all logging of old-growth forests everywhere
- Protect river and lake systems everywhere
- Identify the most important marine hot spots, such as coral reefs, and protect them
- Obtain as complete a map as possible of the Earth's biodiversity
- Make conservation profitable by helping people who live near preserves, supporting them financially and involving them professionally in conservation efforts
- Investigate biodiversity more thoroughly in order to use more species as sources of food and pharmaceuticals and to take advantage of species' ability to restore the environment after it has been damaged
- Develop genetically modified crops that have been thoroughly tested for safety and use them in a regulated way
- Increase the capacity of zoos and botanical gardens to breed endangered species
- Support population planning throughout the world

If these measures can be taken quickly, as a collaboration between governments, scientists, and industries, Wilson believes that it is not too late to reverse the major damage that has been done. It should be possible to feed humanity, if populations can be stabilized by the middle of the 21st century, and if the world's ecospheres are protected and restored.

Modern conservation provides a sobering perspective to those who have believed that science will progress fast enough to solve all the world's problems and ensure a healthy future for humanity. Evolution, genetics, and the other parts of modern biology have demonstrated that human nature is dependent on both genes and the environment. Human lifestyles have already had a deep impact on ecospheres; increasingly and inevitably, these changes will mold the future of the species. One of the most important lessons of modern biology is that the only way to provide a healthy future for humanity is to improve the quality of life for people around the globe—using the tools of genetics and other means—while ensuring that the Earth is a healthy place to live.

Chronology

1651	William Harvey claims that all animals arise from eggs.
1677	Antoni van Leeuwenhoek discovers sperm.
1751	Pierre-Louis Maupertuis studies polydactyly, the inheritance of extra fingers in humans.
	Joseph Adams recognizes the negative hereditary effects of inbreeding.
1802	Jean-Baptiste Lamarck publishes *Research on the Organization of Living Bodies,* in which he claims that species become more perfect and pass on acquired characteristics to their offspring. The hypothesis is overturned by evolution.
1824	Joseph Lister builds a new type of microscope that removes distortion and greatly increases resolution.
1838	Matthias Schleiden discovers that plants are made of cells.
1840	Theodor Schwann discovers that all animal tissues are made of cells.
1856	Gregor Mendel begins experiments on heredity in pea plants.

1857	Joseph von Gerlach discovers a new way of staining cells that reveals their internal structures.
1858	The theory of evolution is made public at a meeting of the Linnean Society in London with the reading of papers by Charles Darwin and Alfred Russel Wallace.
	Rudolf Virchow states the principle of *Omnis cellula e cellula*: every cell derives from another cell—including cancer cells.
1859	Charles Darwin publishes *On the Origin of Species.* The complete first print sells out on the first day.
1865	Gregor Mendel presents his paper "Experiments in Plant Hybridization" in meetings of the Society for the Study of Natural Sciences in Brnø, Moravia. The paper outlines the basic principles of the modern science of genetics. It is published the next year but receives little attention.
1868	Fredrich Miescher isolates DNA from the nuclei of cells; he calls it nuclein.
1871	Francis Galton carries out experiments in rabbits that disprove Darwin's hypothesis of how heredity functions.
1876	Oscar Hertwig observes the fusion of sperm and egg nuclei during fertilization.

1879	Walther Flemming observes the behavior of chromosomes during cell division.
1885	August Weismann states that organisms separate reproductive cells from the rest of their bodies, which helps explain why Lamarck's concept of evolution and inheritance is wrong. He tries and fails to observe Lamarckian inheritance in the laboratory by cutting off the tails of mice for many generations.
1900	Hugo de Vries, Carl Correns, and Erich von Tschermak-Seysenegg independently publish papers that confirm Mendel's principles of heredity in a wide range of plants.
	Archibald Garrod identifies the first disease that is inherited according to Mendelian laws, which means that it is caused by a defective gene.
	Theodor Boveri demonstrates that different chromosomes are responsible for different hereditary characteristics.
1901	Karl Landsteiner identifies the ABO blood groups.
1902	William Bateson popularizes Mendel's work in a book called *Mendel's Principles of Heredity: A Defense.*
1903	Walter Sutton connects chromosome pairs to hereditary behavior, demonstrating that genes are located on chromosomes.

1905	Nettie Stevens and Edmund Wilson independently discover the role of the X and Y chromosomes in determining the sex of animal species.
1906	William Bateson discovers that some characteristics of plants depend on the activity of two genes.
1908	Archibald Garrod shows that humans with an inherited disease are lacking an enzyme (a protein), demonstrating that there is a connection between genes and proteins.
1909	William Bateson coins the term *genetics.*
1910	The Eugenics Record Office is opened at Cold Spring Harbor, New York.
	Thomas Hunt Morgan discovers the first mutations in fruit flies, *Drosophila melanogaster,* bred in the laboratory. This leads to the discovery of hundreds of new genes over the next decades.
1911	Morgan discovers some traits that are passed along in a sex-dependent manner and proposes that this happens because the genes are located on sex chromosomes. He proposes the general hypothesis that traits that are likely to be inherited together are located on the same chromosome.
1913	Alfred Sturtevant constructs the first genetic linkage map, allowing researchers to

pinpoint the physical locations of genes on chromosomes.

1920 Hans Spemann and Hilde Proescholdt Mangold begin a series of experiments in which they transplant embryonic tissue from one species to another. The scientists show that particular groups of cells they called organizers send instructions to neighboring cells, changing their developmental fates.

1921 Erwin Baur, Eugen Fischer, and Fritz Lenz publish a book called *Menschliche Erblichkeitslehre und Rassenhygiene,* which attempts to link genetics to race and is used by eugenicists in the United States and Germany as a justification for declaring that there are inferior races and a motivation for sterilizing and killing social undesirables.

1922 Ronald A. Fisher uses mathematics to show that Mendelian inheritance and evolution are compatible.

1927 Hermann Muller shows that radiation causes mutations in genes that can be passed down through heredity.

1928 Fredrick Griffith discovers that genetic information can be transferred from one bacterium to another, hinting that hereditary information is contained in DNA.

1931 Barbara McClintock shows that as chromosome pairs line up beside each other during the copying of DNA, fragments can break off one chromosome and be inserted into the other in a process called recombination.

Archibald Garrold proposes that diseases can be caused by a person's unique chemistry—in other words, genetic diseases may be linked to defects in enzymes.

1933 Theophilus Painter discovers that staining giant salivary chromosomes in fruit flies reveal regular striped bands.

1934 Calvin Bridges shows that chromosome bands can be used to pinpoint the exact locations of genes.

1935 Nikolai Timofeeff-Ressovsky, K. Zimmer, and Max Delbrück publish a groundbreaking work on the structure of genes that proposes that mutations alter the chemistry and structure of molecules.

1937 George Beadle and Boris Ephrussi show that genes work together in a specific order to produce some features of fruit flies.

1940 George Beadle and Edward Tatum prove that a mutation in a mold destroys an enzyme and that this characteristic is inherited in a Mendelian way, leading to their hypothesis that one gene is related to one enzyme (protein), formally proposed in 1946.

1943 Max Delbruck and Salvador Luria demonstrate evolution in the laboratory by showing that bacteria evolve defenses to viruses through mutations that are acted on by natural selection.

1944 Oswald Avery, Colin MacLeod, and Maclyn McCarty show that genes are made of DNA.

Erwin Schrödinger publishes *What Is Life?*

1948 The American Society for Human Genetics is founded.

1950 Barbara McClintock publishes evidence that genes can move to different positions as chromosomes are copied.

Erwin Chargaff discovers that the proportions of A and T bases in an organism's DNA are identical, as are the proportion of Gs to Cs.

1951 Rosalind Franklin uses X-ray diffraction to obtain images of DNA; the patterns reveal important clues to the building plan of the molecule.

1953 James Watson and Francis Crick publish the double helix model of DNA, which explains both how the molecule can be copied and how mutations might arise.

In the same issue of the journal *Nature,* Rosalind Franklin and Maurice Wilkins publish X-ray studies that support the Watson-Crick model. This launches the field of molecular biology that shows, over the next 20 years, how the information in genes is used to build organisms.

1958	Francis Crick describes the central dogma of molecular biology: DNA creates RNA creates proteins. He challenges the scientific community to figure out the molecules and mechanisms by which this happens.
1959	Jerome Lejeune discovers the first disease due to defects in chromosomes: Down syndrome is caused by the inheritance of an extra chromosome.
	Marshall Nirenberg, Marianne Grunberg-Manago, and Severo Ochoa show that the cell reads DNA in three-letter words to translate the alphabet of DNA into the 20-letter alphabet of proteins.
1961	Sidney Brenner, François Jacob, and Matthew Meselson discover that messenger RNA is the template molecule that carries information from genes into protein form. Crick and Brenner suggest that proteins are made by reading three-letter codons in RNA sequences, which represent three-letter codes in DNA. M. W. Nirenberg and J. H. Matthaei use artificial RNAs to create proteins with specific spellings, helping them learn the complete codon spellings of amino acids.
1965	Leonard Hayflick discovers that human cells raised in laboratory cultures have a limited life span, prompting a search for molecular mechanisms of aging.

1966 Marshall Nirenberg and H. Gobind Khorana work out the complete genetic code—the DNA recipe for every amino acid.

1970 Hamilton Smith and Kent Wilcox isolate the first restriction enzyme, a molecule that cuts DNA at a specific sequence—which will become an essential tool in genetic engineering.

1972 Janet Mertz and Ron Davis use restriction enzymes and DNA-mending molecules called ligases to carry out the first recombination: the creation of an artificial DNA molecule.

Paul Berg creates a new gene in bacteria using genetic engineering.

1973 Stanley Cohen, Annie Chang, Robert Helling, and Herbert Boyer create the first transgenic organism by putting an artificial chromosome into bacteria.

1975 Edward Southern creates Southern blotting, a method to detect a specific DNA sequence in a person's DNA; the method will become crucial to genetic testing and biology in general. Cesar Milsein, Georges Kohler, and Niels Kai Jerne develop a method to make monoclonal antibodies.

1977 Walter Gilbert and Allan Maxam develop a method to determine the sequence of a DNA molecule; Fredrick Sanger and colleagues independently develop another very rapid method for doing so, launching the age of high-throughput DNA sequencing.

Frederick Sanger finishes the first genome, the complete nucleotide sequence of a bacteriophage.

Phillip Sharp and colleagues discover introns, information in the middle of genes which do not contain codes for proteins and must be removed before an RNA can be used to create a protein.

1977 Genentech, the first biotech firm, is founded based on plans to use genetic engineering to make drugs.

1978 Recombinant DNA technology is used to create the first human hormone.

1980 Christiane Nüsslein-Volhard and Eric Wieschaus discover the first patterning genes that influence the development of the fruit fly embryo, bringing together the fields of developmental biology and genetics.

1981 Three laboratories independently discover oncogenes: proteins that lead to cancer if they undergo mutations.

1982 Insulin becomes the first genetically engineered drug.

1983 Walter Gehring's laboratory in Basel and Matthew Scott and Amy Weiner, working at the University of Indiana, independently discover HOX genes: master patterning molecules for the creation of the head-to-tail axis in animals as diverse as flies and humans.

1985 Kary B. Mullis publishes a paper describing the polymerase chain reaction, a method which rapidly and easily copies DNA molecules.

1986 First outbreak of BSE (mad cow disease) among cattle in the United Kingdom

1987 First human genetic map published

1988 The Human Genome Project is launched by the U.S. Department of Energy and the National Institutes of Health, with the aim of determining the complete sequence of human DNA.

1989 Alec Jeffreys discovers regions of DNA that undergo high numbers of mutations. He develops a method of DNA fingerprinting that can match DNA samples to the person they came from and can also be used in establishing paternity and other types of family relationships.

The Human Genome Organization (HUGO) is founded.

1990 W. French Anderson carries out the first human gene replacement therapy to treat an immune system disease in four-year-old Ashanti DeSilva.

1993 The company Monsanto develops and begins to market a genetically engineered strain of tomatoes called Flavr Savr.

The Huntington disease gene is found.

Cynthia Kenyon discovers mutations in *C. elegans* that double the worm's life span.

1994 Mary-Claire King discovers BRCA1, a gene that contributes to susceptibility to breast cancer.

1995 The first confirmed death from Creutzfeldt-Jakob disease, the human form of BSE, is reported in the United Kingdom.

1996 Researchers complete the first genome of a eucaryote, baker's yeast. The completion of the genome of *Methanococcus jannaschii,* an archaeal cell, confirms that archaea are a third branch of life, separate from bacteria and eucaryotes.

Gene therapy trials to use the adenovirus as a vector for healthy genes are approved in the United States.

1997 Ian Wilmut's laboratory at the Roslin Institute produces Dolly the sheep, the first cloned mammal.

1998 Scientists obtain the first complete genome sequence of an animal, the worm *Caenorhabditis elegans.*

1999 Jesse Gelsinger dies in a gene therapy trial, bringing a temporary halt to all viral gene therapy trials in the United States.

2000 The genome of the fruit fly *Drosophila melanogaster* is completed.

Scientists complete a working draft of the human genome. The complete genome is published in 2003.

2002 The mouse genome is completed.

2004 Scientists in Seoul announce the first successful cloning of a human being, a claim which is quickly proven to be false.

2008 Samuel Wood of the California company Stemagen successfully uses his own skin cells to produce clones, which survive five days.

Glossary

allele one variant of a particular gene

amyloid plaque a cluster of protein fragments that accumulates in tissues and does not dissolve, frequently found in association with Alzheimer's and other neurodegenerative diseases

apoptosis a cellular self-destruct program triggered by genes, usually in response to external stimuli

biodiversity coined from biological diversity, referring to the amount of life found in a particular environment

bioremediation a process by which microorganisms restore features of the environment to their original state, for example when there has been contamination by pollutants or radiation

chromatin immunoprecipitation (ChIP) procedure to determine whether a given protein binds to or is localized to a specific DNA sequence in vivo

clade a branch of an evolutionary or family tree containing only organisms that have descended from a specific common ancestor

cloning a method which makes an exact copy of a DNA sequence, a chromosome, or an entire genome

computerized tomography (CT) a method which uses X-rays or another method to scan patient tissues, creating slicelike photographs that are assembled into three-dimensional images by computer

conditional mutagenesis a type of genetic engineering that knocks out or knocks in a gene only in specific tissues, rather than in an entire organism, often in combination with molecules that allow researchers to decide when the alteration takes place

constitutional activation a form of a gene or molecule that is always active, usually because of mutations that remove its ability to be switched off

crystallography a method of turning proteins or other biological molecules into crystals, often the first step in determining a three-dimensional atomic structure of a molecule

DNA (deoxyribose nucleic acid) a molecule made of nucleic acids that forms a double helix in cells, holds a species' genetic information, and encodes RNAs and proteins

DNA microarray (DNA chip) a set of probes made of nucleic acids, usually mounted on a glass slide, used to compare the RNAs made by different types of cells

dominant an allele that determines the phenotype of an organism, even when that organism has a different allele as the second copy of the gene

eugenics strategies and actual programs to influence the gene pool and future evolution of humans by controlling their mating—in some cases, by sterilizing or killing people judged to be unfit

fermentation a process by which cells degrade substances to produce energy, in the absence of oxygen

fitness the degree to which an organism is adapted to its environment

forward genetics a method of discovering gene functions that starts with a phenotype and searches for the molecule that is responsible for it

gastrulation a process that takes place in the early development of the embryo, in which undifferentiated cells form three layers that go on to produce the body's major tissues and organs

gene a region of DNA that encodes a protein

genetically modified organism (GMO) an organism whose genes have been altered through an artificial process in the laboratory

genome the entire set of DNA in an organism or species, usually referring to the DNA in nucleus (of cells that have one)

genotype an organism's complete collection of genes, including both dominant and recessive alleles

germ cell a specialized cell capable of creating a new organism (a sperm or egg cell)

green fluorescent protein (GFP) a protein that releases green fluorescent light when exposed to energy of a particular frequency

heredity the means by which features of a parent organism are passed to its offspring

homologue a DNA sequence, tissue, organ, or other body structure that is the closest evolutionary relative of a similar structure in another organism

HOX gene a gene containing a structure called a homeobox, which usually controls important developmental processes in embryos

intron a sequence in an RNA or gene that does not encode a part of a protein

in vivo in the living body of an animal or plant

knock in a gene that has been artificially added to an organism

knock out a gene that has been artificially removed from an organism

ligase an enzyme that can join other molecules together; in genetic engineering, ligases are used to combine fragments of DNA to make new genes

lobotomy the surgical removal of part of the brain

magnetic resonance imaging (MRI) a method that detects the presence and locations of particular atoms by exposing them to strong magnetic fields

mass spectrometry a method of detecting the composition of substances, such as proteins or other chemical compounds

materialism a philosophy that seeks physical and chemical explanations for phenomena, including mental behavior and states

messenger RNA a molecule made of nucleic acids, based on the information in a gene, used as a template for the production of a protein

metagenomics an approach to DNA sequencing that tries to capture the sequences of all DNA found in a particular environment rather than that of a particular organism or species

miasma theory of disease an ancient idea that illness is caused by the inhalation of bad air

microRNA a tiny RNA molecule produced by cells whose main function seems to be to bind to specific messenger RNAs and prevent their translation into proteins

microtubule a cellular fiber made of protein subunits called tubulin that helps give the cell its shape and structure, participates in cell division, and serves as a highway along which other molecules are delivered

mitotic spindle a structure built during cell division; it is made of microtubules, and its function is to separate chromosomes into two equal sets

monogenic trait a feature of an organism that is determined by the presence of a particular allele of a single gene

mutagen a substance that causes mutations in genes

mutation a change in an organism's DNA sequence caused by a copying error, a mutagen, or some other form of damage

natural selection the process by which the environment gives some members of a species an advantage at having more offspring due to their genetic makeup

nuclear magnetic resonance a method that uses a very strong magnetic field to identify the atoms that make up molecules and plot their positions relative to each other, a common method of investigating the three-dimensional structures of proteins

nucleotide (base) a subunit of DNA and RNA that consists of a base linked to a phosphate group and sugar

ontogeny the stages of an individual organism's development, from fertilization to birth and adulthood

phenotype the complete set of measurable physical and behavioral characteristics of an organism determined by its genes

phylogeny the stages of a species's evolution, from the first cell to its current form

protein a molecule made of amino acids, produced by a cell based on information in its genes

recapitulation Ernst Haeckel's theory that the development of a single organism passes through phases that retrace its evolutionary history

receptor (protein) a molecule in a cell or on its surface that binds to a specific partner molecule, usually leading to a change in its activity, and often ultimately resulting in a change in the set of genes active within a cell

recessive an allele that must be present in two copies in an organism to fully determine its phenotype

restriction enzyme a protein that cuts single-stranded or double-stranded DNA at specific sequences

reverse genetics altering a gene to discover its functions in cells and observe its effects on an organism's phenotype

ribonucleic acid (RNA) a molecule made of nucleotides that is produced by transcribing the information in a DNA sequence

sequence a list of the subunits of a molecule such as DNA or proteins, in the order in which they are attached to each other

small interfering RNA (siRNA) an artificial molecule made of RNA, used to prevent specific messenger RNAs from being translated into proteins

spontaneous generation a disproven theory that held that complex living organisms such as maggots or flies commonly arose on their own without an egg

synchrotron a circular instrument in which electrons or other subatomic particles are accelerated through the use of magnets, often used in biology as a source of high-energy X-rays to study the structures of molecules

systems biology an interdisciplinary field that sees life as the result of complex networks of many interacting elements, usually attempting to study it through computer models

telomerase an enzyme that adds small repeated DNA sequences to telomeres, helping protect chromosomes from being degraded

telomere a region at the ends of chromosomes that does not contain genes but protects chromosomes from being degraded as DNA is copied

tiling array a type of DNA microarray designed to investigate whether RNA molecules have been made from any of the sequences within a particular region of the genome, including segments that are not known to contain genes

tumor suppressor gene a gene that leads to tumors when it becomes defective; the healthy form protects cells from becoming cancerous

variation the diversity within a species, caused by the existence of different forms of genes, and the range of phenotypes that this diversity produces

viral therapy methods that aim to use viruses to deliver healthy forms of genes or other genetic material to cells

vitalism the hypothesis that a special form of energy (often thought to be spiritual or nonmaterial in nature) is necessary to produce life from nonliving substances and that life cannot be explained purely in terms of physical and chemical forces

Further Resources

Books and Articles

Allen, Myles. "A novel view of global warming." *Nature* 433 (January 2005). A climate researcher's critique of Michael Crichton's book *State of Fear,* which cites scientific publications to criticize the way the research community has dealt with the theme of global warming. Allen believes that Crichton distorted the evidence and is manipulating readers' opinions by injecting pseudo-science into a novel.

Asimov, Isaac. *The Beginning and the End.* New York: Pocket Books, 1983. Essays on genetics and other branches of science, with a particular focus on the impact they will have on humans and society in the future, from one of the world's foremost science fiction and popular science authors.

Bodman, Walter, and Robin McKie. *The Book of Man: The Quest to Discover Our Genetic Heritage.* London: Little, Brown and Company, 1994. A well-written account of the major people and themes of human genetics from the late 19th century to the beginning of the Human Genome Project.

Branden, Carl, and John Tooze. *Introduction to Protein Structure,* 2nd ed. New York: Garland Publishing, 1999. A detailed overview of the chemistry and physics of proteins, for university students with some background in both fields.

Brown, Andrew. *In the Beginning Was the Worm.* London: Pocket Books, 2004. The story of an unlikely model organism in biology: the worm *C. elegans,* and the scientists who have used it to understand some of the most fascinating issues in modern biology.

Browne, Janet. *Charles Darwin: The Power of Place.* New York: Knopf, 2002. The second volume of the definitive biography of Charles Darwin.

———. *Charles Darwin: Voyaging.* Princeton, N.J.: Princeton University Press, 1995. The first volume of the definitive biography of Charles Darwin.

Caporale, Lynn Helena. *Darwin in the Genome: Molecular Strategies in Biological Evolution.* New York: McGraw Hill, 2003. A new look at variation and natural selection based on discoveries from the genomes of humans and other species, written by a noted biochemist.

Carlson, Elof Axel. *Mendel's Legacy: The Origin of Classical Genetics.* Cold Spring Harbor, N.Y.: Cold Spring Harbor Laboratory Press, 2004. An excellent, easy-to-read history of genetics from Mendel's work to the 1950s. Carlson explains the relationship between cell biology and genetics especially well.

———. *The Unfit: A History of a Bad Idea.* Cold Spring Harbor, N.Y.: Cold Spring Harbor Laboratory Press, 2001. An in-depth account of eugenics movements across the world.

Carson, Rachel. *Silent Spring.* Cambridge: Riverside Press, 1962. Carson's book on the impact of DDT on birds and the rest of the ecosphere, proposing that chemical pollutants constitute a serious threat to the health of humans and other species. The book played an important role in the rise of environmental movements over the next decades.

Carter, Rita. *Mapping the Mind.* London: Phoenix, 2000. A beautifully written and illustrated book in which science writer Rita Carver explores the state of the art of research into higher levels of brain function. While the book does not deeply explore the genetics of the brain, it gives an excellent overview of what current methods have revealed about how brain tissues participate in perception, behavior, and other functions.

Cavalli-Sforza, L. Luca. "The Human Genome Diversity Project: Past, Present, and Future." *Nature Reviews Genetics* 6 (2005): 333. An overview of the progress and difficulties encountered by the HGDP from its conception to its development as a resource now being used by researchers all over the world.

Cavalli-Sforza, L. Luca, Paolo Menozzi, and Alberto Piazza. *The History and Geography of Human Genes.* Princeton, N.J.: Princeton University Press, 1994. For over three decades Cavalli-Sforza has been interested in using genes (as well as other fields such as linguistics) to study human diversity and solve interesting historical questions like where modern humans evolved and how they spread across the globe. This book is a compilation of what he and many researchers have found.

Chambers, Donald A. *DNA: The Double Helix: Perspective and Prospective at Forty Years.* New York: New York Academy of Sciences, 1995. A collection of historical papers from major figures involved in the discovery of DNA with reminiscences from some of the authors.

Chimpanzee Sequencing and Analysis Consortium, The. "Initial sequence of the chimpanzee genome and comparison with the human genome." *Nature* 437 (2005): 69–87. This article presents an in-depth contrast of the complete DNA sequences of humans and chimpanzees.

Crichton, Michael. *The Andromeda Strain.* New York: Knopf, 1969. Crichton's breakthrough novel about the arrival of an alien microorganism on Earth and how scientists struggle to control it.

———. *Jurassic Park.* New York, Knopf: 1990. A best seller about a group of researchers who have obtained DNA from dinosaurs and brought them back to life by reconstructing their genomes with patches from other organisms.

———. *State of Fear.* New York: HarperCollins Publishers, 2004. Crichton's techno-thriller about a group of ecoterrorists who perpetrate global disasters in order to convince the public of the seriousness of global warming and other environmental threats.

Crick, Francis. *What Mad Pursuit: A Personal View of Scientific Discovery.* New York: Basic Books, 1988. Crick's account of dead ends, setbacks, wild ideas, and finally glory on the road to the discovery of the structure of DNA, with speculations on the future of neurobiology and other fields.

Darwin, Charles. *The Descent of Man.* Amherst, N.Y.: Prometheus, 1998. In this book, originally published 12 years after *On the Origin of Species,* Darwin outlines his ideas on the place of human beings in evolutionary theory.

———. *On the Origin of Species.* Edison, N.J.: Castle Books, 2004. Darwin's first, enormous work on evolution, which examines a huge number of facts while building a case for heredity, variation, and natural selection as the forces that produce new species from existing ones.

———. *The Voyage of the* Beagle. London: Penguin Books, 1989. A scientific adventure story; Darwin's account of his five years as a young naturalist aboard the *Beagle.* He had not yet discovered the principles of evolution but was aware of the need for a scientific theory of life. Readers watch over his shoulder as he tries to make sense of questions that puzzled scientists everywhere in the mid-19th century.

Diamond, Jared. *Guns, Germs, and Steel.* New York: W. W. Norton, 2005. A new look at how Western civilization came to dominate the globe, integrating information from archeology, anthropology, genetics, evolutionary biology and many other sources. Many consider Diamond the first to accurately apply evolutionary principles to the question of why some societies become dominant over others.

Dorus, Steve, Eric J. Vallender, et al. "Accelerated evolution of nervous system genes in the origin of *Homo sapiens.*" *Cell* 119, no. 7 (December 29, 2004): 1,027–1,040. A systematic study of how quickly genes crucial to brain evolution have evolved in primates and mammals. The study compares genes in humans, macaque monkeys, mice, and rats, revealing that brain-related genes have been subject to pressure from natural selection, especially in the branch of primates leading to humans.

Dubrova, Yuri, Valeri Nesterov, et al. "Further evidence for elevated human minisatellite mutation rate in Belarus eight years after the Chernobyl accident." *Mutation Research* 381 (2007): 267–278. A groundbreaking study examining the long-term impact of the Chernobyl accident on the genomes of people living in the region.

Elliott, William H., and Daphne C. Elliott. *Biochemistry and Molecular Biology.* New York: Oxford University Press, 1997. An excellent college-level overview of the biochemistry of the cell.

Fruton, Joseph. *Proteins, Enzymes, Genes: The Interplay of Chemistry and Biology.* New Haven, Conn.: Yale University Press, 1999. A very detailed historical account of the lives and work of the chemists, physicists, and biologists who worked out the major functions of the molecules of life.

Fukuyama, Francis. *Our Posthuman Future: Consequences of the Biotechnology Revolution.* London: Profile Books Ltd., 2003. A book by a leading political economist about the economic and social impact of current developments in genetic engineering and other forms of biotechnology, with keen insights into the ethical dilemmas that they pose.

———. *The End of History and the Last Man.* New York: Maxwell Macmillan International, 1992. A philosophical perspective on the development of societies and forms of government, proposing that Western liberal democracy is the logical end point of human social evolution.

Gilbert, Scott. *Developmental Biology.* Sunderland, Mass.: Sinauer Associates, 1997. An excellent college-level text on all aspects of developmental biology.

Goldsmith, Timothy H., and William F. Zimmermann. *Biology, Evolution, and Human Nature.* New York: Wiley, 2001. Life from the level of genes to human biology and behavior.

Gregory, T. Ryan, ed. *The Evolution of the Genome.* Boston: Elsevier Academic Press, 2005. An advanced-level book presenting the major themes of evolution in the age of genomes, written by leading researchers for graduate students and scientists.

Harper, Peter S. *Practical Genetic Counselling,* 6th ed. London: Hodder Arnold, 2004. An introduction to the basics of genetic counseling, with a review of genetic disorders based on body systems such as the nervous system, the eye, and cardiovascular and respiratory diseases, and a section on privacy issues and other themes under the heading "Genetics and Society."

Henig, Robin Marantz. *A Monk and Two Peas.* London: Weidenfeld & Nicolson, 2000. A popular, easy-to-read account of Gregor Mendel's work and its impact on later science.

Huxley, Aldous. *Brave New World.* New York: Perrenial, 1998. Huxley's anti-utopian novel of the future.

Huxley, J. S. *Man in the Modern World.* London: Chatto & Windus, 1947. Originally published in *The Uniqueness of Man,* 1941, the essay by biologist Julian Huxley, brother of the author of *Brave New World,* supports eugenics movements to improve the human race.

Judson, Horace Freeland. *The Eighth Day of Creation: Makers of the Revolution in Biology.* New York: Simon and Schuster, 1979. A comprehensive history of the science and people behind the creation of molecular biology, from the early 20th century to the 1970s, based on hundreds of hours of interviews Judson conducted with the researchers who created this field.

Koch, Christof. *The Quest for Consciousness: A Neurobiological Approach.* Englewood, Colo.: Roberts & Company, 2004. Koch is trying to establish the biological basis of consciousness and related mental abilities in humans. This fascinating book explores discoveries from the activity of genes to the behavior of modules of the brain and presents scientists' best current knowledge of the relationship between the physical brain and the metaphysical mind.

Kohler, Robert E. *Lords of the Fly: Drosophila Genetics and the Experimental Life.* Chicago: University of Chicago Press, 1994. The story of Thomas Hunt Morgan and his disciples, whose discoveries regarding fruit fly genes dominated genetics in the first half of the 20th century.

Lu, Shi-Jiang, Qiang Feng, et al. "Biological properties and enucleation of red blood cells from human embryonic stem cells." *Blood* (prepublished online August 19, 2008). Describes the successful creation of differentiated blood from embryonic stem cells in cell cultures in the laboratory.

Lutz, Peter L. *The Rise of Experimental Biology: An Illustrated History.* Totowa, N.J.: Human Press, 2002. A very readable, wonderfully illustrated book tracing the history of biology from ancient times to the modern era.

Maddox, Brenda. *Rosalind Franklin: The Dark Lady of DNA.* London: HarperCollins Publishers, 2002. An account of the life and work of Rosalind Franklin, who played a key role in the discovery of DNA's structure but who had trouble fitting in to the scientific culture of London in the 1950s.

Magner, Lois N. *A History of the Life Sciences.* New York: M. Dekker, 1979. An excellent, wide-ranging book on the development of ideas about life from ancient times to the dawn of genetic engineering.

McElheny, Victor K. *Watson and DNA: Making a Scientific Revolution.* Cambridge, Mass.: Perseus, 2003. A retrospective on the work and life of the extraordinary scientific personality James Watson, codiscoverer of the structure of DNA.

Musaro, Antonió, Nadia Rosenthal, et al. "Stem cell-mediated muscle regeneration is enhanced by local isoform of insulin-like growth factor 1." *Proceedings of the National Academy of Sciences* (February 3, 2004): 1,206–1,210. A paper by Nadia Rosenthal's group on the effects of a mutation in the IGF-1 gene that stimulates the regeneration of muscle in mice and extends animals' lifespans.

Plotkin, Mark J. *Tales of a Shaman's Apprentice.* New York: Penguin Books, 1993. A landmark book by a pioneer in the field of ethnobotany. For decades, Plotkin has been working with remote tribes on conservation issues and to record their knowledge of pharmacological and other uses of indigenous plants.

Ptashne, Mark, and Alexander Gann. *Genes and Signals.* Cold Spring Harbor, N.Y.: Cold Spring Harbor Laboratory Press, 2002. A readable and nicely illustrated book presenting a modern view of how genes in bacteria are regulated and what these findings mean for the study of other organisms.

Purves, William K., David Sadava, et al. *Life: The Science of Biology.* Kenndallville, Ind.: Sinauer Associates and W. H. Freeman, 2003. A comprehensive overview of themes from the life sciences. The book is most suited for beginning university students, but most of the chapters will be accessible to younger students and teachers. The illustrations used to demonstrate methods in biology and processes such as embryonic development are very clear and informative.

Sacks, Oliver. *The Island of the Colour-Blind and Cycad Island.* London: Picador, 1996. Neurobiologist Sacks's personal account of his travels to the Pacific islands of Cycad and Pingelap. There he encountered people with an unusual genetic condition that allows them only to see shades of gray; the book contains his reflections on the impact of this condition on island culture.

Scott, Christopher. *Stem Cell Now: From the Experiment That Shook the World to the New Politics of Life.* New York: Pi Press, 2006. An excellent, very readable introduction to stem cells and the role that they are likely to play in the medicine of the future, taking into account political, social, and ethical dimensions of their use.

Stent, Gunther. *Molecular Genetics: An Introductory Narrative.* San Francisco: W. H. Freeman, 1971. A classic book for college-level students about the development of genetics and molecular biology by a researcher and teacher who witnessed it firsthand.

Strachan, Tom, and Andrew P. Read. *Human Molecular Genetics 3.* New York: Garland Publishing, 2004. An excellent college-level textbook giving a comprehensive overview of methods and findings in human genetics in the molecular age.

Tanford, Charles, and Jacqueline Reynolds. *Nature's Robots: A History of Proteins.* New York: Oxford University Press, 2001. A history of biochemical and physical studies of proteins and their functions and the major researchers in the field.

Thorson, James. *Aging in a Changing Society,* 2nd ed. Philadelphia: Brunner/Mazel, 2000. Thorson is professor of gerontology at the University of Nebraska. This book presents facts, trends, and a fascinating social perspective on the rising life expectancy in the modern world.

Tudge, Colin. *In Mendel's Footnotes*. London: Vintage, 2002. An excellent review of ideas and discoveries in genetics from Mendel's day to the 21st century.

———. *The Variety of Life: A Survey and a Celebration of All the Creatures That Have Ever Lived*. New York: Oxford University Press, 2000. A beautifully illustrated tree of life classifying and describing the spectrum of life on Earth.

Vogel, Friedrich, and Arno Motulsky. *Human Genetics,* 3rd ed. New York: Springer-Verlag, 1997. A college-level, in-depth overview of human genetics in the molecular age.

Wang, Eric, Greg Kodama, et al. "Global Landscape of Recent Inferred Darwinian Selection for *Homo sapiens*." *PNAS* 103, no. 1 (2006): 135–140. A study comparing DNA sequences from humans, apes, and rodents. The work reveals human genes that have been subject to natural selection, particularly genes related to brain development and function.

Watson, James D. *The Double Helix*. New York: Atheneum, 1968. Watson's personal account of the discovery of the structure of DNA.

Watson, James D., and Francis Crick. "A Structure for Deoxyribose Nucleic Acid." *Nature* 171 (1953): 737–738. The original article in which Watson and Crick described the structure of DNA and its implications for genetics and evolution.

Wilson, Edward O. *The Future of Life*. London, Abacus: 2002. An easy-to-read overview of the current state of biodiversity throughout the world, with fascinating insights into the impact of human activity on the environment and proposals for coping with overpopulation, species extinctions, and other ecological problems, written by the world's foremost expert on ants and a two-time Pulitzer Prize winner.

Web Sites

There are tens of thousands of Web sites devoted to the topics of molecular biology, genetics, evolution, and the other themes of this book. The selection below provides original articles, teaching materials, multimedia resources, and links to hundreds of other excellent sites.

American Society of Naturalists. "Evolution, Science, and Society: Evolutionary Biology and the National Research Agenda." Available online. URL: http://www.rci.rutgers.edu/~ecolevol/fulldoc.pdf. Accessed April 28, 2009. A document from the American Society of Naturalists and several other organizations, summarizing evolutionary theory and showing how it has contributed to other fields including health, agriculture, and the environmental sciences.

Bradshaw Foundation. "Journey of Mankind—the Peopling of the World." Available online. URL: http://www.bradshawfoundation.com/journey/. Accessed April 28, 2009. An online lecture and film giving an excellent visual demonstration of how and when modern humans likely spread from Africa to populate the globe.

British Broadcasting Corporation. "BBC—Press Office—Richard Dimbledy Lecture 2007: Dr. J. Craig Venter." Available online. URL: http://www.bbc.co.uk/pressoffice/pressreleases/stories/2007/12_december/05/dimbleby.shtml. Accessed April 28, 2009. An interesting lecture from Craig Venter. Venter's was the first individual's genome to be completely sequenced. In the lecture, he describes some anecdotal discoveries about his own genetic code, as an illustration of what can be learned through personal genomics and with a view toward a day of personalized medicine.

California Institute of Technology. "The Caltech Institute Archives." Available online. URL: http://archives.caltech.edu/index.cfm. Accessed April 28, 2009. This site hosts materials tracing the history of one of America's most important scientific institutes since 1891. One highlight is a huge collection of oral histories with firsthand accounts of some of the leading figures who have been at Caltech, including George Beadle, Max Delbrück, and others.

———. "Videos—CNS 120." Available online. URL: http://www.klab.caltech.edu/cns120/wiki/Videos. Accessed April

28, 2009. A lecture series by Christof Koch, who is carrying out research into the neurobiology of consciousness at Caltech. The series, which can be watched online, gives an excellent overview of Koch's work and the way he has approached questions about the evolution and biology of the human mind.

Center for Genetics and Society. "CGS: Detailed Survey Results." Available online. URL: http://www.geneticsandsociety.org/article.php?id=404. Accessed April 28, 2009. This article presents the results of numerous surveys conducted in the United States and elsewhere on topics related to genetics, human cloning, and stem cell research, providing a fascinating view of people's knowledge of basic genetic topics as well as how opinions have changed over the past few years.

Department of Energy, Human Genome Project. "Genetics Legislation." Available online.. URL: http://www.ornl.gov/sci/techresources/Human_Genome/elsi/legislat.shtml. Accessed April 28, 2009. This page presents an overview of legislation regarding human genome information and the protection of personal genetic information. The Web site is a good starting point for teachers and students who want to get an overview of scientific and ethical issues related to human genetics, including information about laws pertaining to genetic testing, patient rights, medical discoveries, etc.

———. "Genetic Disease Information." Available online. URL: http://www.ornl.gov/sci/techresources/Human_Genome/medicine/assist.shtml. Accessed April 28, 2009. A basic, easy-to-understand guide to the facts about known genetic diseases and ethical and legal issues surrounding diagnoses. The site, which was created by the Human Genome Project, provides links to places where the tests are available and information about new approaches to dealing with the diseases.

Dolan DNA Learning Center, Cold Spring Harbor Laboratory. "DNA Interactive." Available online. URL: http://www.dnai.org. Accessed April 28, 2009. A growing collection of multimedia and archival materials including several hours of filmed interviews with leading figures in molecular biology, a timeline of discoveries, an archive on the American eugenics movement, and a wealth of teaching materials on the topics of this book.

———. "Genes to Cognition Online." Available online. URL: http://www.g2conline.org/. Accessed April 28, 2009. This Web site for students, teachers, and the general public (as well as scientists) offers a huge amount of material on the relationship between genes and thinking and a wide range of related topics. A unique feature of the site is a new, dynamic, style of navigation based on "concept mapping," a learner-directed technique for structuring and visualizing information. The DNA Learning Center is currently testing the site in classrooms to explore new ways of teaching and learning about science.

European Bioinformatics Institute (EBI). "2can." Available online. URL: http://www.ebi.ac.uk/2can/home.html. Accessed April 28, 2009. An educational site from the EBI—one of the world's major Internet providers of information about genomes, proteins, molecular structures, and other types of biological data. Many of the tutorials and basic introductions to the themes are accessible to pupils or people with a bit of basic knowledge in biology.

Exploratorium. "Microscope Imaging Station." Available online. URL: http://www.exploratorium.edu/imaging_station/index.php. Accessed April 28, 2009. San Francisco's Exploratorium is an interactive science museum; its Web site has a range of wonderful activities based on biological themes such as development, blood, stem cells, and the brain. There are also videos, desktop wallpapers that can be downloaded for free, and feature articles on current themes from science.

Institute of Human Origins. "Becoming Human: Paleoanthropology, Evolution and Human Origins." Available online. URL: http://www.becominghuman.org/. Accessed April 28, 2009. An attractive site with a focus on paleoanthropology and human origins, with a video documentary that can be watched online or downloaded, classroom resources, and articles on

"How Science Is Done." "The Chromosome Connection," an activity in the Learning Center section of the site, introduces pupils to differences between humans and apes from a molecular perspective.

National Academies Press. "The New Science of Metagenomics: Revealing the Secrets of Our Microbial Planet." Available online. URL: http://books.nap.edu/catalog.php?record_id=11902. Accessed April 28, 2009. A fascinating report on biodiversity, summarizing what has been learned by metagenomics projects to sequence the DNA of microbes that inhabit the human body and various ecospheres. The report provides perspectives for future projects and identifies eight key areas in which metagenomics may be useful.

National Center for Biotechnology Information. "Bookshelf." Available online. URL: http://www.ncbi.nlm.nih.gov/sites/entrz?db=books. Accessed April 28, 2009. A collection of excellent online books ranging from biochemistry and molecular biology to health topics. Most of the works are quite technical, but many include very accessible introductions to the topics. Some highlights are: *Molecular Biology of the Cell, Molecular Cell Biology,* and the *Wormbook.* There are also annual reports on health in the United States from the Centers for Disease Control and Prevention.

National Geographic. "Outpost: Human Origins @ nationalgeographic.com." Available online. URL: http://www.nationalgeographic.com/features/outpost/. Accessed April 28, 2009. A virtual expedition, accompanying human fossil hunter Lee Berger on a search for ancient human remains in Botswana and South Africa.

National Health Museum. "Access Excellence: Genetics Links." Available online. URL: http://www.accessexcellence.org/RC/genetics.php. Accessed April 28, 2009. Links and resources from the "Access Excellence" project of the National Health Museum.

National Public Radio. "Wild Cows Cloned." Available online. URL: http://www.npr.org/templates/story/story.php?storyId=1225049. Accessed April 28, 2009. An audio interview with researcher Robert Lanza, pioneering stem cell researcher (see chapter 1). In this interview Lanza discusses his cloning of an endangered wild cow called the banteng, using cells taken from an animal that had died 23 years earlier. Lanza also appears in an interview from 2006 on more general themes in stem cell research at the following URL: http://www.npr.org/templates/story/story.php?storyId=5204335. The NPR Web site offers a wide range of interviews and broadcasts on biological and medical themes that can be listened to online.

Nobel Foundation. "Video Interviews with Nobel Laureates in Physiology or Medicine." Available online. URL: http://nobelprize.org/nobel_prizes/medicine/video_interviews.html. Accessed April 28, 2009. Video interviews with laureates from the past four decades, many of whom have been molecular biologists or researchers from related fields. Follow links to interviews with winners of other prizes, Nobel lectures, and other resources.

Patricia Piccinini. "Speculative Fabulations for Technoculture's Generations: Taking Care of Unexpected Country," by Donna Haraway. Available online. URL: http://www.patriciapiccinini.net/. Accessed April 28, 2009. This article is a critical review of several years of Piccinini's sculptures, and can be found under the essays link on the artist's Web site. Piccinini imagines what genetic hybrids of humans and other animals might look like in the future and creates very real-looking representations of them.

Public Broadcasting Service (PBS). "American Experience: Jesse James." Available online. URL: http://www.pbs.org/wgbh/amex/james/index.html. Accessed April 28, 2009. This is the home page of a PBS documentary centered on the life of Jesse James, including the transcript of the broadcast, a wide range of images, and other supplementary materials.

Research Collaboratory for Structural Bioinformatics. "RCSB Protein Data Bank." Available online. URL: http://www.rcsb.org. Accessed April 28, 2009. This site provides "a variety of tools and resources for studying the structures of biological macromolecules and their relationships to sequence, function, and disease." There is a multimedia tutorial on how to use the tools and databases. One special feature is the "Molecule of the Month," with beautiful illustrations by David Goodsell.

Science Friday. "Science Friday Archives: Cancer Update with Robert Weinberg." Available online. URL: http://www.sciencefriday.com/program/archives/200710123. Accessed April 28, 2009. Science Friday is heard live every Friday on National Public Radio stations. This link points to the podcast of a program originally broadcast on Oct. 12, 2007, with Robert Weinberg, renowned cancer researcher at MIT. Weinberg's group had just discovered that small molecules called microRNAs regulate the production of proteins in tumor cells, a finding with significant implications for diagnosis and therapy.

Scientific American. "The First Human Cloned Embryo." Available online. URL: http://www.sciam.com/article.cfm?id=the-first-human-cloned-em. Accessed April 28, 2009. An article by Robert Lanza and his colleagues describing the methods they used to make the first clones of human embryonic stem cells. The article explains the methods used and the researchers' hopes for how this type of work may change medicine in the future.

TalkOrigins. "The Talk Origins Archive." Available online. URL: http://www.talkdesign.org. Accessed April 28, 2009. A Web site devoted to "assessing the claims of the Intelligent Design movement from the perspective of mainstream science; addressing the wider political, cultural, philosophical, moral, religious, and educational issues that have inspired the ID movement; and providing an archive of materials that critically examine the scientific claims of the ID movement." (A subsection of the site deals specifically with the origins of humans: http://www.talkorigins.org/faqs/homs.)

Tech Museum of Innovation, San Jose, California. "Understanding Genetics: Human Health and the Genome." Available online. URL: http://www.thetech.org/genetics. Accessed April

28, 2009. An excellent collection of news and feature stories on scientific discoveries and ethical issues surrounding genetics.

University of California, San Francisco. "Live Long and Prosper: A Conversation About Aging with Cynthia Kenyon—Science Café—UCSF." Available online. URL: http://www.ucsf.edu/science-cafe/conversations/kenyon. Accessed April 28, 2009. UCSF's Science Café features interviews with university scientists about their work—at a level easy to understand for the general public. This link is devoted to the research of Cynthia Kenyon, director of the Larry L. Hillblom Center for the Biology of Aging and a pioneer in the biology of aging in the worm *C. elegans* and other model organisms. Visitors to the Web site can read the article or listen to it in mp3 format. Kenyon gives an overview of her own work for non-scientists at the home page of her own lab: http://kenyonlab.ucsf.edu/html/non-scientist_overview.html.

University of California, Santa Cruz. "UCSC Genome Bioinformatics." Available online. URL: http://genome.ucsc.edu/bestlinks.html. Accessed April 28, 2009. A portal to high-quality resources for the study of molecules and genomes, from UCSC and other sources. The map of the BRC2A gene presented in chapter 2 was obtained using this site.

University of Cambridge. "The Complete Works of Charles Darwin Online." Available online. URL: http://darwin-online.org.uk. Accessed April 28, 2009. An online version of Darwin's complete publications, 20,000 private papers, and hundreds of supplementary works.

University of Utah, Genetic Science Learning Center. "Learn Genetics." Available online. URL: http://learn.genetics.utah.edu/. Accessed April 28, 2009. An excellent Web site introducing the basics of genetic science, including a "Biotechniques Virtual Laboratory," special features on the genetics and neurobiology of addiction, stem cells, and molecular genealogy, and podcasts on the genetics of perception and aging.

University of Washington. "Gene Tests Home Page." Available online. URL: http://www.genetests.org/. Accessed April 28, 2009. A site mainly aimed at professionals interested in obtaining up-to-date information on new links between genes and disease, with links to articles on the latest research, laboratories carrying out studies of particular diseases, and ongoing clinical trials. But the site also has a wealth of information on the genetic tests that are available, links to doctors and clinics who provide specific tests, a lexicon of technical terms, and educational resources.

University of Washington Television. "UWTV Program: Genomic Views of Human History." Available online. URL: http://www.uwtv.org/programs/displayevent.aspx?rID=2493. Accessed April 28, 2009. A lecture from Mary-Claire King (see chapter 4) that can be watched online. The theme is how new tools of genomic analysis are being used to investigate the genes of modern humans, shedding light on historical puzzles such as ancient migrations and the settlement of the globe.

Vega Science Trust. "Scientists at Vega". Available online. URL: http://www.vega.org.uk/video/internal/15. Accessed April 28, 2009. Filmed interviews with some of the great figures in 20th-century and current science, including Max Perutz, Kurt Wüthrich, Aaron Klug, Fred Sanger, John Sulston, Bert Sakmann, Christiane Nüsslein-Volhard, etc.

Wisconsin Medical Society. "Wisconsin Medical Society." Available online. URL: http://www.wisconsinmedicalsociety.org/savant_syndrome. Accessed April 28, 2009. A site devoted to the work of Darold Treffert, a psychiatrist who is likely the world's foremost expert on people with savant syndrome. It includes an overview of the field, references to the literature, and written and video portraits of over 20 savants.

Index

Italic page numbers indicate illustrations or maps.

L

M

N